STP 1458

Landscape Ecology and Wildlife Habitat Evaluation: Critical Information for Ecological Risk Assessment, Land-Use Management Activities, and Biodiversity Enhancement

Lawrence Kapustka, Hector Galbraith, Matthew Luxon, and Gregory Biddinger, editors

ASTM Stock Number: STP1458

ASTM International
100 Barr Harbor Drive
PO Box C700
West Conshohocken, PA 19428-2959

Printed in the U.S.A.

Library of Congress Cataloging-in-Publication Data

Landscape ecology and wildlife habitat evaluation : critical information for ecological risk assessment, land-use management activities, and biodiversity enhancement/ Lawrence Kapustka... [et al.].

p. cm. — (STP ; 1458)

Selected papers presented at the symposium "Landscape ecology and wildlife habitat evaluation" held in Kansas City, Missouri, on 7–9 April 2003.

Includes bibliographical references and index.

ISBN (invalid) 080313476

1. Ecological risk assessment—Congresses. 2. Land use—Environmental aspects—Congresses. 3. Habitat (Ecology)—Congresses. 4. Landscape ecology—Congresses. I. Kapustka, Lawrence. II. ASTM special technical publication ; 1458.

QH541.15.R57L36 2004
333.95′14—dc22 2004049022

Peer Review Policy

Each paper published in this volume was evaluated by two peer reviewers and at least one editor. The authors addressed all of the reviewers' comments to the satisfaction of both the technical editor(s) and the ASTM International Committee on Publications.

To make technical information available as quickly as possible, the peer-reviewed papers in this publication were prepared "camera-ready" as submitted by the authors.

The quality of the papers in this publication reflects not only the obvious efforts of the authors and the technical editor(s), but also the work of the peer reviewers. In keeping with long-standing publication practices, ASTM International maintains the anonymity of the peer reviewers. The ASTM International Committee on Publications acknowledges with appreciation their dedication and contribution of time and effort on behalf of ASTM International.

Printed in Bridgeport, NJ
June 2004

Foreword

This publication, *Landscape Ecology and Wildlife Habitat Evaluation: Critical Information for Ecological Risk Assessment, Land-Use Management Activities, and Biodiversity Enhancement,* contains selected papers presented at the symposium of the same name held in Kansas City, Missouri, on 7–9 April 2003. The symposium was sponsored by Committee E-47 on Biological Effects and Environmental Fate. The symposium chairmen and co-editors were Lawrence Kapustka, Hector Galbraith, Matthew Luxon, and Gregory Biddinger.

Contents

Overview

This book contains a collection of papers that were derived from papers presented at a symposium *on Landscape Ecology and Wildlife Habitat Evaluation: Critical Information for Ecological Risk Assessment, Land-Use Management Activities, and Biodiversity Enhancement Practices* that was held 7–9 April 2003 in Kansas City, Missouri. The purpose of the symposium was to bring together scientists with diverse interests in landscape ecology, ecological risk assessment, and environmental management. It was designed to explore contemporary knowledge of theoretical and applied ecology, especially embodied in landscape ecology and population dynamics, especially as they relate to characterizing environmental risks to wildlife and requirements of environmental managers addressing current situations and predicting consequences of actions.

Land-use patterns have been described as the most critical aspect affecting wildlife populations and regional biodiversity. Environmental contamination by chemicals often ranks fairly low in terms of factors limiting wildlife populations. Regulatory and legislative efforts have begun to promote "brownfield development" as an alternative to expansion into uncontaminated areas and with less stringent cleanup standards. Indeed, until recently, many areas which have low to moderate levels of chemical contamination were nevertheless subjected to intrusive remediation efforts; the consequence being substantial destruction of existing wildlife habitat and low potential for enhancing better quality habitat at the affected site. Nevertheless, current practices in Ecological Risk Assessment generally do a poor job of considering biological and physical factors as most focus entirely or nearly so on chemical effects. Therefore, the essential tool used to characterize sites does poorly in weighing the merits of alternative remediation options.

The opening session of the symposium provided three perspectives that drew upon the applied discipline of landscape ecology, approaches used to characterize wildlife habitat, and challenges of environmental management of biological resources from a global corporate perspective. The series of papers that followed, explored theoretical aspects of landscape ecology, population dynamics affected by landscape conditions, and tools and approaches in various stages of development that can be used in assessing environmental risks over different temporal and spatial scales. Finally, several presentations covered real-world applications of different tools and approaches.

The symposium was sponsored by the ASTM Committee E47 on Biological Effects and Environmental Fate. Financial assistance was provided by the American Chemistry Council and the U.S. Army Center for Health Promotion and Preventive Medicine (USACHPPM) Health Effects Research Program. The Subcommittee E47.02 on Terrestrial Assessment and Toxicology anticipates development of two or more Standard Guides covering materials covered in this symposium.

Lawrence Kapustka
Ecological Planning and Toxicology Incorporated
Corvalis, OR
Symposium Chairman and Editor

Hector Galbraith
Galbraith Environmental Sciences
Boulder, CO
Symposium Chairman and Editor

Matthew Luxon
Winward Environmental LLC
Seattle, WA
Symposium Chairman and Editor

Gregory R. Biddinger
Exxon Mobil Refining & Supply Company
Fairfax, VA
Symposium Chairman and Editor

Session I

Virginia H. Dale,[1] *Patrick J. Mulholland,*[1] *Lisa M. Olsen,*[1] *Jack W. Feminella,*[2] *Kelly O. Maloney,*[2] *David C. White,*[3] *Aaron Peacock,*[3] *and Thomas Foster*[4]

Selecting a Suite of Ecological Indicators for Resource Management

---

REFERENCE: Dale, V. H., Mulholland, P. J., Olsen, L. M., Feminella, J. W., Maloney, K. O., White, D. C., Peacock, A., and Foster, T., **"Selecting a Suite of Ecological Indicators for Resource Management,"** *Landscape Ecology and Wildlife Habitat Evaluation: Critical Information for Ecological Risk Assessment, Land-Use Management Activities and Biodiversity Enhancement Practices, ASTM STP 1458*, L. A. Kapustka, H. Galbraith, M. Luxon, and G. R. Biddinger, Eds., ASTM International, West Conshohocken, PA, 2004.

ABSTRACT: We discuss the use of ecological indicators as a natural resource management tool, focusing on the development and implementation of a procedure for selecting and monitoring indicators. Criteria and steps for the selection of ecological indicators are presented. The development and implementation of indicators useful for management are applied to Fort Benning, Georgia, where military training, controlled fires (to improve habitat for the endangered red cockaded woodpecker), and timber thinning are common management practices. A suite of indicators is examined that provides information about understory vegetation, soil microorganisms, landscape patterns, and stream chemistry and benthic macroinvertebrate populations and communities. For example, plants that are geophytes are the predominant life form in disturbed areas, and some understory species are more common in disturbed sites than in reference areas. The set of landscape metrics selected (based upon ability to measure changes through time or to differentiate between land cover classes) included percent cover, total edge (with border), number of patches, mean patch area, patch area range, coefficient of variation of patch area, perimeter/area ratio, Euclidean nearest neighbor distance, and clumpiness. Landscape metrics indicate that the forest area (particularly that of pine) has declined greatly since 1827, the date of our first estimates of land cover (based on witness tree data). Altered management practices in the 1990s may have resulted in further changes to the Fort Benning landscape. Storm sediment concentration profiles indicate that the more

---

[1]Environmental Sciences Division, Oak Ridge National Laboratory, Oak Ridge, TN 37831.
[2]Department of Biological Sciences, Auburn University, Auburn, AL 36849-5407.
[3]Center for Biomarker Analysis, University of Tennessee, Knoxville, TN 37932-2575.
[4]Panamerican Consultants, Inc., 4711 Milgen Road, Columbus, GA 31907.

highly disturbed catchments had much greater rates of erosion and sediment transport to streams than less disturbed catchments. Disturbance also resulted in lower richness of EPT (i.e., number of taxa within the aquatic insect orders Ephemeroptera, Plecoptera, and Trichoptera) than in reference streams but similar total richness of invertebrate species. Each indicator provides information about the ecological system at different temporal and spatial scales.
KEYWORDS: disturbance, forests, indicators, resource management

Introduction

The questions that our work addresses are on a local resource management level. What are the best indicators to be measuring? How can those metrics be properly interpreted? Because of its proactive mode of management, this effort focuses on lands owned and managed by the Department of Defense of the United States. We first examine criteria that are suitable for indicators and then consider steps of selection of indicators. A suite of indicators is proposed, and a case study dealing with potential indicators at Fort Benning, Georgia is presented. Overall, the paper provides insights into the value of indicators, how they are selected, and how they can be used.

Criteria for Selecting Ecological Indicators

Criteria for selecting ecological indicators were developed based on the goal of capturing the complexities of the ecological system but remaining simple enough to be effectively and routinely monitored (Dale and Beyeler 2001):

• *Be easily measured.* The indicator should be easy to understand, simple to apply, and provide information that is relevant, scientifically sound, easily documented, and cost-effective (Lorenz et al. 1999).

• *Be sensitive to stresses of the system.* Ecological indicators should react to anthropogenic stresses placed on the ecological system, while also having limited and documented sensitivity to natural variation (Karr 1991).

• *Respond to stress in a predictable manner.* The response of the indicator should be decisive and predictable even if the indicator responds to the stress by a gradual change. Ideally, there is some threshold level at which the observed response is lower than the level of concern of the impact.

• *Be anticipatory: signify an impending change in key characteristics of the ecological system.* Change in the indicator should be measurable even before substantial change in the ecological system occurs.

• *Predict changes that can be averted by management actions.* The value of the indicator for management depends on its relationship to changes in human actions.

• *Be integrative: together with the full suite of indicators, provide a measure of coverage of the key gradients across the ecological systems (e.g., soils, vegetation types, temperature, etc.).* The full suite of indicators for a site should provide a synchronized perspective of the key attributes of major environmental gradients. These gradients may relate to time, space, soil properties, elevation, or any other factor that is important to the ecological system (e.g, see Figure 1).

• *Have a known response to natural disturbances, anthropogenic stresses, and ecological changes over time.* The indicator should have a definitive reaction to both natural disturbance and to anthropogenic stresses in the system. As ecological conditions change in a system (e.g., via succession), the response of the indicator should be predictable. This criterion most often pertains to metrics that have been extensively studied and have a clearly established pattern of response.
• *Have low variability in response.* Indicators that have a small range in response to particular stresses allow for change in the response value to be distinguished from background variability.

Selecting Ecological Indicators

Identification of the key criteria for ecological indicators sets the stage for a seven-step procedure for selecting indicators. These steps are discussed in view of land use decisions on military lands but are applicable to resource issues on other public and private lands.

Hierarchical Overlap of Suite of Ecological Indicators Over Time

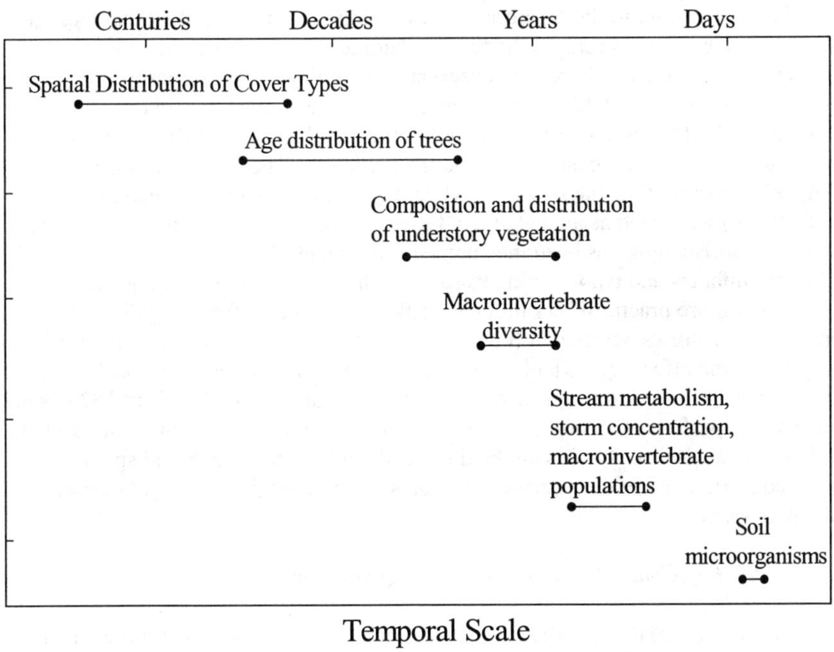

Figure 1 — *A suite of indicators can be depicted across time*

Step 1: Identify Goals for the System.

The first step in problem solving is to define the issue and develop clear goals and objectives. Often, goals are a compromise among the concerns of interested parties. Sometimes objectives change as adherence to one target compromises another. The more complex the nature of the problem, the more important it becomes to establish clear goals and objectives within the spatial and temporal parameters of the system. The selection of ecological indicators is complex in the sense that many factors are involved, feedbacks are common, and diverse groups of stakeholders have different perspectives, value systems, and intentions.

For spatial analysis, it is useful to consider both the immediate area of interest and a broader perspective. The area contained inside the socio-politically delineated boundary can be referred to as the *focal area*, for it is the area of immediate concern to the resource manager. In dealing with ecological management issues, situations often arise when it is useful to look outside of the focal area to a *context area*. Both the focal and context areas can be defined by ecological, social, or political concerns influencing system characteristics.

For the same reason that it is important to consider spatial context when assessing management options, it is also important to consider temporal context. Management areas are defined by past, present, and future social, political, and ecological influences. *Focal time* can be used to refer to the temporal context being considered in the focal area, and *context time* can be used to refer to the temporal context of the entire situation.

As an example, the focal area of conservation planning at Fort Benning is defined by the boundaries of the installation (a political unit), but the context area extends throughout much of the Southeast along the fall line that bisects Fort Benning and differentiates between the Coastal Plain and the Piedmont. One focal time for Fort Benning is the current time back to 1974 when the red cockaded woodpecker (*Picoides borealis,* RCW) was listed as an endangered species. Another focal time might be the last century, for Fort Benning has been the "home of the infantry" since 1918 and is now the site of major infantry and tank training exercises. The context time must consider the intensive agriculture practiced by European settlers since the1800s and by Native Americans for centuries before that time (Kane and Keeton 1998; Foster et al. 2003). To better quantify the effects of agriculture before military activity began at Fort Benning, a vegetation map has been created based on witness tree surveys conducted in 1827 as part of land surveys performed in order to distribute the land (Olsen et al. 2001; Black et al. 2002; Foster et al. 2003). By viewing land use and land cover in the broad spatial and temporal context, meeting the management goals can be considered in light of these broader perspectives.

Step 2: Identify Key Characteristics of the Ecological System

Characteristics are the specific functional, compositional, and structural elements that, when combined, define the ecological system. All ecological systems have elements of composition and structure that arise though ecological processes. The characteristic conditions of an area depend on sustaining key ecological functions that, in turn, produce additional compositional and structural elements. If the linkages between underlying processes, composition, and structural elements are broken, then sustainability is jeopardized and restoration may be difficult and complex.

Key characteristics include the physical features that allow species, ecosystems, or landscapes to occur. For example, at Fort Knox, Kentucky, locations of threatened calcareous habitats of rare species can be predicted based on a combination of soils, geology, and slope (Mann et al. 1999). This edaphic-based approach has also been used to identify locations of Henslow's sparrow (*Ammodranmus henslowii*) habitat at Fort Knox and sites at Fort McCoy, Wisconsin, that can support wild lupine (*Lupinus perennis*), the sole host plant for the larvae of the endangered Karner blue butterfly (*Lycaecides melissa samuelis*) (Dale et al. 2000).

Identification of the key ecological characteristics of a system also involves attention to social, economic, and political features of a site. Combinations of social, economic, political, and ecological concerns, such as laws and regulations, peoples' values, regional economics, and ecological conditions, determine the importance of a characteristic. The Southern Appalachian Assessment (SAA) provides an example of multiple agencies working together to identify key characteristics of a large area (USDA 1996). The first step in this identification process was to determine the major concerns about the system emanating from social, economic, and ecological perspectives of the eight-state region. The assessment focused on terrestrial, aquatic, atmospheric, and social/cultural/economic conditions. Thus, the assessment was concerned with the condition of the natural resources as well as how people use the resources and their expectations. Because the SAA covers such a large area and such broad topics, a list of key terrestrial characteristics was developed for categories of forest health, wildlife and plant species, and important habitats. Aquatic characteristics include water quality, aquatic species, and habitats. The influences on ecological conditions of historical disturbances, land uses, and social and political forces were also considered, and both local environments and landscape perspectives were evaluated.

Once the important characteristics of a system are identified, the typical range of variation in those characteristics can be established within the focal and context areas and times. This information on the range of terrestrial, aquatic, atmospheric, and social/cultural/economic conditions provided the bulk of the five-volume Southern Appalachian Assessment (USDA 1996). The variability in these characteristics can be presented with regard to changes over time, environmental gradients in the area, or different levels of anthropogenic influences.

In their consideration of key characteristics, military natural resource managers have focused on endangered species and systematic inventories of vascular plant and wildlife. For example, the Army has instituted the Land Condition-Trend Analysis (LCTA) program as a standardized way to measure, analyze, and report data from inventory plots on plant communities, habitat, disturbances, impacts of military training, soil erosion potential, allowable uses, and restoration needs (Diersing et al. 1992). The purpose of that program was both to characterize the vegetation and to monitor change and detect trends in natural resources (Bern 1995). Sample plots were established in a stratified random manner using satellite imagery. Because the military testing and training typically result in intense, local, and broadly spaced impacts, the LCTA plots often do not capture the spatial distribution of the effects. For example, at Yuma Proving Ground, Arizona, about 60 to 70% of the plots had no land use over the period 1991 to 1993 even though the actually land use was more extensive (Bern 1995). Therefore, the LCTA approach needs to be supplemented by a scheme designed to focus on discerning impacts and to integrate over broad spatial scales. Yet to relate the characteristics to the impacts, the stress also needs to be identified.

Step 3: Identify Key Stresses

Stress to an ecological system is typically defined as any anthropogenic action that results in degradation (e.g., less biodiversity, reduced primary productivity, or lowered resilience to disturbances) (Odum et al. 1979; Barret and Rosenberg 1981; Odum 1985; Mageau et al. 1995). Stress can be classified into four categories: physical manipulations, changes in disturbance regimes, introduction of invasive species, and chemical changes [a slight revision of Rapport and Whitford's (1999) categories that use "stress" for anthropogenic activities]. Physical manipulations include human activities that can change soil conditions or construction of structures. Human activities may also cause fragmentation or eliminate critical habitats for some species.

Changes in disturbance intensity, frequency, duration, and extent can have major impacts on ecological systems (Dale et al. 1998). Disturbances are considered to be those events that are not typical of a system. For example, fires within a fire-moderated system, such as the lodgepole pine (*Pinus contorta*) forest of the western United States, would not be a disturbance to the system (even though individual organisms are impacted) (Fahey and Knight 1986). It is the absence of such fires that may cause a disturbance, for fires are an integral part of establishment and development of community structure of these forests. Thus, disturbances must be considered with regard to the life history of the major organisms in the community.

The introduction of invasive species is a major problem in many ecological systems. Often these introductions are nonnative species that do not have predators or competitors within the new system and thus become out of control. These introduced species can physically override the presence of other organisms and replace them quickly. There are numerous examples of such replacements (Westbrooks 1998). Occasionally invasive species may take over because of the elimination of some physical or biological constraints that may have been in the system in the past. *Lonicera maackii* (Rupr.) Herder (Amur honeysuckle), a large invasive shrub introduced into the United States in the late 19th century, has naturalized in at least 24 eastern states. It is abundant in habitats ranging from disturbed open sites to forest edges and interiors. *Lonicera maackii* negatively impacts native species, especially tree seedlings and forest herbs. Open, disturbed forests (e.g., Fort Campbell, Kentucky, where training can open forest canopies) are especially susceptible to colonization (e.g., Deering and VanKat 1998).

Chemical changes in the environment typically occur as a direct result of human activities. Point sources of toxins that result from spills or groundwater movements are a common cause of such a chemical change. Air pollution can also cause widespread and non-point source solution changes in systems.

Stress can be depicted as a gradient or a threshold such as intensity of impact, duration of event, or frequency of impact. Stresses are ultimately what most management plans are for, both preventively and retrospectively. Often, changes in characteristics of a system result directly from one or more stresses. Typically, stresses interact and may exacerbate conditions for biotic survival or maintenance (Paine et al. 1998). Multiple stresses may be simultaneously analyzed or considered one at a time, depending on the goal of the analysis.

The stresses on military installations fit into the four categories of physical manipulations, changes in disturbance regimes, introduction of invasive species, and chemical changes. The training and testing typical of most installations creates a diversity of physical stresses ranging from soil erosion to vegetation removal. Alterations to fire frequency and intensity are the most common form of changing disturbance regimes. In some cases (such as Eglin Air Force Base on the Florida Panhandle), a prior landowner controlled fires, and the Department of Defense is now reinstituting a regular fire regime. The introduction of invasive species is a common problem on most installations. At Fort McCoy, Wisconsin, the leafy spurge (*Euphorbia esula*) threatens to encroach into oak savannas and outcompete the wild lupine. Kudzu (*Pueraria thunbergiana*) is present on most military installations in the Southeast where it literally overgrows anything in its path. Chemical changes on most installations occur as point sources in areas devoted to intense military activities (e.g., painting of aircraft). Usually, these sites are considered sacrifice areas in terms of conservation goals. However, chemical control of introduced species or along roadsides can also affect ecosystem management.

Step 4: Determine How Stresses May Affect Key Characteristics of the Ecological System

Once the process of selecting potential issues and identifying ecological characteristics and stresses within the context and focal systems is completed, the indicator selection process moves into the more specific stage of indicator selection. The process of developing and evaluating landscape-based ecological indicators is large and complicated, varies by region, and requires conceptual and causal links between stresses and the resulting ecological change (Brooks et al. 1998). Each concern that has been determined through the issue identification process needs to be analyzed in order to identify associated stresses, the cause of those stresses, the scope of those stresses on the management area, and the resulting changes in the characteristics of the management area.

Stresses are important to an ecological system in that they can disrupt composition, structure, or function. To the extent that these changes alter key characteristics of a system, the effect is significant. For example, insects or pathogens can increase tree mortality, reduce growth, and eventually change species composition and habitat patterns. Yet stresses that disrupt rare communities may be of the greatest concern to composition. For example, in the Southern Appalachians, 84% of the federally listed species occur in 31 rare communities and streamside habitats (USDA 1996), which means that management for endangered species can concentrate on select sites. However, there are considerable challenges to managing large tracts of land on the basis of a few endangered species.

Matrices that relate stresses to key ecological characteristics may be the best way to depict the effect that human activity may have on a system. For example, matrices containing the ways that military use can affect different types of vegetation at Fort

McCoy, Wisconsin have been developed (Dale et al. 2002b). The focus is on vegetation structure of the ground layer and the shrubs and trees because the wild lupine on which the larvae of the endangered Karner blue butterfly exclusively feeds occurs in the ground layer, and the shrub and tree layers provide the oak savanna system in which the lupine thrives. Such a matrix brings attention to those characteristics that are likely to change under current stresses and, thus, provides a way to identify indicators.

In much the same way that the spatial and temporal scales of the focal and context areas need to be defined, so too do the spatial and temporal scales of the individual stresses. As a result, stress effects may be limited to certain places or times. For example, ozone damage to sensitive trees may be greater at higher elevations where sufficient moisture is available from cloud cover to prevent stomata closure and allow more ozone to be absorbed. As a temporal example, some organisms are only susceptible to stress during their dispersal phase, while stresses at other times have little effect. For example, tank activity at Fort McCoy, Wisconsin actually enhances the presence of wild lupine upon which the endangered Karner blue butterfly ovipost (Smith et al. 2001). Yet, tank activity during the larvae stages can kill the insect.

Step 5: Select Indicators

The selected indicators should reflect the criteria (discussed earlier) and identify stress effects on key characteristics of the system. In general, these criteria call for indicators that are sensitive to the identified stressors in the system, sophisticated enough to capture the ecological system complexities, and responsive to identified stressors in such a way that they can be easily measured and monitored. Knowing how the stresses affect the key characteristics of the ecological system assists in the selection of indicators.

The selection of indicators is best made in a hierarchical manner. The selection process is initiated by considering the entire area of interest. For most military applications, this perspective would entail the installation as the focal site and the present as the focal time. However, the larger spatial and temporal context should also be considered. Thus, examination of the major physical gradients across the landscape or region should consider topography, soils, geology, land-use history, disturbance history, patterns of water (streams, lakes, and wetlands), and human use (roads, trails, buildings, and training and testing sites). Often the vegetation type, size, or density reflects the combination of these physical forces and serves as a useful indicator of their strength. For example, at Fort Stewart, Georgia, the amount of hardwood ingrowth into longleaf pine (*Pinus palustris*) stands indicates the time since the last growing-season fire. Thus, the pattern of vegetation types, such as hardwood ingrowth, or other land covers should be evaluated to see if it portrays features of the landscape that are indicative of stresses at the site and that may affect the ecological properties of the site. At Arnold Air Force Base in Tennessee, the high degree of forest fragmentation is indicative of past timber-harvesting practices and may portend effects on neotropical migrants (Robinson et al.1995).

Ideally the suite of indicators should represent key information about structure, function, and composition. Yet the complexity of the relationship between structure function, and composition only hints at the intricacy of the ecological system on which it is based. Often it is easier to measure structural features that can convey information about the composition or functioning of the system than to measure composition or function. Sometimes measures from one scale can provide information relevant to another scale. For example, the size of the largest patch of a habitat often restricts the species or trophic levels of animals that are able to be supported based solely on their minimal territory size (Dale et al. 1994). Analysis of patch size for Henslow's sparrow at Fort Riley, Kansas indicates that the largest patch on the installation supports a declining population (the population's finite rate of increase is less than one) (Dale et al. 2000).

After the landscape is analyzed, the ecosystem and the species levels should be investigated. This process of considering characteristics of the system and potential indicators in a spatially hierarchical fashion needs to apply to each gradient of importance at the site. Placing the information on a spatial or temporal axis provides a means to check that information at all spatial scales. Alternatively, it is important to include indicators that encapsulate the diversity of responses over time (so that one is not just measuring immediate responses of the system). All major gradients are included in the analysis. We have focused on spatial and temporal scales, but it is also useful to consider the representativeness of indices across major physical gradients (soils, geology, land use, etc.).

Step 6: Test Potential Indicators Against Criteria

A crucial aspect for legitimizing the selection procedures for ecological indicators is the establishment of a scientifically sound method of monitoring system change. Each of the potential indicators needs to be tested to determine if it effectively measures the system characteristics of interest and meets the other criteria for indicators. This test should follow scientific procedures (e.g., theory and hypothesis development, hypothesis testing with control comparison, statistically significant results, etc.). The working hypotheses should reflect how specific indicators measure changes in key characteristics under stress. Experiments should be designed to compare measures of the indicators and key characteristics with and without stress events. For example, the condition of these indicators both before, during, and after documented stresses can then be compared with similar data collected in control sites. Based on the results of the tests for each potential indicator, the final set of ecological indicators can then be selected that is believed to be the most effective combination of indicators for monitoring the characteristics of interest to the management planners. The statistical analysis of such indicators is a basic aspect of most statistical text books.

Step 7: Select Final Indicators and Apply Them to the Decision-Making Process

The final ecological indicators are selected based on the test in Step 6. Then, management can implement monitoring of the suite of selected indicators. Long-term monitoring is an essential part of all environmental management programs, with adjustment of management activities based on indicator information and its relationship to overall management goals. The process of linking management to monitoring is part of adaptive management that views management actions as experiments and accumulates knowledge to achieve continual learning (Holling 1978; Walters 1986).

Often the application of measuring indicators or of adding refinements to measures can occur very quickly. This implementation aspect is especially rapid on Department of

Defense installations where the mentality is to act. For example, after we had used soil, geology, and slope to identify the sites at Fort McCoy, Wisconsin, that the wild lupine could occupy (Dale et al. 2000), the environmental site manager modified his monitoring program for wild lupine to focus only on areas that the analysis indicated could support the plant. This modification allowed the monitoring program to focus on those sites of greatest importance.

Case Study

The objective of this case study is to identify indicators that signal ecological change in intensely and lightly used ecological systems at Fort Benning. Currently, military training, controlled fires (to improve habitat for the endangered red cockaded woodpecker), and timber thinning are common management practices on the installation. All of Fort Benning has experienced some anthropogenic changes either from past farming, logging, absence of burning, or military testing. Because the intent is that these indicators become a part of the ongoing monitoring system at the installation, the indicators should be feasible for the installation staff to measure and interpret. The focus is on Fort Benning, but the goal is to develop an approach to identify indicators that would be useful at several military installations. Because some of these effects may be long-term or may occur after a lag time, early indications of both current and future change need to be identified. The intent of this identification of indicators is to improve managers' ability to manage activities that are likely to be damaging and to prevent long-term, negative effects. Therefore, a suite of variables is needed to measure changes in ecological conditions. The suite that we are examining includes measures of terrestrial understory and overstory vegetation, soil microbial biomass and community composition, landscape patterns, and instream physiochemical and biotic water quality conditions. Because of the limited space in this publication, for further details we direct the reader to the project web site:
(http://www.esd.ornl.gov/programs/SERDP/research_projects.html#conservation).

The analyses of vegetation data collected from sites at Fort Benning with five discrete land-use histories showed high variability in species diversity and lack of distinctiveness of understory cover and led us to consider life form and plant families as indicators of military use (Dale et al. 2002a). Life form successfully distinguished between plots based on military use. For example, phanerophyte species (trees and shrubs) were the most frequent life form encountered in sites that experienced infantry foot traffic training. Analysis of soils collected from each transect revealed that depth of the A layer of soil was significantly higher in reference and infantry foot traffic training areas which may explain the life form distributions. In addition, the diversity of plant families and, in particular, the presence of grasses and composites were indicative of training and remediation history. These results are supported by prior analysis of life form distribution subsequent to other disturbances (Adams et al. 1987; McIntyre et al. 1995; Stohlgren et al. 1999) and demonstrate the ability of life form and plant families to distinguish between military uses in longleaf pine forests.

The soil microbial community of a longleaf pine ecosystem at Fort Benning also responds to military traffic (Peacock et al. 2001). Using the soil microbial biomass and community composition as ecological indicators, reproducible changes showed increasing traffic decreases soil viable biomass, biomarkers for microeukaryotes and Gram-negative bacteria, while increasing the proportions of aerobic Gram-positive bacterial and actinomycete biomarkers. Our results indicate that as a soil is remediated it does not escalate through states of succession in the same way as it descends following military use. We propose to explore this hysteresis between disturbance and recovery process as a predictor of the resilience of the microbial community to repeated disturbance/recovery cycles.

The landscape metrics for Fort Benning were calculated and analyzed, and an assessment was made of the accuracy of the land cover estimates obtained from remote sensing as compared to *in situ* observations of land cover (Olsen et al. 2001). Metrics at the class and landscape level were compiled and analyzed to determine which were the best indicators of ecological change at Fort Benning. A set of metrics was selected, based upon change through time or ability to differentiate between land cover classes. We found the most useful metrics for depicting changes in land cover and distinguishing between land cover classes at Fort Benning were percent cover, total edge (with border), number of patches, mean patch area, patch area range, coefficient of variation of patch area, perimeter/area ratio, Euclidean nearest neighbor distance, and clumpiness. An accuracy assessment was performed of the 1999 land cover classification that was created using a July 1999 Landsat ETM image as compared to a 0.5-m digital color orthophoto of Fort Benning taken in 1999. The overall accuracy was found to be 85.6 for the 30-m resolution data (meaning that 85.6% of the test sites were correctly classified).

Landscape metrics indicate that the forest pattern (particularly that of pine) has declined greatly since 1827 (e.g., the area of pine forest declined from 78% to 34% of the current installation). Altered management practices in the 1990s may have resulted in changes to the landscape at Fort Benning. Several trends, such as an increase in non-forested and barren lands in riparian buffers were slowed or reversed in the last decade. Pine forest, on the other hand, appears to have been increasing in the last ten years. Improved monitoring techniques coupled with an aggressive management strategy for perpetuating pine forest at Fort Benning may have resulted in an increase in pine populations and a decrease in hardwood invasion. This management strategy includes harvesting timber and burning to establish and maintain viable pine communities. While it appears that the percentage of non-forest land has been slowly increasing, the number of non-forest patches has increased tremendously in the last decade. In other words, the non-forest land has become more fragmented over time. Consequently, the size of these patches has decreased significantly.

We are evaluating the efficacy of several stream chemistry and biology parameters as indicators of disturbance associated with military training and natural resource management activities at Fort Benning. This work is based on the idea that stream ecosystems are sensitive to disturbances within their catchments because many disturbances alter the patterns of runoff, drainage water chemistry, and inputs of biologically important materials to receiving streams. In addition, stream ecosystems are important components of the landscape and indicators of disturbance to stream biological communities and biogeochemical processes are an important part of any assessment of ecosystem health. Our research uses a disturbance gradient approach in which 1st- to 3rd-order streams draining catchments with strongly contrasting disturbance levels have been selected for study. These catchments are distinguished by percent bare ground for some have little disturbance and others have widespread erosion caused by regular tank traffic. The inclusion of several reference streams in our study design provides data on the range of values for physicochemical and biological parameters expected for catchments showing minimal level of disturbance. Data from streams along the disturbance gradient are being compared to evaluate the suitability and sensitivity of specific disturbance indicators. The potential aquatic indicators at Fort Benning have been narrowed to:

- Suspended sediment concentrations (both baseflow and storms) and baseflow (PO_4, DOC) and stormflow (NH_4, NO_3, and PO_4) nutrient concentrations (indicator of erosion and biogeochemical status)
- Diurnal dissolved oxygen profiles (indicator of in-stream metabolism)
- Streambed organic matter content (indicator of food or habitat), and sediment movement dynamics (indicator of in-stream habitat stability or quality)
- Macroinvertebrate populations and communities, including EPT richness, Shannon diversity, biotic tolerance indices, and Bray-Curtis similarity of disturbed and reference streams (indicator of biological response)

For example, storm sediment concentration profiles show that streams in highly disturbed catchments had much higher rates of erosion and sediment transport than streams in less disturbed catchments.

The effects of historical land use / disturbance on stream macroinvertebrates are also being examined. Using remotely sensed imagery from 1974 and 1999, we used the GIS extension ATtILA to estimate areal percentage of 1) bare ground on slopes >3%, 2) successional stage of vegetation (early-regeneration forested land) on slopes >3%, and 3) road density (km road/km^2 catchment) for each catchment. These three land use variables were then combined to derive a disturbance index (DI), which was used to rank and compare each catchment's historic and contemporary disturbance level. With these data we are examining the degree to which current measures of biotic water quality relate to historical vs. contemporary disturbance conditions. Preliminary analysis indicated that percent silt in the streambed was positively correlated with levels of historical (1974) land use among the catchments. Moreover, relative abundance of macroinvertebrate functional feeding groups also was related to historical land use. Disturbance also resulted in lower richness of EPT (i.e., number of taxa within the aquatic insect orders Ephemeroptera, Plecoptera, and Trichoptera) than in reference streams but similar total richness of invertebrate species. These data indicate 1) a legacy of environmental disturbance in Fort Benning catchments that spans at least 25 years, and 2) knowledge of historical land use conditions may be critical in interpreting contemporary water quality conditions.

Conclusions

Ecological indicators offer a means to measure the effects of resource management. A key challenge is dealing with the complexity of ecological systems. Criteria and procedures for selecting indicators offer a way to deal with this complexity. The Department of Defense is developing ways to implement the use of ecological indicators for ecosystem monitoring and management. The next step is implementing indicators into resource-management practices.

Acknowledgments

Suzanne Beyeler and Jordan Smith provided assistance. The project was supported by a contract from the Strategic Environmental Research and Development Program (SERDP) Ecosystem Management Program (SEMP) to Oak Ridge National Laboratory (ORNL). Oak Ridge National Laboratory is managed by the University of Tennessee-Battelle LLC. for the U.S. Department of Energy under contract DE-AC05-00OR22725.

References

Adams, A. B., Dale, V. H., Kruckeberg, A. R., and Smith, E., 1987, "Plant Survival, Growth Form and Regeneration Following the May 18, 1980, Eruption of Mount St. Helens, Washington," *Northwest Science*, Vol. 61, pp. 160-170.

Barrett, G. W. and Rosenberg, R., 1981, *Stress Effects on Natural Ecosystems*, John Wiley & Sons, New York.

Bern, C. M., 1995, *Land Condition-Trend Analysis Installation Report: Yuma Proving Ground Arizona 1991-1992*, The Center for Ecological Management of Military Lands: Colorado State University, Fort Collins, Colorado.

Black, B. A., Foster, H. T., and Abrams, M. D., 2002, "Combining Environmentally Dependent and Independent Analysis of Witness Tree Data in East-Central Alabama," *Canadian Journal of Forest Research*, Vol. 32, No. 11, pp. 2060-2075.

Brooks, R. B., O'Connell, T. J., Wardrop, D. H., and Jackson, L. E., 1998, "Towards a Regional Index of Biological Integrity: The Example of the Forested Riparian Ecosystem," *Environmental Monitoring and Assessment*, Vol. 51, pp. 131-143.

Dale, V. H., and Beyeler, S. C., 2001, "Challenges in the Development and Use of Ecological Indicators," *Ecological Indicators*, Vol. 1, No. 1, pp. 3-10.

Dale, V. H., Offerman, H., Pearson, S., and O'Neill, R. V., 1994, "Effects of Forest Fragmentation on Neotropical Fauna," *Conservation Biology*, Vol. 8, pp. 1027-1036.

Dale, V. H., Lugo, A., MacMahon, J., and Pickett, S., 1998, "Ecosystem Management in the Context of Large, Infrequent Disturbances," *Ecosystems*, Vol. 1, pp. 546-557.

Dale, V. H., King, A. W., Mann, L. K., and Ashwood, T. L., "Contributions of Spatial Information and Models to Management of Rare and Declining Species," *Spatial Information for Land Management*, M. J. Hill and R. J. Aspinall, Eds., Gordon and Breach Science Publishers, Lausanne, Switzerland, 2000, pp. 159-172.

Dale, V.H., Beyeler, S.C., and Jackson, B. 2002a. "Understory Indicators of Anthropogenic Disturbance in Longleaf Pine Forests at Fort Benning, Georgia, USA," *Ecological Indicators*, Vol. 1, No. 3, pp. 155-170.

Dale, V. H., Fortes, D. T., and Ashwood, T. L., "A Landscape Transition Matrix Approach for Land Management," *Integrating Landscape Ecology into Natural Resource Management*, J. Liu and W. Taylor, Eds., Cambridge University Press, Cambridge, 2002b, pp. 265-293.

Deering, R. H., and VanKat, J. L., 1998, "Forest Colonization and Developmental Growth of the Invasive Shrub *Loricera maackii*," *American Midland Naturalist*, Vol. 141, pp. 43-45.

Diersing, V. E., Shaw, R. B., and Tazik, D. J., 1992, "U.S. Army Land Condition Trend Analysis (LCTA) Program," *Environmental Management*, Vol. 16, pp. 405-414.

Foster, T., Black, B., and Abrams, M., In press, "A Witness Tree Analysis of the Effects of Native Americans on the Pre-European Settlement Forests in East-Central Alabama," *Human Ecology*.

Hawkins, V., and Selman P., 2002, "Landscape Scale Planning: Exploring Alternative Land Use Scenarios," *Landscape and Urban Planning*, Vol. 60, No. 4, pp. 211-224.

Kane, S., and Keeton, R., 1998, *Fort Benning: the Land and the People*, Southeast Archeological Center, National Park Service, Tallahassee, Florida.

Karr, J. R., 1991. "Biological Integrity: A Long Neglected Aspect of Water Resource Management," *Ecological Applications*, Vol. 1, pp. 66-84.

Mageau, M.T., Costanza, R., and Ulanowicz, R. E., 1995, "The Development and Initial Testing of a Quantitative Assessment of Ecosystem Health," *Ecosystem Health*, Vol. 1, pp. 201–213.

Mann, L. K., King, A. W., Dale, V. H., Hargrove, W. W., Washington-Allen, R., Pounds, L., and Ashwood, T. A., 1999, "The Role of Soil Classification in Geographic Information System Modeling of Habitat Pattern: Threatened Calcareous Ecosystems," *Ecosystems*, Vol. 2, pp. 524-538.

McCoy, E. D., and Mushinsky, H. R., 2002, "Measuring the Success of Wildlife Community Restoration," *Ecological Applications*, Vol. 12, No. 6, pp. 1861-1871.

McIntyre, S., Lavorel, S., and Tremont, R.M., 1995, "Plant Life-History Attributes - their Relationship to Disturbance Responses in Herbaceous Vegetation," *Journal of Ecology*, Vol. 83, No. 1, pp. 31-44.

Odum, E. P., 1985, "Trends Expected in Stressed Ecosystems," *BioScience*, Vol. 35, pp. 419-422.

Odum, E. P., Finn, J. T., and Franz, E., 1979, "Perturbation Theory and the Subsidy–Stress Gradient," *BioScience*, Vol. 29, pp. 349–352.

Olsen, L. M., Dale, V. H., and Foster, T., 2001, "Landscape Patterns as Indicators of Ecological Change at Fort Benning, GA." 2001 ESRI Users Conference, SanDiego, California.

Paine, R. T., Tegner, M.J., and Johnson, A. E., 1998, "Compounded Perturbations Yield Ecological Surprises: Everything Else is Business as Usual," *Ecosystems*, Vol. 1, pp. 535-545.

Peacock, A. D., MacNaughton, S. J., Cantu, J. M., Dale, V. H., and White, D. C., 2002, "Soil Microbial Biomass and Community Composition Along an Anthropogenic Disturbance Gradient within a Longleaf Pine Habitat," *Ecological Indicators*, Vol. 1, No. 2, pp. 113-122.

Robinson, S. K., Thompson III, F. R., Donovan, T. M., Whitehead, D. R., and Faaborg, J., 1995, "Regional Forest Fragmentation and the Nesting Success of Migratory Birds," *Science*, Vol. 267 pp. 1987-1990.

Smith, M. A., Turner, M. G., and Rusch, D. H., 2001, "The Effect of Military Training Activity on Eastern Lupine and the Karner Blue Butterfly on Fort McCoy, Wisconsin," *Environmental Management*, Vol. 29, pp.102-115.

Stohlgren, T. J., Schell, L. D., and Vanden Heuvel, B., 1999, "How Grazing and Soil Quality Affect Native and Exotic Plant Diversity in Rocky Mountain Grasslands," *Ecological Applications*, Vol. 9, No. 1, pp. 45-64.

USDA. 1996. *The Southern Appalachian Assessment: Summary Report*, U. S. Department of Agriculture, Washington, D.C.

Westbrooks R. G. 1998. *Invasive Plants: Changing the Landscape of America*, Federal Interagency Committee for the Management of Noxious and Exotic Weeds, Washington, D.C.

David G. Richards[1]

Integrating Mineral Development and Biodiversity Conservation into Regional Land-Use Planning

---

REFERENCE: Richards, D. G., "***Integrating Mineral Development and Biodiversity Conservation into Regional Land-Use Planning,***" *Landscape Ecology and Wildlife Habitat Evaluation: Critical Information for Ecological Risk Assessment, Land-Use management Activities, and Biodiversity Enhancement Practices, ASTM STP 1458*, L. A. Kapustka, H. Galbraith, M. Luxon and G. R. Biddinger, Ed(s)., ASTM International, West Conshohocken, PA, 2004.

ABSTRACT: A major independent multi-stakeholder analysis of how the mining industry can maximize its role in the transition to sustainable patterns of development – the Mining, Minerals and Sustainable Development (MMSD) project – concluded in 2001. Prominent among the recommendations in the MMSD report were the need for the mining industry to improve its performance in biodiversity assessment and management, and the need for all parties to commit to better models for decision-making processes in land use and access.

Mining is a temporary use of land, but history teaches us that the net effect of mining in a landscape is usually negative for biodiversity. There are benefits to human society in health, wealth and education, but society increasingly demands that environmental values be protected without compromising economic and social foundations. These expectations are captured in the concept of sustainable development.

Often, the most prospective areas for future mines will also be those with the greatest biodiversity value and with the greatest need for poverty alleviation. Many governments lack the capacity, will or resources to reconcile these conflicting needs equitably. Corruption in government and oppression of local populations have accompanied some mine developments.

Leading companies in the mining industry believe that these negative experiences are not inevitable, that better decisions on land use and access can be achieved and that sustainable benefits can be delivered through mineral development. One key to achieving these outcomes is the regional landscape-scale analysis of projects and conservation priorities, supported by fair, transparent and consistent decision-making processes.

Rio Tinto is a large diversified mining company which played a leading role in the actions leading to the commissioning of the MMSD project and participated fully in it. Examples from recent projects in Rio Tinto, illustrating aspects of regional planning and conservation actions, are presented in support of the case outlined above.

KEYWORDS: mineral development, biodiversity conservation, regional land-use planning

Introduction

The signs of mining seem to be a permanent feature of some landscapes. In reality the duration of mining activities – extraction and processing – tends to be relatively short. It is the failure to return mined lands to other uses that creates the impression that mining's environmental impacts are, inevitably, permanent. For example, there is no mining for metals currently being carried out in Cornwall, UK, one of the homes of underground mining traditions and expertise. The last time there was a significant mining industry there was the end of the 19th

---

[1] Principal Advisor Environment, Health, Safety & Environment Department, Rio Tinto plc, 6 St James's Square, London SW1Y 4LD, UK.

century, although there have been several short and unsuccessful revivals since then. The mined landscapes are essentially unrestored since the 1880s, but their current beneficial use as important assets in attracting tourists to the area is relatively recent. They have always been historically important but now their dereliction has become picturesque. It has taken over 100 years for a beneficial use to emerge.

Of course there are many examples of continuous or sequential episodic mining activities being carried out in an area for more than a century – Bingham Canyon in Utah, is one. But equally there are cases where a mine has been developed and rehabilitated in less than ten years, with no obvious lasting impact on the region's landscape or environment. One such example is that of Flambeau, Wisconsin, which is described in more detail later in this paper.

Mining, unlike agriculture, is not necessarily a permanent part of the geography and economy of many landscapes. It is a temporary use of the land. Equally it is not like agriculture in its specific economic power. It is localized and can yield great wealth out of small areas. An illustration of this is that the value of Rio Tinto's mineral output in 2002, \$10.8 Bn, was generated from a disturbed area of just over 1400 km^2. To generate this revenue from agriculture, even using intensive high-yield methods as practiced in the USA and western Europe, would require the use of an estimated 180 000 km^2. This is a factor of over 120. It might be argued that the relative impacts of agriculture on environmental values such as water quality, water availability and biodiversity are less than those of mining, though this is a debatable point itself. Even if true, I would propose that this is nowhere near enough to outweigh the economic benefits of mining.

The challenges are to ensure that the environmental footprint of mining remains as small as possible by preventing pollution and to use the financial benefits wisely so that sustainable improvements in livelihoods can be created. The legacy of the mining industry contains too many examples of failure to achieve one or both of these objectives.

In this paper, recent projects and initiatives aimed, *inter alia*, at resolving the sources of the historical conflict between mining and conservation are reviewed; a better framework for achieving this reconciliation through land-use decision-making processes is discussed; and examples from Rio Tinto's recent experience, in which multiple uses and concerns have been factored into development projects, are presented.

The Business Case for Improvement

In the latter part of the 20th century the leading companies in the mining industry made great advances in setting and achieving higher standards of performance, to the extent that they did not feel they deserved the poor reputation the industry had acquired from its legacy. Bad reputation hurts the bottom line when the lack of trust it produces causes neighboring communities to protest against new mines, or investors to choose to put their money elsewhere, or regulators to feel predisposed against a permit application. These changes in outlook came to be expressed in terms of the license to operate, and acquiring and maintaining one remains a strong element of the business case for high standards of social, environmental, health and safety performance.

Another aspect of the business case is the need for continuing access to land for exploration and mining. Individual mineral deposits are finite and non-renewable, mineral commodities sustain our lifestyles, and the growing world population will use minerals to secure social and economic development. The notion of responsible mining – commitment to shared high values and delivery of better social and environmental performance – as a condition of access to land has been increasingly accepted by leading mining companies. A growing component of the business case in the future will probably be the access to premium commodity markets for only those mining companies which meet conditions, based on environmental and social performance, imposed by customers.

So, in the late 1990s the conditions were right for leaders in the mining and metals industry to seek to re-negotiate their relationship with the rest of society. A number of obstacles stood in the way of this process, including a lack of trust of mining on the part of many constituencies, both governmental and non-governmental. The industry was also not organized in a way which facilitated engagement with other stakeholders at regional and global scales. The regulatory framework within which individual companies operated around the world was variable in terms of standards and enforcement, and very competitive market conditions encouraged many companies to seek short term benefit by accepting low standards of performance. It was clear from these influences that the multi-stakeholder consultation and analysis which would be a necessary part of changing the status quo would not be entirely comfortable for the industry. Nevertheless, something had to change.

The Global Mining Initiative (GMI)

The prime movers for starting the process of change were the leaders – presidents, chairmen and CEOs – of nine large mining companies with global asset portfolios. Meeting in late 1997 they realized the need to address the issue of the trust deficit existing between mining and other constituencies, and to bring some of the industry's critics into the process of setting the direction for the industry of the future. The trust deficit was never a one-way phenomenon. For their part, governments and non-governmental organisations (NGOs) would be encouraged to confront some of their prejudices about mining and to contribute their ideas to an objective debate in which no single constituency would be sure of winning all the arguments.

The conceptual framework within which this debate took place is sustainable development (SD), in which economic development, social justice and environmental integrity can be achieved in a sustainable process. Despite the high profile achievements of the 1992 UN Conference on Environment and Development, SD was still not a mainstream concept in 1997 for many in government, industry and society as a whole, yet all indications pointed to it becoming the framework within which legislative change would drive behavior change in society. The traditional arguments made by industry - that it created primary wealth which could be used to improve livelihoods - and those made by critics - that profits were made at the expense of environmental damage and social oppression – were viewed not as the territory of perpetual confrontation but instead as the starting points for a challenging process of reconciliation of different societal needs.

The industry leaders decided that the process to debate these issues needed to be independent if it was to be attractive to external organizations and if its conclusions were to be credible. The nine companies launched their project in 1998 as the Global Mining Initiative, and identified three parallel "tracks" for the work program.

- **An Independent Analytical Process**

After a successful scoping document, this was commissioned by the World Business Council for Sustainable Development in April 2000 on behalf of the GMI, and was entitled the Mining Minerals and Sustainable Development (MMSD) project.

- **An Industry Engagement Process**

Through this, the original nine companies sought to bring a larger part of the mining industry into the process and to involve them in a series of actions culminating in an international conference on mining and sustainable development to be held in 2002.

- **Industry Association Management**

The lack of an effective global voice was seen by many as contributing to the poor reputation and weak influence of the industry on its acceptance by the public and by regulators. A single global trade organization with the mission of carrying forward the industry's commitment to making an effective contribution to SD – the International Council on Mining and Metals (ICMM) – was formed in 2001.

The GMI was always intended by its sponsors to be a time-limited initiative, promoting the actions listed above and bringing the results to a wider public debate, first at the conference, Resourcing the Future, which was held in Toronto in May 2002, and also at the World Summit on Sustainable Development (WSSD) in Johannesburg in September of the same year. The initiative achieved its aims and ceased to exist in 2002.

The Mining, Minerals and Sustainable Development (MMSD) Project

The MMSD was an independent multi-stakeholder analysis of the issues surrounding mining and sustainable development, organized at global and regional levels. It was independent in that the research and analysis for the global project was carried out by an independent NGO, the International Institute for Environment and Development (IIED), and by other independent centers of expertise in four regional projects in Australia, Southern Africa, North America and South America.

It was a multi-stakeholder process in that the views of an enormous number of individuals and organisations were sought and incorporated into the scope of work for the project. The global analytical work was grouped into eight "challenges" facing the mining industry, and the regional projects each established their own priority areas for analysis. Working groups were convened by the IIED to generate debate and discussion, contribute and analyze case studies and to attempt to produce consensus on the issues. A total of over 20 international workshops was held in the period from April 2000 to November 2001, involving over 700 people.

The integrity and quality of the work done by MMSD was assured by a strong governance structure. The IIED Work Group, led by a Project Manager, was accountable to a Project Co-ordinator who, in turn, managed the interactions with the two other groups involved in the project. The Sponsors' Group was composed not only of the mining companies who were the initiators of the work, plus the additional companies they had brought into the project, but also charitable foundations, intergovernmental organizations and NGOs. The quality control function was provided by an Assurance Group made up of eminent individuals from many backgrounds, balanced across the main stakeholder groups and regions.

The MMSD project produced its final report – Breaking New Ground – in May 2002 (Mining, Minerals and Sustainable Development Project 2002), and the findings were debated at the GMI conference later that month. The final chapter of the report was called 'An Agenda for Change' and brought together the conclusions and recommendations from the analysis of the eight challenges defined at the outset. Many of the calls for action were directed at the mining industry, where higher standards of environmental and social performance were demanded in order for the operations of mining companies to be greater contributors to sustainable development.

The report also acknowledged some fundamental things about the industry – its products are essential for modern life and the unique properties of many minerals and metals underpin the social and economic development which follows the initial task of poverty alleviation. The view that mining companies make unreasonable profits at the expense of people and the environment was also exposed as a myth – returns have been very low over at least the last 20 years, and investors look elsewhere for high yields. Finally, and most importantly for the analysis of land access issues, mines are located only where there are mineral deposits, and these are distributed very unevenly over the Earth's surface. Mines are not transferable economic opportunities in the sense that many industrial development options are.

Another area in which the MMSD drew out the complexity of the decision-making processes surrounding mineral development was in the area of stakeholder consultation and prior informed consent. There is a marked gradation in the rights of stakeholders across the spectrum from traditional owners whose rights would be affected by proposed new projects to concerned individuals in developed countries who object to some aspect of the development and commercialization of resources. Governments came in for a lot of criticism for not creating the laws and enforcement regimes which would produce consistently better outcomes from mining projects. No-one walked away from the MMSD project with their prejudices intact and without a list of areas for improvement in their actions.

The success of the MMSD process in delivering a fair and balanced analysis of the issues and of the GMI in showing leadership in addressing the complex problems the industry faced can be measured by the remark contributed to the GMI conference by Kofi Annan, UN Secretary-General. He said that the conference had "mobilized an unprecedented coalition for change" (ICMM 2002a).

Biodiversity and Regional Land-Use Planning

What has this got to do with the subject of the ASTM Symposium – Landscape Ecology and Wildlife Habitat Evaluation – at which this paper was presented? Biodiversity was the subject of one of the working groups convened by the MMSD Work Group, and much of the work of this group centred on issues surrounding decisions on access to land. The group held two workshops in 2002 and the progress made inside and outside these was judged by many to have been one of the most successful in the whole project.

At the first workshop the lack of trust between mining companies, biodiversity conservation NGOs, social development organizations, protected area managers, indigenous peoples' groups and intergovernmental organizations was palpable. All formal contributions were essentially defensive of positions already well-known to all others present. The breakout sessions started to unpick this tangle, and a surprising amount of consensus on the issues around land access began to emerge. Perhaps there should not have been such surprise – at a workshop held in Gland, Switzerland in September 2000 on World Heritage and Mining, protected area managers and mining companies found they shared many common principles and ideas for the evaluation of future World Heritage sites (International Council on Metals and the Environment 2001).

For the second workshop several papers were commissioned as thought starters, and these continued the process of articulating a shared vision of the accommodation of conservation objectives and development planning, including mining projects, in land-use plans at a regional scale (Mining. Minerals and Sustainable Development Project 2002b). The rationale for this shared vision follows in the next section, taken from an unpublished paper prepared by the author for the GMI conference. Inevitably, the perspective of this vision comes from someone working in the mining industry but, while the words may not be exactly the same, the views expressed by protected area managers have been remarkably similar. Our viewpoints may differ, but it is clear that we are surveying the same scene.

Mining and Conservation – Shared Vision?

What would a landscape look like if biodiversity conservation objectives were reconciled with development projects at a regional scale? First, there would be core areas in which no development, including mining, takes place. These would be the unique landscape features, the refuges of endangered species, a representative sample of all habitats, the fragile ecosystems and the sacred sites. The rest of the landscape would be managed for multiple uses and multiple objectives.

Development would only be considered under strict conditions where biodiversity and other values are highest, and the cost of meeting those conditions would act as a filter to render marginal projects unviable. Strict conditions might include

- longer baseline survey timeframes
- stricter emissions standards
- smaller footprint for the operation
- higher levels of engineering and other security
- limitations on access and other infrastructure
- requirements to offset unavoidable impacts by investments elsewhere in the region
- bonds to cover clean-up and closure costs identified in technically sound and updated closure plans

Where biodiversity and other values are less the presumption in the planning process would move in favor of development, with gradations in between. Some of the strict conditions listed above would be relaxed to reflect the reduction in risk. Less strict conditions should not imply lower standards of environmental and social performance, but neither should attractive projects be destroyed by requirements to include prohibitive protection and mitigation costs if the risks of significant adverse impacts are low.

This configuration would produce something like the UNESCO Man and Biosphere (MAB) Reserve (UNESCO 2003) concept applied at a regional landscape scale. MAB reserves consist of core areas surrounded by buffer zones and transition zones, such that the potential impacts of human activities on the areas of greatest conservation value are minimised. Its adoption implies several things, not all of which are assured in many cases. It assumes that there are conservation objectives for the biodiversity and ecosystems outside the core areas, so that these can be used to influence development proposals. This concept is in line with the definition of Protected Areas in the UN Convention on Biological Diversity (CBD): a geographically defined area which is designated or regulated and managed to achieve specific conservation objectives (Secretariat of the Convention on Biological Diversity 2001). The ideas set out above are likely to deliver better outcomes for conservation and development only if there is an adequate and accessible base of biodiversity and other data to inform decisions on land use, and if there are sufficient resources and capacity in governments to set up and implement the planning and decision-making processes and if corruption and oppression do not subvert the process.

For the mining industry it implies that operations can be carried out within the range of predicted, designed and permitted environmental and social impacts, when these impacts have been reconciled with benefits in an inclusive process designed to produce equitable outcomes. Past performance in parts of industries such as mining and oil and gas has not given rise to confidence that these outcomes can be delivered consistently. One challenge for industry is to find effective ways of raising standards across the whole spectrum of its components, including junior companies, state-owned companies and artesanal and small-scale mining. Unless companies with high standards of performance are recognized and rewarded, and those with low standards are held accountable and penalized, the incentive to invest in improvement is weakened.

Where conservation objectives are set for areas of land, and used in land-use and management decisions, it will always be necessary to compile and manage information on the designation and status of these areas. A system to perform this function must be international in reach, either as a central database or an effective network of consistent databases held at a national or regional level. Designation of protected areas in the system should be based on high-resolution ecosystem-based information including information on demographics and land-use practices in the area. The system must have consistent criteria for the definition of core zones from which specific types of development are excluded. Outside these zones the process for reconciling conservation objectives with other aims should be backed by consistent evaluation criteria. The system must be actively managed, identifying changes in the planned outcomes of decisions and having the capacity or influence to recover the position as far as possible.

The notion that human activity including economic development must be reconciled with biodiversity conservation and ecosystem management if outcomes are to be successful is not new for protected area managers and the broader conservation community. It forms the basis of the UNESCO MAB system and is recognized in IUCN – World Conservation Union categories V and VI (IUCN – World Conservation Union 1994). It is also not new that these objectives are best reconciled at the scale of landscapes and regions. Much of the original thinking in the recent literature of protected area management has stressed the need for clusters, mosaics, corridors, buffer zones and other significant linkages of land under management for different objectives (McNeely 1995; Stolton and Dudley 1999 Carey et al. 2000).

The difference envisaged by the MMSD biodiversity working group is that mineral development projects can and should be considered as possible ways in which sustainable development is achieved without threatening conservation.

What is Stopping Progress Towards this Vision?

Assuming that the mining industry would like decisions on land access to be made in a more equitable, inclusive and sustainable way, that the conservation movement would like protected areas to be more effective in securing conservation objectives, and that governments and intergovernmental bodies would like conservation goals to be reconciled with development aspirations, what is stopping the pursuit of a largely shared vision?

Lack of trust on all sides has already been identified as the greatest barrier. The transition from the GMI to the ICMM as the leading body representing the mining industry's engagement with sustainable development issues was marked by the Toronto Declaration, a statement made by the CEOs of leading mining companies at the end of the GMI conference in response to the challenges set out by the MMSD report. The ICMM commitment was: *"In partnership with IUCN-The World Conservation Union and others, seek to resolve the questions associated with protected areas and mining"* (ICMM 2002a). A Task Force on Biodiversity and Mining was formed from representatives of member companies and associations, and discussions with the IUCN led to the announcement, at the World Summit on Sustainable Development (WSSD) in Johannesburg, of a relationship between the two organizations (ICMM 2002b).

Work since then has developed a joint terms of reference and a work program for the dialogue between IUCN and ICMM. A landscape approach to the analysis and resolution of mining and biodiversity issues is explicit in the relationship. One of the strategic objectives of the ICMM task force is: *"to contribute to the development and adoption of integrated approaches to land use and access, based on sound science and the principles of sustainable development".*

The program for the dialogue contains work *"undertaken on developing a longer term and broader program of work focused on land use planning, biodiversity conservation and mining"* (IUCN 2002). A specific task for 2003 is *"A discussion paper aimed at developing integrated and transparent approaches to land-use planning, biodiversity conservation and mining, including 'no-go' areas, with due regard to the precautionary principle, participation of local communities, indigenous groups and other key constituencies and the principle of science-based decision making".*

In October 2002, the ICMM Council of chairmen, CEOs and presidents passed a resolution committing the ICMM to: *"work in partnership with IUCN and others to develop integrated and transparent approaches to land-use planning, biodiversity conservation and mining, including 'no-go' areas, based upon the principles of sound science".*

Much of the work in 2003 was directed at achieving significant progress in some areas so that these could be reported at the World Parks Congress (WPC) in September 2003, but the ICMM recognizes that the development of better ways of making inclusive and integrated decisions on land use will take much longer to achieve. The IUCN reviewed how the dialogue is progressing after the WPC, but the ICMM is fully committed to a long process of co-operation with the conservation sector. The Terms of Reference of the IUCN-ICMM dialogue envisage the convening of a widely based consultation group to take forward the issues of mining's presence in the landscape. Organizations such as the UNESCO World Heritage Committee and the World Bank Group will be invited to attend, as well as representatives of development organizations, other parts of the UN family and governments.

There are many initiatives under way to carry forward the full range of conclusions of the MMSD project (Culverwell et.al. 2003). These will require the engagement of a wide spectrum of participants, not least the governments who can create the legal, social and fiscal environment within which change can happen and be encouraged.

Rio Tinto's Experiences

Rio Tinto is a large diverse mining company with over 90 operations spread over more than 20 countries. It produces a wide range of commodities, from aluminum to zircon, and its businesses are grouped in six global product groups – Industrial Minerals, Diamonds, Iron Ore, Energy Minerals, Aluminum and Copper. Exploration and Technology are also organized globally. Standards and policies in the areas of External Affairs and Health, Safety & Environment are developed centrally and their implementation is assured by global programms (Rio Tinto 2001, 2002, 2003).

It has been a leader in the process of change taking place in the mining industry over the past five years, building on internal progress made in the preceding 10 – 20 years. Sir Robert Wilson, as chairman of Rio Tinto, was chairman of the GMI, co-chair of the Sponsors' Group of the MMSD, chairman of the GMI conference, and the first chairman of the ICMM Council.

Opportunities to put into practice new approaches to the planning and evaluation of new mines do not come along very frequently, even in a large group like Rio Tinto. Nevertheless, there have been several examples of where the analysis of alternatives and the establishment of the Licence to Operate have been true to the vision of mining as an activity integrated into the social, economic and environmental landscape.

Flambeau, Wisconsin

One of these, the Flambeau mine in Wisconsin, has already been mentioned. Flambeau was developed and operated by the Kennecott Minerals Corporation, a subsidiary of Rio Tinto since 1989. The ore deposit was discovered in 1968 and consisted of a relatively small tonnage of high-grade copper, gold and silver ore lying close to the surface. The site is adjacent to the Flambeau River in an area of Wisconsin without mining history (FIG. 1). The initial project concept in the early 1970s was established, according to prevailing practice, almost entirely on the basis of technical and economic considerations. Operations would consist of an open pit mine, ore concentrator and tailings dam, with the pit reclaimed after mining as a lake.

FIG. 1 Flambeau mine site before mining (foreground) and Flambeau River (upper background)

FIG. 2 Flambeau mine during operation (1996)

Local community reaction to this was negative in almost every respect, with concern over protection of the Flambeau River and local ecosystem a major issue. This led the company to re-evaluate project options. Permit applications were resubmitted in the mid-1980s, with a revised project concept. Ore concentration would be carried out at a remote existing location, removing the need for a tailings disposal facility and reducing the footprint to 181 acres. The pit would be filled with waste rock after mining and the land returned to a mixture of habitats – grassland, wetland and woodland. Ultimately, some of the site buildings would be maintained as a base for sustainable economic activities after mine closure. Environmental protection commitments were also strengthened, including state-of-the-art water treatment facilities and the use of impermeable liners to prevent contamination.

A Local Agreement and Conditional Use Permit were negotiated with three local government bodies, and included provisions to maximize economic benefits to local businesses and minimize disturbance and risk to water supplies. State permits to operate were granted in 1991 and production lasted from May 1993 to August 1997 (FIG. 2). No lost time injuries were incurred throughout the project's life and there were no environmental incidents of any significance. Site rehabilitation took place throughout 1998 and 1999 and was completed in 2001 (FIG 3). Vegetation monitoring will continue for four years and groundwater monitoring for a period of 40 years (Kennecott Minerals Corporation, 2003).

FIG. 3 Flambeau mine reclaimed after mining (2002)

The retained buildings and rail spur have been leased by a local government agency and business to provide sustainable economic benefits. Recreation trails have been established on the site and are well-used by local people. The local communities received approximately $10 million of funds from taxes paid to the state by the mining company, and were able to obtain matched funding. As a result, 500 jobs were either created or retained in the local county. One of the most significant and lasting signs of the mine's presence is a new library costing $1.3 million, which was paid for by direct contributions from the mine and out of taxes paid by the company.

Throughout the life of the mine over 80% of direct employment was local, and training provide by the mine has raised skill levels in the area so that future employment opportunities are enhanced.

This example illustrates how the initial experience of having the license to operate withheld by local community groups produced an appropriate response from the company. Given the sensitivity of the site and the concerns of local people, different project options were selected despite their higher costs. This is an example of where mining under tighter controls can make it possible for the economic benefit to be realized without unacceptable social and environmental costs. Following mining the land has been returned to an enhanced mixture of scientific, commercial and recreational uses.

Not all project economics will be robust enough able to sustain the sort of radical re-think that happened at Flambeau but that, surely, is the point of responsible mining in sensitive settings.

Diavik, NWT Canada

The Diavik diamond deposit is located under a large tundra lake, Lac de Gras, situated 100 km north of the tree line in the arctic region of the Northwest Territories of Canada (FIG 4). It was discovered in 1994 and is owned by Rio Tinto (60%) and Aber Diamond Corporation (40%). Social and environmental baseline studies were carried out between 1994 and 1997; the environmental assessment was concluded and approved in 1999; permits and licenses were obtained and the mine came into production in 2003 (Diavik Diamond Mines Inc. 2003).

Two distinguishing features of the Diavik project assessment were its regional scope and the breadth and depth of community consultation accompanying it (Diavik Diamond Mines Inc. 1999). Although the project footprint will be small, with less than 0.5% of the area of Lac de Gras being used for mining, the assessment of environmental and social effects was carried out at two scales – local (approx. 30 km x 20 km) and regional (approx. 110 km x 90 km) (FIG 5). Eight aboriginal communities claim traditional land-use ties to the project region, and concerns over possible environmental effects were expressed from as far away as the Arctic Ocean 520 km to the north, where the Coppermine River draining Lac de Gras meets the sea.

To address these concerns and to fully engage with community stakeholders, over 300 meetings were held with communities over the period 1994 –1999. These enabled community concerns to help develop and refine project plans. Another new diamond mine, Ekati, opened in 1998, and the Diavik assessment was careful to consider cumulative effects of both projects.

Tundra ecosystems are typically fragile and show low biological productivity in the extreme weather conditions. Water quality and fishery protection were the main concerns in the aquatic realm. Terrestrially, effects on the migration of caribou have both biological and socio-economic relevance. Consultation, followed by the implementation of appropriate mitigation measures, led to these issues being addressed to the satisfaction of stakeholders. The approval by Canadian federal government agencies in 1999 stated that *"with the implementation of all of the mitigation measures identified in the comprehensive study report, the Diavik Diamonds Project is not likely to cause significant adverse environmental effects* " and *"The ... project is important, not only for the Northwest Territories, but for all of Canada. Northerners stand to realize very significant direct benefits from job creation and business opportunities"* (Canadian Environmental Assessment Agency 1999)."

FIG. 4 Diavik project site in 1999 looking southwest. The project is centred on East Island in Lac De Gras

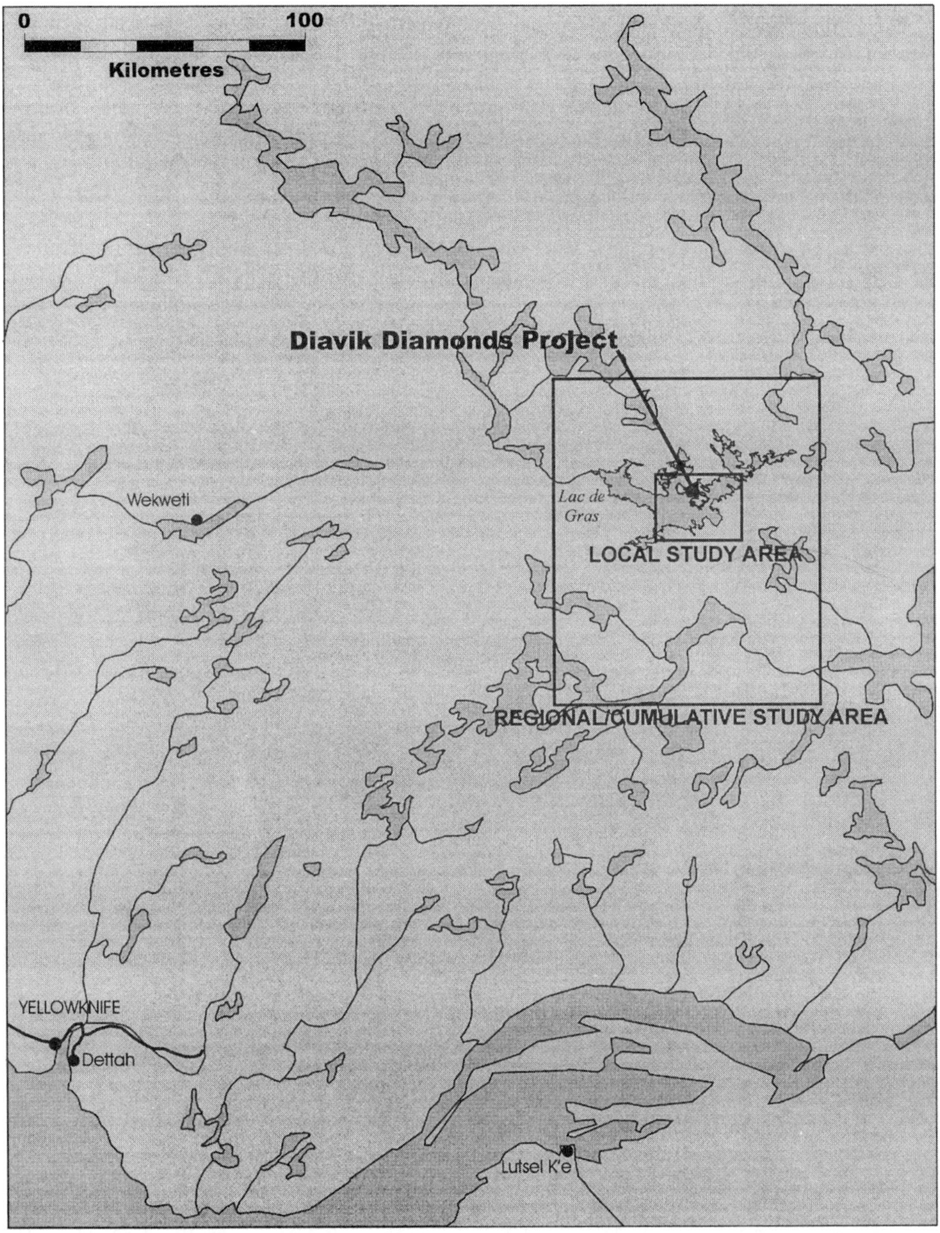

FIG. 5 The wildlife baseline study areas covered both regional and local scales

The commitment to managing social and environmental effects at the regional scale is being carried through to the operating phase of the mine. Diavik has established a socio-economic agreement with the government and aboriginal groups to provide jobs and business opportunities to northern and aboriginal residents and businesses, particularly those from neighbouring communities. The planned distribution of the income from the project has also been agreed and a significant proportion will go to the region and its inhabitants (Ellis 2000). The environmental monitoring program agreed with government and with local communities also covers the collection of physical, chemical and biological data over a broad area.

As it enters the production phase (FIG 6), Diavik has excellent relationships with regulators and with local communities, and has a sound understanding of the steps it will need to take to ensure that outcomes of the mine's presence meet the expectations of all stakeholders.

FIG. 6 Project site in 2002 showing construction of dikes to keep lake water from the mining pits

QMM, Madagascar

Exploration in 1986 discovered potentially economic deposits of heavy mineral (ilmenite and rutile) sands in a fossil dune complex near Fort Dauphin in south east Madagascar (FIG 7). The area has little mining tradition – mica is mined nearby but there are no modern commercial mines – and the area is blighted by extreme poverty. The evaluation of the deposit has been carried out by QIT Madagascar Minerals (QMM), which is 80% owned by Rio Tinto and 20% by the Government of Madagascar.

The mineral deposits underlie a complex littoral forest ecosystem with many important variations in biological diversity, forest structure, endemic and endangered species. For many years these forests have been exploited in an unsustainable way by local people for food, building materials, charcoal and fuelwood. The original forest is present in small remnants and the progressive losses have been documented over the project life so far (FIG 8). Regionally the picture is little better. Charcoal production and other uses of timber are driving deforestation of the foothills behind Fort Dauphin, encroaching on the Andohahela National Park.

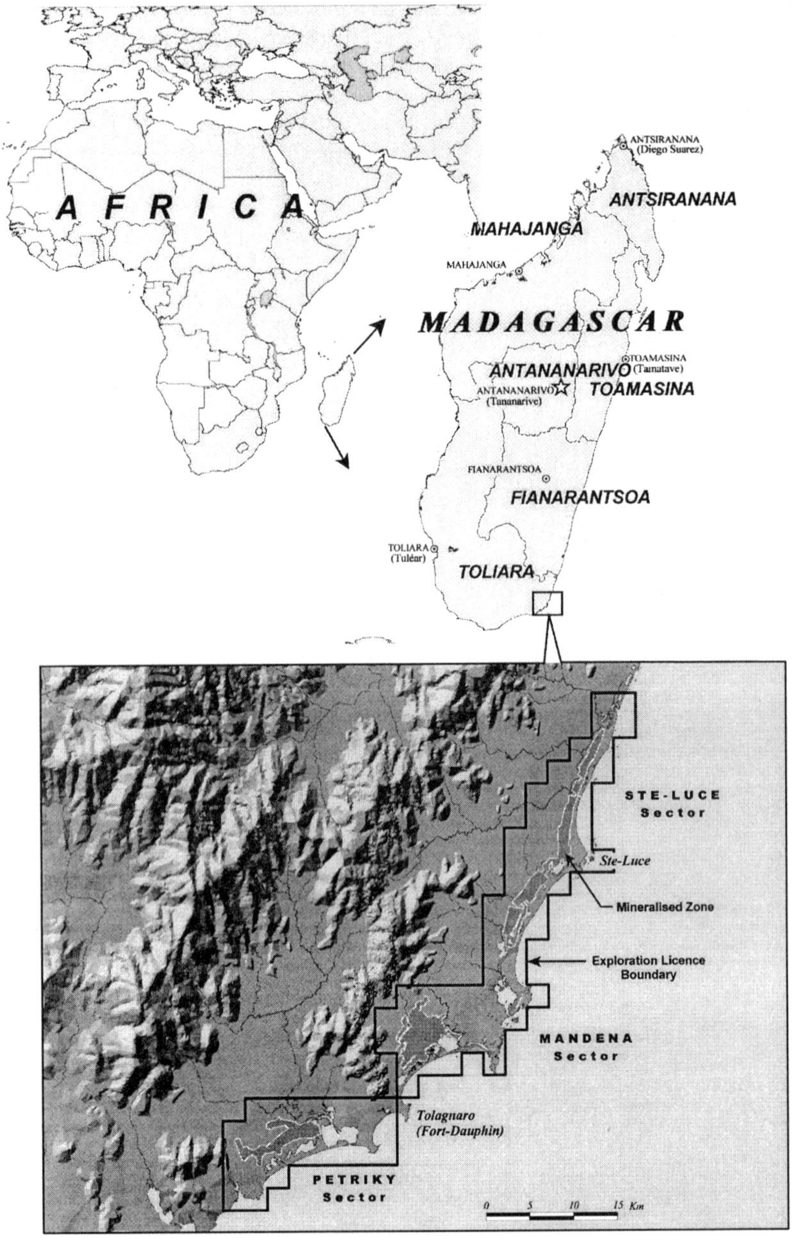

FIG. 7 The location of the deposit

After several years in Madagascar, coupled with advice received both nationally and internationally, it became abundantly clear to QMM that successful resolution of the complex social and environmental issues facing the project and region would require a more intense and continuous effort than traditional project assessment methods. The response was to establish a fully-staffed social and environmental program which has achieved the following things:

- establishment of an Ecological Research Station in 1996 in the first proposed mining area, Mandena, with smaller stations in the other areas;
- basic research and data collection on the social environment
- extensive consultations with villagers to determine their concerns and how to address them;
- basic research on the terrestrial and aquatic ecosystems and interactions between them;
- basic research on the interactions between social and natural environments;
- establishment of conservation zones, with the co-operation of all stakeholders, to protect key ecosystems;
- investigation and testing of restoration techniques for degraded ecosystems;
- development of sustainable plantations for fuelwood and charcoal supply;
- full-scale rehabilitation trials.

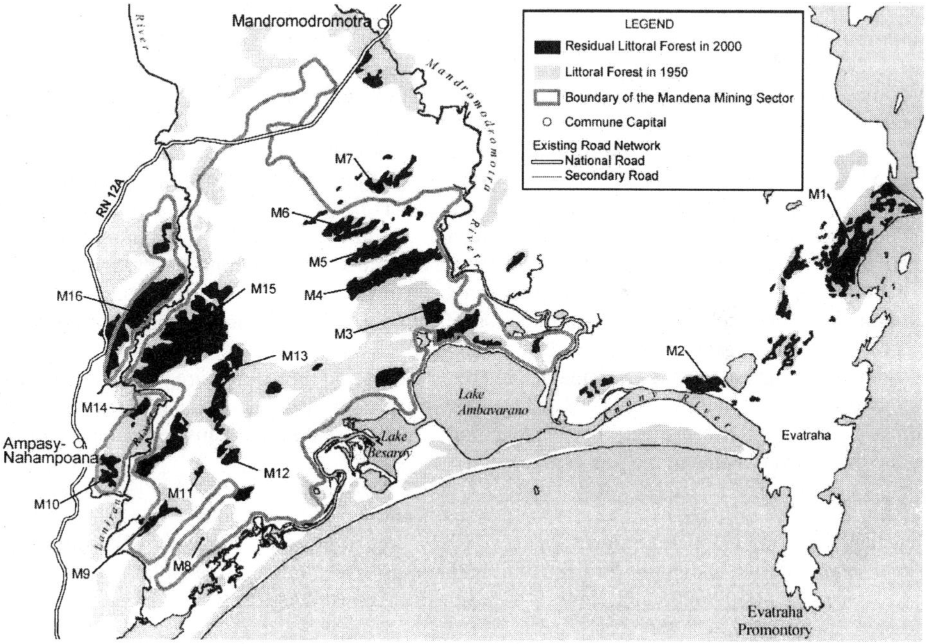

FIG. 8 Evolution of Mandena Forest Cover 1950-2000

The planning and evaluation of this work has been advised by a panel of international as well as national Malagasy experts, who are free to consult with stakeholders of their choice and who advise QMM on all aspects of the project (QIT Madagascar Minerals 2001).

One key aspect of the project evaluation has been the recognition that regional development has to go hand in hand with project development in order to prevent economic distortions in the area. These might provoke migration into the area and lead to adverse secondary effects on local communities and natural resources.

QMM has supported the initiation of a Regional Development Framework by the Government of Madagascar, with the support of donor agencies. It has also begun the process of reaching agreements with government departments and local communes to establish sustainable resource management practices in areas near the proposed operations. QMM is now supporting a similar approach to natural resource management at a wider regional level.

The project has always seen its potential for success and its threats in terms of the landscape in which it is located. Using the economic power of the project to catalyse change in natural resource use patterns is a bold aspiration, and the project is by no means certain to go ahead. If it does, the measure of its success will be the extent to which social and economic development can advance without the current rate of depletion of biological resources continuing.

Conclusion

In the area of landscape management and regional land-use planning, the commitment of leading elements of both conservation and mining sectors to an analysis of multiple uses and multiple benefits at the regional scale is very encouraging, despite the bad experiences of the past. The MMSD project challenged both constituencies to work together to achieve better outcomes for development and conservation. Concrete examples show what can be achieved in practice.

The conservation community must be able to see past the legacy of mining and come to recognize mining development projects as part of the solution to its problems rather than only as a threat to its aims. This will require it to trust that the mining industry is sincere in its commitments and capable of delivering the outcomes it promises. Equally, the mining industry, when planning and implementing management action, must move beyond its own archive of cautionary tales where trust has been squandered. It must earn the trust of the conservation community, so that the exclusion of activities such as mining is not the only approach offered for the protection and conservation of sensitive areas.

Society's progress towards sustainable patterns of development will require all constituencies to find ways of reconciling their needs and expectations with those of others, and trust will be earned by how well we manage these reconciliations. Transparency, inclusiveness and equity are the essential elements of the process.

References

Canadian Environmental Assessment Agency, 1999, Press Release: Minister of Environment sends Diavik Diamonds Project to Regulatory Phase, Online, Available: www.ceaa-acee.gc.ca/0009/0003/0025/rel991103_e.htm, 19 August 2003.

Carey, C., Dudley, N., and Stolton, S., 2000, *Squandering Paradise – the importance and vulnerability of the world's protected areas.* WWF – World Wide Fund for Nature (Formerly World Wildlife Fund) International, Gland, Switzerland.

Culverwell, M., Lee, B., and Koziell, I., 2003, *Towards and Improved Governance Agenda for the Extractive Sector,* Royal Institute for International Affairs, London, UK.

Diavik Diamond Mines Inc., 1999, Diavik Diamonds Project Environmental Assessment Executive Summary, DDMI, Yellowknife, NWT, Canada.

Diavik Diamond Mines Inc., 2003, The Diavik Diamonds Project, Online, Available: http://www.diavik.ca/html/diavik_diamonds_project.html, DDMI, Yellowknife, NWT, Canada, 19 August 2003.

Ellis, R., 2000, *The Diavik Diamonds Project Distribution of the Project Resource Income,* DDMI, Yellowknife, NWT, Canada.

International Council on Metals and the Environment (ICME), 2001, *Proceedings of the Technical Workshop on World Heritage and Mining, Gland, Switzerland, 21-23 September 2000,* ICME, Ottawa, Canada.

International Council on Mining and Metals (ICMM), 2002a, Message to Closing Session of the Global Mining Initiative Conference, Toronto, 15 May 2002, in *ICMM Newsletter,* Vol. 1, No. 4, ICMM, London, UK.

International Council on Mining and Metals (ICMM), 2002b, *ICMM Toronto Declaration,* ICMM, London, UK.

International Council on Mining and Metals (ICMM), 2002c, *News Release – Mining Industry and IUCN – The World Conservation Union Announce Partnership on Mining and Biodiversity,* ICMM, London, UK, 31st August 2002.

IUCN – World Conservation Union Commission on National Parks and Protected Areas and World Conservation Monitoring Centre, 1994, *Guidelines for Protected Area management categories,* IUCN, Gland, Switzerland.

IUCN – World Conservation Union, 2003, Draft Terms of Reference for the dialogue between IUCN/ICMM, Online, Available: http://www.iucn.org/info_and_news/press/miniucnicmmtor.pdf, IUCN, Gland, Switzerland, 19 August 2003.

Kennecott Minerals Corporation (KMC), 2003, Flambeau Mine, Online, Available: http://www.kennecottminerals.com/mines/flambeau.html, KMC, Salt Lake City, Utah, 19 August 2003.

McNeely, J. (Ed.), 1995, *Expanding Partnerships in Conservation*, Island Press Washington, DC for IUCN.

Mining, Minerals and Sustainable Development (MMSD) Project, 2002a, *Breaking New Ground – Mining, Minerals and Sustainable Development*, Earthscan, London, UK for IIED and WBCSD.

Mining, Minerals and Sustainable Development (MMSD) Project, 2002b, Minutes of the 2nd MMSD "Mining and Biodiversity" Workshop, in e-book attachment to: *Breaking New Ground – Mining, Minerals and Sustainable Development*, Earthscan, London, UK for IIED and WBCSD.

QIT Madagascar Minerals (QMM) S.A., 2001, *Ilmenite Project – Social and Environmental Impact Assessment – May 2001,* QMM, Montreal, Canada.

Rio Tinto, 2001, Data Book 2000, Rio Tinto plc, London UK.

Rio Tinto, 2002, *Social and environment review highlights 2002,* Rio Tinto plc, London, UK.

Rio Tinto, 2003, *The way we work*, Rio Tinto plc, London, UK.

Secretariat of the Convention on Biological Diversity, 2001, *Handbook of the Convention on Biological Diversity,* Earthscan, London, UK.

Stolton, S. and Dudley, N. (Eds.) 1999, *Partnerships for Protection – New Strategies for Planning and Management for Protected Areas*, Earthscan, London, UK.

UNESCO Man & Biosphere Programme (MAB), 2003, Online, Available: http://www.unesco.org/mab/wnbr.htm, UNESCO, Paris, France, 19 August 2003.

Session II

Alan R. Johnson,[1] *Craig R. Allen,*[2] *and Kristi A. N. Simpson*[1]

Estimating Functional Connectivity of Wildlife Habitat and Its Relevance to Ecological Risk Assessment

---

REFERENCE: Johnson, A. R., Allen, C. R., and Simpson, K. A. N., **"Estimating Functional Connectivity of Wildlife Habitat and Its Relevance to Ecological Risk Assessment,"** *Landscape Ecology and Wildlife Habitat Evaluation: Critical Information for Ecological Risk Assessment, Land-Use Management Activities, and Biodiversity Enhancement Practices, ASTM STP 1458*, L. A. Kapustka, H. Galbraith, M. Luxon, and G. R. Biddinger, Eds., ASTM International, West Conshohocken, PA, 2004.

ABSTRACT: Habitat fragmentation is a major threat to the viability of wildlife populations and the maintenance of biodiversity. Fragmentation relates to the sub-division of habitat into disjunct patches. Usually coincident with fragmentation *per se* is loss of habitat, a reduction in the size of the remnant patches, and increasing distance between patches. Natural and anthropogenic processes leading to habitat fragmentation occur at many spatial scales, and their impacts on wildlife depend on the scales at which species interact with the landscape. The concept of functional connectivity captures this organism-based view of the relative ease of movement or degree of exchange between physically disjunct habitat patches. Functional connectivity of a given habitat arrangement for a given wildlife species depends on details of the organism's life history and behavioral ecology, but, for broad categories of species, quantities such as home range size and dispersal distance scale allometrically with body mass. These relationships can be incorporated into spatial analyses of functional connectivity, which can be quantified by indices or displayed graphically in maps. We review indices and GIS-based approaches to estimating functional connectivity, presenting examples from the literature and our own work on mammalian distributions. Such analyses can be readily incorporated within an ecological risk framework. Estimates of functional connectivity may be useful in a screening-level assessment of the impact of habitat fragmentation relative to other stressors, and may be crucial in detailed population modeling and viability analysis.

KEYWORDS: fragmentation, functional connectivity, dispersal, metapopulation, population viability analysis

---

[1] Assistant professor and graduate research assistant, respectively, Department of Environmental Toxicology, Clemson University, PO Box 709, Pendleton, SC 29670

[2] Unit leader, USGS South Carolina Cooperative Fish and Wildlife Research Unit, G27 Lehotsky Hall, Clemson University, Clemson, SC 29634

Introduction

Wildlife populations are subject to multiple stressors. Among the leading threats to the viability of wildlife populations are the loss and fragmentation of habitat. Habitat loss can be defined as any process that results in a decrease in the area of available habitat for the species. Habitat fragmentation is any process that increases the partitioning of available habitat into spatially disjunct patches. Usually accompanying habitat fragmentation *per se* is an overall loss of habitat, a decrease in the average patch size and an increase in the average inter-patch distance (Andrén 1994). Many activities, such as conversion of land to agricultural or urban uses, road construction, timber harvesting, wetland drainage, or dam construction contribute to habitat fragmentation. As wildlife habitat is fragmented, there is a concomitant increase in other land cover types in the intervening space between patches, collectively referred to as the matrix.

Habitat fragmentation has multiple effects on wildlife populations, most of them tending to decrease population viability and increase the probability of local extinctions. As the total amount of habitat declines, the carrying capacity or maximal population size decreases. As inter-patch distances increase, local populations within a patch can become increasingly isolated. The subdivision of the population into smaller local populations by itself increases the vulnerability of the local populations to random extinction events due to demographic stochasticity or localized catastrophes (Holsinger 2000). Small local populations may suffer genetic consequences, such as inbreeding depression or erosion of genetic diversity, which threaten long-term persistence (Dudash and Fenster 2000; Sherwin and Moritz 2000). If extinction of a local population does occur, isolated habitat patches are less likely to be recolonized. The scale at which these effects are observed depend upon the inter-patch distances relative to the vagility of the wildlife species under consideration. The ability of an organism to traverse a given distance between habitat patches may also be significantly influenced by the nature of the intervening matrix, such as the presence of movement corridors (Johnson et al. 1992; Beier and Noss 1998).

Wildlife populations are frequently exposed to other stressors in conjunction with habitat fragmentation. For instance, conversion of land for intensive agriculture involves not only substantial loss and fragmentation of original habitats, but potential effects of pesticides and fertilizers, changes in drainage patterns, soil erosion and sedimentation in surface waters, and encroachment of agricultural pests and weeds. Each of these can affect ecological systems at various scales and levels of organization, leading to cumulative effects on wildlife populations (Freemark 1995). The term "chemically induced habitat fragmentation" has been applied to situations in which pesticides or other chemicals disrupt interactions between local populations (Nabhan and Buchmann 1996). Ecological risk assessments need to account for interaction of such multiple stressors.

Habitat fragmentation may result in a disruption of the functional connectivity of the landscape for the species under consideration, where functional connectivity is a measure of the facilitation of ecologically important fluxes (in this case, movement of organisms) across the landscape (Forman 1995; Taylor et al. 1993). Clearly, movement of organisms across a landscape may be influenced by the amount and spatial arrangement of habitat. However, properties of the organism, such as its dispersal ability, behavior at habitat/matrix boundaries, and mortality risk within the matrix, can also influence functional connectivity. It is important to distinguish between structural connectivity, which based on spatial contiguity of habitat within the landscape, and functional connectivity which depends on the organism's interaction with the landscape structure (Forman 1995; Tischendorf and Fahrig 2000). A structurally connected habitat may be functionally fragmented if, for instance, long narrow strips of habitat connecting large patches do not function as actual movement corridors (Simberloff et al. 1992; Beier and Noss 1998). Conversely, a structurally fragmented collection of disjunct habitat patches may be functionally connected for an organism that is able to disperse long distances across the intervening matrix.

Quantifying animal movements in complex landscapes is a daunting task. Measurements in the field require intensive sampling (Tischendorf and Fahrig 2000), and even computer simulations require numerous assumptions about often poorly characterized aspects of animal behavior. However, a reasonable approximation of functional connectivity for most wildlife species may be estimable based on dispersal distances, which can be predicted from allometric scaling relationships with home range or body size (Sutherland et al. 2000; Bowman et al. 2002). Sutherland et al. (2000) report median and maximum natal dispersal distances that each span approximately four orders of magnitude for a suite of avian and mammalian species. This large variation in dispersal ability of organisms is likely the primary determinant of functional fragmentation experienced in a given arrangement of spatially disjunct habitat patches.

In this paper we review the major approaches that have been proposed for analyzing functional connectivity in landscapes. We present representative examples of several of these analytical approaches applied to a range of animal species and landscapes. Finally, we consider the relevance of such analyses to the assessment of risk for wildlife populations, and how such analyses can be incorporated into the overall framework for ecological risk assessment.

Quantifying Functional Connectivity

Graph theory can be applied to represent functional connectivity between habitat patches (Keitt et al. 1997; Urban and Keitt 2001; Bunn et al. 2000). A mathematical graph is composed of point-like "vertices", some of which are joined by lines called "edges" (terminology varies, we follow that of Gross and Yellen 1999). Each vertex has a degree, defined as the number of incident edges (a self-loop, that is an edge that connects a vertex to itself, contributes 2 to the degree). A walk is an ordered sequence of vertices and edges, $(v_0, e_1, v_1, \ldots, v_{n-1}, e_n, v_n)$. A path is a walk in which no edge or vertex is repeated (except possibly the initial and final vertex). A connected graph is one in which every pair of vertices is connected by at least one walk. A component of a graph is a maximal connected subgraph. A connected graph, therefore, has a single component.

In the context of fragmented landscapes, we define a landscape graph G, where each vertex represents a habitat patch, and an edge joins two vertices if and only if organismal movement functionally connects the corresponding patches. Various approaches can be used to define connections between patches. One special case is represented by percolation theory, in which the landscape is viewed as a "lattice," or regular array of grid cells as in a raster-based GIS. Each grid cell is either habitat or matrix, and each habitat cell is connected to the other habitat patches in its local neighborhood. Often the neighborhood is defined as immediately adjacent cells, in which case functional connectivity defaults to spatial contiguity. The most notable feature of percolation-based landscape models it the abrupt emergence of long-range connectivity at a critical threshold (Gardner et al. 1987; With and Crist 1995). Percolation theory has played a substantial role in generating neutral models (sensu Caswell 1976) of spatially heterogeneous landscapes (With and King 1997).

In the more general case, vertices represent habitat patches that can vary in size, shape and spatial arrangement. Edges are placed depending on the ability of the organism to disperse between the corresponding patches. Various methods can be used to define edges. The simplest is a threshold distance, assuming that dispersal is possible up to some fixed distance (e.g., Keitt et al. 1997; Van Langevelde 2000). Alternatively, edges can be defined based on a function that defines the probability of dispersal as a function of distance. Typically, a probability that declines exponentially with distance is assumed, although other functional forms could be applied. Analysis can proceed by producing graphs in which edges are randomly added according to the specified probability (each such graph representing one stochastic realization from an ensemble of possible graphs), or by placing edges indicating all possible connections in one graph, with the edges weighted by the probability. Inter-patch distances can be defined in several ways: Euclidean distance (centroid-to-centroid, or edge-to-edge) or least-cost distance which accounts for the non-uniform nature of the matrix. The choice of distance measure will depend upon the ecological situation and data availability.

Corresponding to the representation as a graph, various matrices can be defined to represent connectivity between patches. The simplest is the adjacency matrix, **A**, for which each element a_{ij} equals the number of edges, if any, joining vertex i to vertex j (Gross and Yellen 1999). If a probability of dispersal is associated with an edge, we can define a matrix **P** in which each element p_{ij} is the probability of dispersal to patch i from patch j. If dispersal probability declines exponentially with distance, $p_{ij} = \exp(-K\, d_{ij})$, where d_{ij} is the inter-patch distance and 1/K is the average dispersal distance. A matrix **C** can be formed, where each element c_{ij} is proportional to the contribution of colonists from patch j to patch i, computed as $p_{ij}A_jO_j$, where A_j is the area of patch j and O_j represents the occupancy of patch j by the species ($O_j = 1$ if the species is present in patch j, $O_j = 0$ otherwise). Finally, Hanski and Ovaskainen (2000) have proposed a landscape matrix **M**, consisting of elements $m_{ij} = p_{ij}A_iA_j$, with properties related to metapopulation persistence.

A number of measures or indices of functional connectivity have been suggested in the ecological literature, or can be adapted from graph theory. Many of these are summarized in Table 1, although this list is not necessarily exhaustive. For the sake of clarity, we have adopted a uniform notation in the equations given in Table 1, which sometimes differs from the notation used elsewhere in the literature. The meaning of each symbol in our notation is summarized in Table 2. In some cases, variant definitions for the same index exist in the literature. Since not all variants and technical details can be included in a tabular summary, the references listed in Table 1 should be consulted regarding the precise definition and application of each index.

TABLE 1 – *Indices related to functional connectivity*

Quantity	**Definition**	**References**
(dis)continuity index	$\ln \sum_i A_i / P_i$	Vogelman 1995 Bélanger and Grenier 2002
correlation length	$\frac{\sum_i A_i R_i}{\sum_i A_i}$ *where* $R_i = \frac{1}{N_{patchcells}} \sum_k \sqrt{(x_k - \langle x \rangle)^2 + (y_k - \langle y \rangle)^2}$	Keitt et al. 1997
patch cohesion	$\left(1 - \frac{\sum_i N_{perimcells}}{\sum_i N_{perimcells}\sqrt{N_{patchcells}}}\right)\left(1 - \frac{1}{\sqrt{N_{mapcells}}}\right)^{-1}$	Schumaker 1996
index of isolation	$\left(\sum_j A_j / d_{ij}^2\right)^{-1}$	Whitcomb et al. 1981
average patch isolation	$\sum_i I_i \Big/ N_{patches}$ *where* $I_i = \frac{d_{NN}}{A_i^c A_{NN}^b}$	Moilanen and Nieminen 2002
mean patch proximity	$\sum_i PX_i \Big/ N_{patches}$ *where* $PX_i = \sum_j A_j / d_{ij}$	Gustafson and Parker 1992
average patch connectivity	$\sum_i C_i \Big/ N_{patches}$ *where* $C_i = \sum_j A_j e^{-\alpha d_{ij}}$	Vos et al. 2001
metapopulation capacity	leading eigenvalue of **M**	Hanski and Ovaskainen 2000 Ovaskainen and Hanski 2001
dispersion	$\sum_i \sum_j s_{ij}$	van Langevelde 2000
vertex-connectivity	minimum number of vertices that must be removed from G to yield a disconnected graph	Gross and Yellen 1999
order (size) of the largest component	number of vertices (edges) in the largest component of G	Palmer 1985
average vertex degree	$\sum_i k_i / N_{vertices} = 2N_{edges} / N_{vertices}$	Gross and Yellen 1999
connectance	proportion of nonzero elements in the adjacency matrix	Gardner and Ashby 1970
characteristic path length	average shortest path length between all pairs of vertices	Watts and Strogatz 1998 Watts 1999

TABLE 2 – *Symbols used in equations for functional connectivity indices*

Symbol	Referent
A_i	area of patch *i*
A_{NN}	area of the nearest neighbor patch
b, c	constants
d_{ij}	distance between patches *i* and *j*
d_{NN}	distance to the nearest neighbor patch
G	landscape graph as described in the text
i, j	subscripts used to refer to individual patches or corresponding vertices
k	subscript used to refer to an individual cell in a raster representation
k_i	degree of the *i*th vertex in a landscape graph
$\mathbf{M}$	landscape matrix as described by Hanski and Ovaskainen 2000
N_{edges}	number of edges in a landscape graph
$N_{mapcells}$	number of cells (all patch + all non-patch cells) in a raster map
$N_{patches}$	number of patches
$N_{patchcells}$	number of cells in a patch (perimeter + interior cells) in a raster representation
$N_{perimcells}$	number of cells on the perimeter of a patch in a raster representation
$N_{vertices}$	number of vertices in a landscape graph
P_i	perimeter of patch *i*
R_i	characteristic radius of patch *i*, also known as radius of gyration
s_{ij}	reciprocal of cumulative edge length of the shortest path from vertex *i* to *j*
x_k	*x* coordinate of the *k*th cell in a raster representation of a patch
$\langle x \rangle$	*x* coordinate of the patch centroid
y_k	*y* coordinate of the *k*th cell in a raster representation of a patch
$\langle y \rangle$	*y* coordinate of the patch centroid
K	migration or dispersal parameter

Each of these indices captures the spatial heterogeneity of a fragmented landscape in a slightly different way. The first three entries in Table 1 (i.e., the continuity or discontinuity index, the correlation length, and patch cohesion) are functions only of the size and shape of habitat fragments, independent of their spatial arrangement. The next four entries (i.e., index of isolation, average patch isolation, mean patch proximity, and average patch connectivity) depend on patch size and inter-patch distances, differing primarily on whether the dependence on distance is inversely linear, quadratic or exponential. Metapopulation capacity is a fragmentation-related quantity with a strong basis in population theory. The landscape matrix on which it is based depends on patch areas and inter-patch distances. Dispersion depends only on inter-patch distances. The remaining entries (i.e., vertex-connectivity, largest component, average vertex degree, connectance, and characteristic path length) explicitly depend only on the topological properties of the landscape graph. However, factors such as patch size and shape, as well as inter-patch distances and properties of the matrix may be implicitly represented, depending on the procedure used to define vertices and edges in the landscape graph.

Examples

One of the first attempts to quantify functional connectivity based on a landscape graph was the work of Keitt et al. (1997). They examined the fragmented coniferous habitat for the Mexican Spotted Owl (*Strix occidentalis lucida*) in the southwestern United States. Since the actual dispersal distance for this subspecies is poorly known, they examined functional connectivity across a range of potential dispersal distances. Using correlation length as an index, they demonstrated a dramatic jump in connectivity as dispersal increased past a threshold value of about 40 km. Further, by individually removing each patch from the habitat map and re-computing the correlation length, they were able to assess the contribution of each patch to overall functional connectivity. As might be expected, large patches tended to contribute more connectivity than small patches. However, a few patches were identified that had a much greater impact on connectivity than would be expected based on size alone. These patches served as stepping stones or bridges between otherwise disjunct networks of habitat patches. This methodology provides a systematic procedure for identifying critical habitat patches that contribute to functional connectivity due to either size or location.

A simple GIS-based approach to the analysis of functional connectivity is provided by our work on improving models of mammalian species distribution for the South Carolina Gap Analysis Program (SC-GAP). One of the primary objectives of SC-GAP is to develop predictive models of vertebrate species distributions based on land cover data derived from remotely-sensed digital imagery (Landsat-TM). As a first approximation, habitat relationship models are developed from existing field data and expert judgement, which associate each species with land cover types that represent suitable habitat. However, the size and spatial arrangement of patches can affect their ability to support a viable population. An isolated patch that is too small to support a minimal viable population size should be unoccupied. Such a patch might well be occupied if it is near a large patch or part of a network of small patches that collectively support a viable population. We applied this logic to all native mammal species in South Carolina, determining a minimum critical area (MCA) using estimates of home range size and minimum viable population size (Allen et al. 2001), and estimates of dispersal distance to determine functional connectivity. Empirical home range and dispersal data were used when available, otherwise they were estimated from data on similar species or from allometric equations.

The eastern harvest mouse (*Reithrodontomys humulis*) provides an example. This species is associated with dry scrub/shrub thicket, grasslands/pasture, cultivated land, marshes, bays and pocosins. In the piedmont region of South Carolina suitable habitat for this species is very patchily distributed. We estimate the minimum critical area necessary to support a viable population to be 20 ha. Many of the patches of potential habitat identified from classified Landsat imagery are considerably smaller than this threshold. To assess the functional connectivity of the landscape, we buffered a distance of 115 m around each habitat patch. This buffer represents the predicted median dispersal distance based on the allometric equation of Sutherland et al. (2000). Figure 1 illustrates the results of incorporating MCA and functional connectivity into habitat mapping.

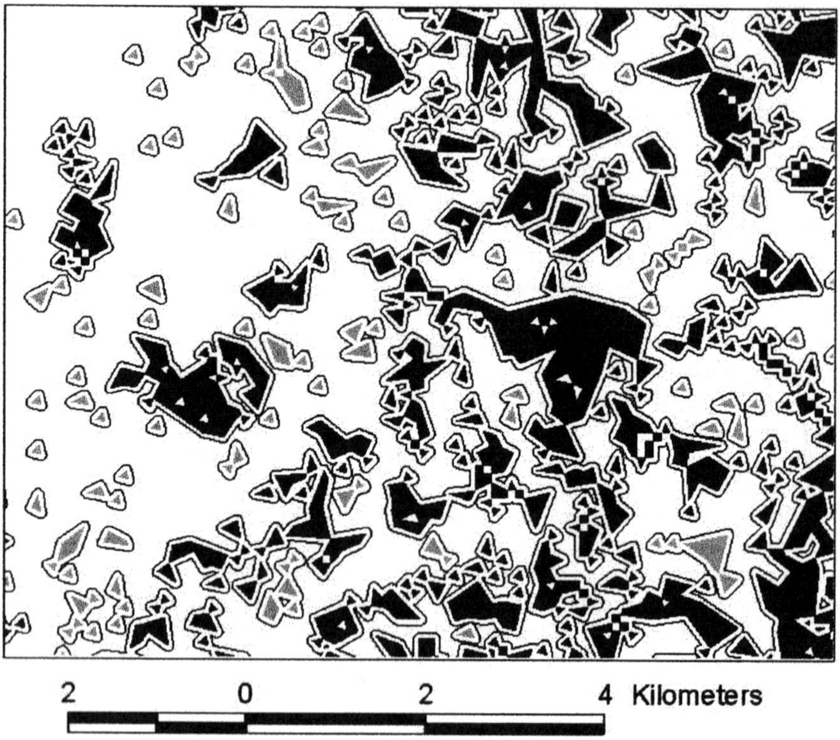

Figure 1 – *Map of potential distribution for the eastern harvest mouse (Reithrodontomys humulis) in a portion of Oconee County, South Carolina. Black = individual patches or functionally connected clusters of patches that meet the minimum critical area (MCA) threshold; Gray = isolated patches or clusters below MCA. Buffers outlining patches illustrate the dispersal distance (115 m) used in assessing functional connectivity.*

Others have applied GIS methods to examine functional connectivity of various landscapes for various species. D'Eon et al. (2002) evaluated the connectivity for a variety of forest-dwelling wildlife species in British Columbia. Functional connectivity of habitat patches in this landscape appeared to be high for most of the species evaluated. The primary exception was the northern flying squirrel (*Glaucomys sabrinus*), the least vagile of the species considered. GIS was also used in the rule-based assessment by Schadt et al. (2002) of habitat suitability and functional connectivity for the Eurasian lynx (*Lynx lynx*). They used detailed land cover data to define habitat patches, and a cost-path analysis was used to evaluate connectivity between patches. Fragmented forest areas separated by less than 1 km were considered suitable for home ranges and functionally belonged to the same patch. Dispersal up to 100 km or more between patches is possible, but depends on the nature of the matrix, and major roads and rivers form effective barriers. Applying such rules to a map of land cover for Germany, the authors found a network of 10 areas large enough to sustain a viable population of lynxes.

Schumaker (1996) examined the functional connectivity of old-growth forests in the Pacific Northwest, potential habitat for the Northern Spotted Owl (*Strix occidentalis caurina*). Although Schumaker considered several indices of functional connectivity, including one of his own derivation, his standard for comparison was how well these indices correlated with simulated dispersal success. In the simulations, the landscape was represented as a grid of habitat and non-habitat (matrix) cells. A collection of dispersers, originating from habitat cells, moved according to a random walk that could be biased in the forward direction by a "linearity parameter." With each step of the walk, there was a constant probability of mortality. Dispersers continued searching until they either died or located an unoccupied territory (habitat cell).

Although Schumaker (1996) modeled dispersal, he did not model population dynamics within habitat patches. An example of a metapopulation model based on measured patch characteristics and population dynamics is the work of Verboom et al. (1991) on the European nuthatch (*Sitta europaea*) in forest fragments of the Dutch agricultural landscape. The nuthatch dynamics fit a simple metapopulation model that included patch size and inter-patch distances as parameters. Van Langevelde (2000) analyzed functional connectivity in European nuthatch population using a graph theoretic approach similar to that of Keitt et al. (1997), and again found that connectivity could change dramatically with relatively small changes in the threshold distance representing dispersal.

We are not aware of any work with vertebrate populations exposed to toxicants that incorporates functional connectivity or a metapopulation approach. However, Sherratt et al. (1999) present an interesting example of odonate (damselfly) populations in ponds in an English agricultural landscape. They developed GIS coverages identifying every pond above 10 m^2 in Cheshire and County Durham (16 383 ponds in total). Field observations quantified the density of various odonate species around selected ponds. It was observed that, in ponds situated in potato fields, odonate observations dropped to zero the day after a fungicide-aphicide application, while other ponds in the study area were unaffected. From mark-recapture data, it was possible to quantify dispersal for the two most abundant species, *Ischnura elegans* and *Coenagrion puella*. Recovery of odonate densities in the pesticide affected ponds depended upon re-colonization from surrounding ponds, and could be predicted based on the spatial arrangement of ponds and the dispersal capabilities of the damselflies.

Incorporation into Ecological Risk Assessment

Habitat fragmentation and the associated loss of functional connectivity can be viewed as one of multiple stressors within an ecological risk assessment framework. Indeed, consideration of habitat alteration is often included, at least as a general stressor category, in ecological risk assessment. Explicit recognition or consideration of changes in functional connectivity resulting from habitat alteration, however, is much less common. We argue that a more systematic treatment of functional connectivity is possible, and likely to lead to more accurate and credible assessments of risks to wildlife.

Incorporating loss of functional connectivity as a stressor in ecological risk assessment requires the derivation of appropriate stressor-response relationships (U.S. EPA 1998). Although ecologists have emphasized the importance of functional connectivity for the health and persistence of wildlife populations, effective quantification of the effects of a given reduction in connectivity has been rare. However, the tools for developing quantitative relationships are now available, and should be used.

Quantification of the stressor can be accomplished by use of the measures and indices set forth in Table 1. It is likely that many of these quantities will be highly correlated with one another, so the risk assessor may be well advised to investigate a suite of measures initially, dropping those that prove redundant. In general, measures that incorporate details such as patch size and inter-patch distance will likely be better predictors of functional connectivity than topological invariants such as connectance, but the appropriate choice may also be dictated by data availability.

Changes in functional connectivity can be expected to elicit responses at several levels of biological organization. At the population level, loss of functional connectivity effectively isolates sub-populations, making them more susceptible to inbreeding depression and localized extinction due to stochastic events. If fragmentation reaches a sufficient magnitude and spatial extent, it can threaten the persistence of the entire regional population. Processes leading to habitat fragmentation seldom, if ever, affect the functional connectivity for only one species. As population dynamics and persistence in the landscape of multiple species are affected, the species composition of the community changes. These changes in species composition can also affect ecological processes, particularly if keystone or ecological dominant species are affected (Boswell et al. 1998).

Quantifying the full range of ecological responses to changes in functional connectivity is a formidable task. However, modeling tools are available that allow at least some of the population level effects to be characterized. Akçakaya and Regan (2002) review existing metapopulation models and their applicability to ecological risk assessment. The simplest approach models patch occupancy, based on an incidence function (e.g., Hanski and Gilpin 1991) or a state transition approach (e.g., Verboom et al. 1991). At an intermediate level of complexity are spatially structured demographic models (e.g., Root 1998). At the most detailed end of the spectrum are individual-based models (e.g., Gaff et al. 2000). Development of appropriate modeling software continues, including efforts to link metapopulation models with models describing changes in habitat suitability and patch geometry over time (Akçakaya 2001).

Spromberg et al. (1998) argue that the functional connectivity of metapopulations in space has several important implications for ecological risk assessment and risk management. Impacts on one local population can cause ecologically significant changes in other populations not directly exposed to a toxicant, an effect they call "action at a distance." Furthermore, because populations in uncontaminated sites can be indirectly affected, these uncontaminated sites cannot serve the traditional role of reference sites. The spatial configuration of the patches is critical to the dynamics of the system and the overall impact of a toxicant. In some cases, if cleanup of the contaminated area is not feasible, the best alternative may be to isolate the contaminated patch, allowing the formerly connected patches to regain more typical population dynamics. At the very least, an awareness of the importance of dispersal and functional connections between fragmented wildlife populations should lead us to look beyond the boundaries of a contaminated site, and to consider the landscape context.

Summary

In this paper, we have examined the concept of functional connectivity for wildlife species in spatially heterogeneous landscapes. When habitat is fragmented into patches, the dynamics of populations depend upon the ability of organisms to move between patches. In many cases, the metapopulation paradigm is appropriate, in which patches host local populations that fluctuate and undergo local extinctions, but colonization of patches by dispersing organisms can allow the population to persist regionally. The ability of the metapopulation to persist, or of a local population to recover from a catastrophic event (e.g., toxicant exposure), depends crucially on the functional connections provided by dispersing individuals. We have considered a variety of quantitative indices, GIS and modeling approaches that can be applied to analyze the functional connectivity of specific landscape configurations. The issue of functional connectivity is relevant to many of the situations faced by ecological risk assessors, and quantitative tools are available that can be incorporated into the risk assessment process. As With (1999) observed, preserving functional connectivity for populations in the landscape may not be sufficient, but it certainly is a necessary component of wildlife management.

Acknowledgments

We are grateful for the comments of J. H. Rodgers, Jr. on an earlier version of this manuscript. This research was supported by grant from the USGS Gap Analysis Program grant to C. R. Allen and A. R. Johnson, and by USDA grant CSREES SC-1700174 to A. R. Johnson. The South Carolina Cooperative Fish and Wildlife Research Unit is jointly supported by a cooperative agreement among the USGS/BRD, the South Carolina Department of Natural Resources, the Wildlife Management Institute, and Clemson University.

References

Akçakaya, H. R., 2001, "Linking Population-Level Risk Assessment with Landscape and Habitat Models," *Science of the Total Environment*, Vol. 274, No. 1-3, pp. 283-291.

Akçakaya, H. R. and Regan H. M., 2002, "Population Models – Metapopulations," *Ecological Modeling in Risk Assessment: Chemical Effects on Populations, Ecosystems, and Landscapes*, R. A. Pastorek, S. M. Bartell, S. Ferson and L. R. Ginzburg, Eds., Lewis Publishers, Boca Raton, pp. 83-95.

Allen, C. R., Pearlstine, L. G., and Kitchens, W. M., 2001, "Modeling Viable Mammal Populations in Gap Analyses," *Biological Conservation*, Vol. 99, No. 2, pp. 135-144.

Andrén, J., 1994, "Effects of Habitat Fragmentation on Birds and Mammals in Landscapes with Different Proportions of Suitable Habitat: A Review," *Oikos*, Vol. 71, No. 3, pp. 355-366.

Beier, P. and Noss, R. F., 1998, "Do Habitat Corridors Provide Connectivity?," *Conservation Biology*, Vol. 12, No. 6, pp. 1241-1252.

Bélanger, L. and Grenier, M., 2002, "Agriculture Intensification and Forest Fragmentation in the St. Lawrence Valley, Québec, Canada," *Landscape Ecology*, Vol. 17, No. 6, pp. 495-507.

Boswell, G. P., Britton, N. F. and Franks, N. R., 1998, "Habitat Fragmentation, Percolation Theory and the Conservation of a Keystone Species," *Proceedings of the Royal Society of London, Series B – Biological Sciences*, Vol. 265, pp. 1921-1925.

Bowman J., Jaeger, A. G., and Fahrig, L., 2002, "Dispersal Distance of Mammals is Proportional to Home Range Size," *Ecology*, Vol. 83, No. 7, pp. 2049–2055.

Bunn, A. G., Urban, D. L., and Keitt, T. H., 2000, "Landscape Connectivity: A Conservation Application of Graph Theory," *Journal of Environmental Management*, Vol. 59, No. 4, pp. 265-278

Caswell, H., 1976, "Community Structure: A Neutral Model Analysis," *Ecological Monographs*, Vol. 46, No. 3, pp. 327-352.

D'Eon, R. G., Glenn, S. M., Parfitt, I. and Fortin, M.-J., 2002, "Landscape Connectivity as a Function of Scale and Organism Vagility in a Real Forested Landscape," *Conservation Ecology*, Vol. 6, No. 2, article 10 [online] URL: http://www.consecol.org/vol6/iss2/art10

Dudash, M. R. and Fenster, C. B., 2000, "Inbreeding and Outbreeding Depression in Fragmented Populations," *Genetics, Demography and Viability of Fragmented Populations*, A. G. Young and G. M. Clarke, Eds., Cambridge University Press, Cambridge, UK, pp. 35-53.

Freemark, K., 1995, "Assessing Effects of Agriculture on Terrestrial Wildlife: Developing a Hierarchical Approach for the US EPA," *Landscape and Urban Planning*, Vol. 31, No.1-3, pp. 99-115.

Forman, R. T. T., 1995, *Land Mosaics: The Ecology of Landscapes and Regions*, Cambridge University Press, Cambridge, UK.

Gaff, H., DeAngelis, D. L., Gross, L. J., Salinas, R. and Shorrosh, M., 2000, "A Dynamic Landscape Model for Fish in the Everglades and its Application to Restoration," *Ecological Modelling*, Vol. 127, No. 1, pp. 33-52.

Gardner, M. R. and Ashby, W. R., 1970, "Connectance of Large Dynamic (Cybernetic) Systems: Critical Values for Stability," *Nature*, Vol. 288, No. 5273, p. 784.

Gardner, R. H., Milne, B. T., Turner, M. G., and O'Neill, R. V., 1987, "Neutral Models for the Analysis of Broad-Scale Landscape Pattern," *Landscape Ecology*, Vol. 1, No. 1, pp. 19-28.

Gross, J. and Yellen, J., 1999, *Graph Theory and Its Applications*, CRC Press, Boca Raton.

Gustafson, E. J. and Parker, G. R., 1992, "Relationships between Landcover Proportion and Indices of Landscape Spatial Pattern," *Landscape Ecology*, Vol. 7, No. 2, pp. 101-110.

Hanski, I. and Ovaskainen, O., 2000, "The Metapopulation Capacity of a Fragmented Landscape," *Nature*, Vol. 404, No. 6779, pp. 755-758.

Holsinger, K. E., 2000, "Demography and Extinction in Small Populations," *Genetics, Demography and Viability of Fragmented Populations*, A. G. Young and G. M. Clarke, Eds., Cambridge University Press, Cambridge, UK, pp. 55-74.

Johnson, A. R., Wiens, J. A., Milne, B. T., and Crist, T. O., 1992, "Animal Movements and Population Dynamics in Heterogeneous Landscapes," *Landscape Ecology*, Vol. 7, No. 1, pp. 63-75.

Keitt, T. H., Urban, D. L., and Milne, B. T., 1997, "Detecting Critical Scales in Fragmented Landscapes," *Conservation Ecology*, Vol. 1, No. 1, article 4 [online] URL http://www.consecol.org/vol1/iss1/art4

Moilanen, A. and Nieminen, M., 2002, "Simple Connectivity Measures in Spatial Ecology," *Ecology*, Vol. 83, No. 4, pp. 1131-1145.

Nabhan, G. P. and Buchmann, S. L., 1996, "Pesticide Disruption of Interactions Between Rare Plants and Their Pollinators: Chemically Induced Habitat Fragmentation in the United States-Mexico Borderlands?," *Comments in Toxicology*, Vol. 5, No. 4-5, pp. 475-486.

Ovaskainen, O. and Hanski, I., 2001, "Spatially Structured Metapopulation Models: Global and Local Assessment of Metapopulation Capacity," *Theoretical Population Biology*, Vol. 60, No. 4, pp. 281-302.

Palmer, E. M., 1985, *Graphical Evolution: An Introduction to the Theory of Random Graphs*, John Wiley and Sons, New York.

Root, K. V., 1998, "Evaluating the Effects of Habitat Quality, Connectivity, and Catastrophes on a Threatened Species," *Ecological Applications*, Vol. 8, No. 3, pp. 854-865.

Schadt, S., Knauer, F., Kaczensky, P., Revilla, E., Wiegand, T. and Trepl, L., 2002, "Rule-Based Assessment of Suitable Habitat and Patch Connectivity for the Eurasian Lynx," *Ecological Applications*, Vol. 12, No. 5, pp. 1469-1483.

Schumaker, N. H., 1996, "Using Landscape Indices to Predict Habitat Connectivity," *Ecology*, Vol. 77, No. 4, pp. 1210-1225.

Sherratt, T. N., Thomas, C. J., Conrad, K. F., Willson, K. H. and Harvey, I. F., 1999, "Landscape Approaches in Ecotoxicology," *Aspects of Applied Biology*, Vol. 53, pp. 227-234.

Sherwin, W. B. and Moritz, C. "Managing and Monitoring Genetic Erosion," *Genetics, Demography and Viability of Fragmented Populations*, A. G. Young and G. M. Clarke, Eds., Cambridge University Press, Cambridge, UK, pp. 9-34.

Simberloff, D., Farr, J. A., Cox, J., and Mehlman, D.W., 1992, "Movement Corridors: Conservation Bargains or Poor Investments?," *Conservation Biology*, Vol. 6, No. 4, pp. 493-504.

Spromberg, J. A., John, B. M. and Landis, W. G., 1998, "Metapopulation Dynamics: Indirect Effects and Multiple Distinct Outcomes in Ecological Risk Assessment," *Environmental Toxicology and Chemistry*, Vol. 17, No. 8, pp. 1640-1649.

Sutherland, G. D., Harestad, A. S., Price, K., and Lertzman, K. P., 2000, "Scaling of Natal Dispersal Distances in Terrestrial Birds and Mammals," *Conservation Ecology* Vol. 4, No. 1, article 16 [online] URL http://consecol.org/vol4/iss1/art16.

Taylor, P.D., Fahrig, L., Henein, K., and Merriam, G., 1993, "Connectivity is a Vital Element of Landscape Structure," *Oikos*, Vol. 79, No. 3, pp. 571-572.

Tischendorf, L. and Fahrig, L., 2000, "On the Usage and Measurement of Landscape Connectivity," *Oikos*, Vol. 90, No. 1, pp.7-19.

U.S. EPA, 1998, "*Guidelines for Ecological Risk Assessment*," USEPA EPA/630/R095/002F. 01 Apr 1998. U.S. Environmental Protection Agency, Risk Assessment Forum, Washington, DC, 175 pp.

Urban, D.L. and Keitt, T.H., 2001, "Landscape Connectivity: A Graph-Theoretic Perspective," *Ecology*, Vol. 82, No. 5, pp. 1205-1218.

Van Langevelde, F., 2000, "Scale of Habitat Connectivity and Colonization in Fragmented Nuthatch Populations," *Ecography*, Vol. 23, No. 5, pp. 614-622.

Verboom, J., Schotman, A., Opdam, P. and Metz, J. A., 1991, "European Nuthatch Metapopulations in a Fragmented Agricultural Landscape," *Oikos*, Vol. 61, No. 2, pp. 149-156.

Vogelman, J.E., 1995, "Assessment of Forest Fragmentation in Southern New England Using Remote Sensing and Geographic Information Systems Technology." *Conservation Biology*, Vol. 9, No. 2, pp. 439-449.

Vos, C. C., Verboom, J., Opdam, P. F. M., and Ter Braak, C. J. F., 2001, "Toward Ecologically Scaled Landscape Indices," *American Naturalist*, Vol. 183, No. 1, pp. 24-41.

Watts, D. J., 1999, *Small Worlds: The Dynamics of Networks between Randomness and Order*, Princeton University Press, Princeton.

Watts, D. J. and Strogatz, S. H., 1998, "Collective Dynamics of 'Small-World' Networks," *Nature*, Vol. 393, No. 6684, pp. 440-442.

With, K. A., 1999, "Is Landscape Connectivity Necessary and Sufficient for Wildlife Management?," *Forest Fragmentation: Wildlife and Management Implications*, J. A. Rocelle, L. A. Lehmann and J. Wisniewski, Eds., Brill, Leiden, pp. 97-115.

With, K. A. and Crist, T. O., 1995, "Critical Thresholds in Species' Responses to Landscape Structure," *Ecology*, Vol. 76, No. 8, pp. 2446-2459.

With, K. A. and King, A. W., 1997, "The Use and Misuse of Neutral Landscape Models in Ecology," *Oikos*, Vol. 79, No. 2, pp. 219-229.

Gerald J. Niemi,[1] *JoAnn M. Hanowski*,[2] *Nick Danz*,[2] *Robert Howe*,[3] *Malcolm Jones*,[2] *James Lind*[2], and *David M. Mladenoff*[4]

Hierarchical Scales in Landscape Responses by Forest Birds

REFERENCE: Niemi, G. J., Hanowski, J. M., Danz, N., Howe, R., Jones, M., Lind, J., and Mladenoff, D., **"Hierarchical Scales in Landscape Responses by Forest Birds,"** *Landscape Ecology and Wildlife Habitat Evaluation: Critical Information for Ecological Risk Assessment, Land-Use Management Activities, and Biodiversity Enhancement Practices, ASTM STP 1458,* L.A. Kapustka, H. Galbraith, M. Luxon, and G.R. Biddinger, Eds., ASTM International, West Conshohocken, PA, 2004.

ABSTRACT: Selection of habitat by birds is manifest at different geographical scales. Most bird communities in forested ecosystems of the northern hemisphere are comprised of migratory species that represent more than 70 % of the species and individuals within a forest patch. Historically from the 1950s to 1970s most studies were focused on the response of forest birds at the patch or forest stand scale. Since the 1980s, field studies have determined both microhabitat needs (e.g., individual trees or species) or the importance of entire landscapes in which populations occur. Advances in computation power, remote sensing, geographic information systems, and multivariate analytical techniques have greatly enhanced our understanding of bird habitat associations at these multiple geographic scales. Based on results for over 50 species, we illustrate the responses of forest birds in the Great Lakes region at three spatial scales: microhabitat, forest patch, and landscape. Management opportunities are easiest to implement at the forest patch scale, but cognizance of natural disturbance regimes, basic life history needs, and landscape context can enhance opportunities for conserving native forest bird assemblages.

KEYWORDS: birds, landscapes, microhabitat, habitat, forests, scale, modeling

[1]Professor and Director, Department of Biology and Natural Resources Research Institute, University of Minnesota Duluth, 5013 Miller Trunk Hwy, MN 55811-1442.

[2]Senior Research Fellow, Junior Scientist, Research Associate, and Assistant Scientist respectively, Natural Resources Research Institute, University of Minnesota Duluth, 5013 Miller Trunk Hwy, MN 55811-1442.

[3]Professor and Director, Department of Natural and Applied Sciences, Cofrin Center for Biodiversity, University of Wisconsin, Green Bay, WI 54311-7001.

[4]Professor, Department of Forest Ecology and Management, University of Wisconsin, 1630 Linden Drive, Madison, WI 53706.

Introduction

During the past 20 years there has been a tremendous surge of information on the importance of scale to plant and animal populations. The basic needs of organisms (e.g., food, cover, and reproductive sites) span scales from square metres to thousands of hectares. Patterns of distribution may reflect these multi-scale needs or they may emerge indirectly from population dynamics and species interactions over many generations (Wiens 1989). As landscapes across the planet have changed with human-induced and natural disturbances, the effects of these changes have increasingly been recognized, especially as our technology such as remote sensing, geographic information systems, and computer capabilities has increased. Over this same time period, the fields of landscape ecology and conservation biology have emerged as critical disciplines to help us understand, assess, manage, and enhance wildlife populations.

Birds as a taxonomic group have been used extensively in the theoretical and applied analysis of landscape ecology and wildlife management. Their characteristic ubiquity across the landscape, their ease of observation relative to other vertebrate groups, and the growing interest in them among the public are reasons for their extensive role in scientific research. In addition, much of the concern about landscape level change and the potential impact on biological diversity was stimulated by early studies of birds. Robbins et al. (1989a) and Terborgh (1989) were among the first scientists to raise concern about the decline of forest birds in the eastern United States, attributed largely to the fragmentation of forest habitats. Similarly, negative population trends were beginning to emerge for a wide variety of forest birds in standard monitoring programs such as the North American Breeding Bird Survey (Robbins et al. 1989b) and counts of migrating birds crossing the Gulf of Mexico from South America to North America (Gauthreaux 1992). At the same time, there has been worldwide attention and concern about loss of tropical rain forests and the concomitant loss in biological diversity. Loss of tropical rain forests which support a large proportion of the world's biological diversity remains a serious global concern, but the fragmentation of forests in North America over the past 200 years also has been very extensive. Forest loss and landscape change continue to be major threats to many forest birds in the United States and throughout the world (Robinson et al. 1995; Norris and Pain 2002).

Today we have vastly improved knowledge on the scales of habitat selection and population dynamics for many birds species (Wiens 1989). Yet, applications of this knowledge in practical management activities is still lacking. The primary goal of this paper is to provide brief examples for responses of birds at different spatial scales. This information can then be incorporated into emerging spatial models of forest landscape change (e.g., He and Mladenoff 1999) to predict local or regional changes in bird communities (Niemi et al. 1998).

Individual bird species appear to select their habitat at different geographical scales. The three most common scales defined by avian ecologists are: 1) microhabitat - square meters, 2) habitat, stand, or patch - 10s of hectares, and 3) landscape - 100s or 1000s of hectares. Our objective was to quantify response of breeding birds in the western Great Lakes region. We used data from an extensive study of forest birds from 1991 to 1996 (www.nrri.umn.edu/mnbirds). Additional details can be found in Hanowski and Niemi

(1995) and Niemi et al. (1998). We summarize variation in responses by species of forest birds at the microhabitat, patch, and landscape scales. We conclude with a brief summary of how this information could be incorporated into landscape and wildlife modeling for the purpose of improving decision-making in land use management for biodiversity. This summary will be from a breeding bird context which is generally the most critical period of time when production and replenishment of the population occurs. In forests of the western Great Lakes, over 70 % of the bird species and individuals are either short-distance or long-distance migrants and, hence, they spend approximately 3 to 6 months within these forest systems.

Microhabitat Scale

During the breeding season birds need adequate food resources, suitable nest sites, and appropriate cover for themselves or their young. Bird species vary considerably in their breadth of response to food or needs for nest sites or cover. For example, species such as the bay-breasted and Cape May warbler follow spruce budworm (*Choristoneura fumiferana*) outbreaks during the breeding season. Similarly, black-billed and yellow-billed cuckoos are specialists during the breeding season on forest tent caterpillars (*Malacosoma americanum)*. These species have highly specialized diets and select breeding habitat largely according to the abundance of specific resources. However, for most bird species we know little about the specificity of their food habits, perhaps because they are, in fact, rather generalized in their food preferences. Many foliage gleaners and ground feeders, for example, appear to be opportunistic in their consumption of invertebrate food resources, although the substrates or behaviors used during feeding are characteristic of different species.

Suitable nest sites are a critical microhabitat component for most breeding birds in forest landscapes (Connor et al. 1976). These microhabitats vary by species, but may include nest cavities, platforms, or tree species that have appropriate physiognomies capable of supporting specific nest structures. Cavity nesting species are particularly prominent in forest landscapes and can be subdivided into primary or secondary cavity nesters. Primary cavity nesting species (e.g., woodpeckers) excavate their own holes, while secondary cavity nesting species [e.g., many owl species, black-capped chickadee (*Parus atricapillus*), and nuthatches] are dependent on the primary cavity nesters for hole excavation. To examine the importance of forest cover type or patch-scale features for forest birds, we tested the strength of habitat associations for 89 bird species in the western Great Lakes national forests (Chequamegon, Chippewa, and Superior). We used an analysis of variance test to evaluate the response of species to patch size habitat. In these tests, the magnitude of the F-value indicated how important this scale of habitat was to the various bird species. Larger values indicate stronger selection for specific patch types (Table 1).

We found that all four of the primary cavity excavators [Northern flicker (*Colaptes auratus*), pileated woodpecker (*Dryocopus pileatus*), hairy woodpecker (*Picoides villosus*), and downy woodpecker (*Picoides pubescens*)] were weak responders to patch-scale habitat. More common species in western Great Lakes forests such as red-eyed vireo (*Vireo olivaceous*), ovenbird (*Seiurus aurocapillus*), and pine warbler (*Dendroica *pinus*) (a pine habitat specialist) exhibited strong selection for specific patch-scale habitat. Among the three most common secondary cavity nesters, the great crested flycatcher (*Myiarchus crinitus*) and white-breasted nuthatch (*Sitta carolinensis*) also did not show strong patch-scale preferences, but the red-breasted nuthatch (*Sitta canadensis*) showed a relatively strong patch-scale preference. The latter association is probably due to strong affinities for conifer-dominated forests. The lack of strong habitat associations for woodpecker species and secondary cavity nesters is likely due to strong microhabitat selection for suitable cavity trees within patches. Hence, habitat analysis at the patch (forest cover type) scale fails to measure the critical features of habitat selection for these species.

TABLE 1–*Rank in F-values (highest to lowest) among 11 bird species and F-values from an analysis of variance comparing the strength of differences in habitat use in three national forests (NF) in the western Great Lakes region. Only the 11 selected examples for 89 species analyzed in all three forests from 1991 to 1996 are shown.*

Species	Rank	Mean	Chequamegon NF	Chippewa NF	Superior NF
Red-eyed Vireo	7	16.7	13.6	30.5	6.1
Ovenbird	8	16.1	14.1	26.2	8.0
Pine Warbler	9	15.6	12.8	30.5	3.6
Red-breasted Nuthatch	34	4.7	7.1	3.9	3.0
Great Crested Flycatcher	62	2.6	3.8	2.5	1.4
Black-capped Chickadee	66	2.2	2.9	2.4	1.4
Northern Flicker	71	2.1	0.9	3.7	1.7
White-breasted Nuthatch	78	1.5	2.4	1.1	1.0
Pileated Woodpecker	80	1.4	2.4	1.4	0.5
Hairy Woodpecker	81	1.4	2.3	1.2	0.7
Downy Woodpecker	89	1.0	1.3	0.9	0.8

Management for primary cavity nesters needs to consider specific tree species and conditions that are most suitable for cavity excavation: silvicultural techniques that maintain these attributes will be appropriate prescriptions for maintaining populations of these birds (Connor et al. 1976). It is beyond the scope here to present these in detail, but a rich literature is available on various microhabitat requirements on a species-specific basis (e.g., North American Bird Series). Besides managing for appropriate tree species, obvious examples of microhabitat management include providing suitable nest sites such as platforms for ospreys (*Pandion haliaetus*) or houses for secondary cavity nesters. In forested landscapes of North America, nest boxes are not commonly used. Locally, nest boxes are set out for use by waterfowl (e.g., wood duck, *Aix sponsa*), owls, and for some passerines, but these are exceptions. In Europe, especially in the United Kingdom, Netherlands, and Finland, nest boxes are widely used to enhance breeding productivity for tits (e.g., *Parus spp.*) and owls. These programs are largely carried out by volunteers and are necessary due to the lack of natural cavities because of intensive forest management. McKenney (1994) has shown that investment in such programs is not economically effective relative to providing for natural cavities (e.g., old trees) through appropriate forest management.

Management to enhance microhabitat for food or cover resources has been applied in limited ways. Certainly management programs for ruffed grouse (*Bonasa umbellus*) by manipulation of forest cutting patterns increases both food and cover resources [e.g., aspen (*Populus spp.*) management]. Most recently, ecosystem management activities have promoted the maintenance of natural features of the forest landscape and thereby maintain the appropriate food, nesting, and cover resources required by native wildlife species (Niemi et al. 1998). These activities require an understanding of the range of natural microhabitat variation within the forest system of interest and involve attempts to mimic or create these conditions via management. The ability to mimic natural disturbance regimes or the range of natural variation has been a hotly debated topic and the management possibilities are only beginning to be explored.

Patch or Habitat Scale

The patch or habitat scale generally consists of small patches of several hectares to large patches of over 100 hectares. Patches usually consist of homogenous units of forest vegetation [e.g., northern hardwood, regenerating, aspen-birch (*Betula spp.)*, lowland white cedar (*Thuja occidentalis*)] that are defined or classified using a variety of criteria depending on the agency or management authority using the data. Currently and historically, this is the most common scale considered in bird studies of forests and for management of wildlife. A plethora of studies have been published over the past 50 + years on habitat selection by forest birds (Wiens 1989). Because this is one of the basic units of habitat selection, it is important to understand the breadth of habitats selected by individual bird species.

Three species with contrasting selection of forest patches within the Chequamegon NF, Wisconsin had significant ($P < 0.05$) habitat associations based on their frequency of occurrence within one or more of the eleven patch types. The hairy woodpecker shows the weakest habitat association. It is most frequently found in hemlock (*Tsuga canadensis*) forest but not more frequently than in pine (*Pinus spp.*), maple (*Acer spp.*), or in pole-sized lowland conifer forest patches. The red-eyed vireo is most common in deciduous forest types such as red oak (*Quercus rubra*), maple, and aspen-fir (*Abies balsamea*), but also in hemlock forest. The hemlock forest type in the Chequamegon NF region is often mixed with other deciduous tree species [e.g., sugar maple (*Acer saccharum*)]. The red-eyed vireo is significantly less frequent in other conifer-dominated forests (e.g., pine and lowland conifer) and open habitats. In contrast, the red-breasted nuthatch is primarily associated with coniferous trees which is reflected in its selection of forest patches. The species most frequently occurred in saw-sized pine and pole-sized lowland conifer. It is significantly more abundant in these patches than in deciduous-forest types such as red oak, maple, and ash (*Fraxinus spp.*) -elm (*Ulmus spp.*).

Breeding bird species vary considerably in their degree of affinity with different forest patch types (Tables 1 and 2). Some species such as woodpeckers

TABLE 2–*Mean frequencies (proportion of stands species was observed) and P-values for bird species with significant (P<0.05) differences in habitat use based on analysis of variance among 11 patch types (n in parentheses) in the Chequamegon National Forest, Wisconsin, 1992-1996. Letters that are different indicate significant differences in frequency among patch types based on Tukey's multiple comparison test.*

Patch type	Hairy Woodpecker	Red-breasted Nuthatch	Red-eyed Vireo
Upland Brush (8)	0.00 b	0.00 c	0.05 f
Red Oak (7)	0.00 b	0.06 bc	0.83 abc
Maple Pole (24)	0.04 ab	0.01 bc	0.86 ab
Maple Saw (11)	0.03 ab	0.03 bc	0.90 a
Pine Pole (11)	0.03 ab	0.12 abc	0.53 cdef
Pine Saw (7)	0.01 ab	0.19 a	0.53 edf
Hemlock (8)	0.12 a	0.15 a	0.79 abcd
Aspen-Fir (12)	0.10 b	0.14 ab	0.72 abcd
Lowland Conifer Pole (9)	0.04 ab	0.07 abc	0.18 ef
Lowland Conifer Saw (4)	0.00 b	0.18 a	0.55 bcdef
Ash-Elm (6)	0.00 ab	0.05 abc	0.56 bcde
Overall P-value	0.02	0.00	0.00

may be responding to very broad cover types but their selection of specific patches may depend on the availability of specific tree species and condition. Species with less-specific nesting requirements such as the red-eyed vireo may select habitat primarily based on food, nesting, or cover requirements provided by deciduous trees or shrubs. The red-eyed vireo nests primarily in deciduous trees or shrubs that are less common in conifer-dominated forests.

Landscape Scale

During the past 15 years the greatest advance in understanding responses by birds to forest habitats has been at the landscape scale. The primary factors associated with these advances are the theoretical and applied development of metapopulation dynamics (e.g., source and sink habitat relationships, Pulliam 1988) and technology; most specifically high speed computers capable of storing vast amounts of information, geographic information systems, geographic positioning systems, and multivariate statistical techniques (i.e., logistic regression and classification and regression trees). Metapopulation theory has laid the foundation to view the entire population as a network of subpopulations; each interacting to varying degrees. Some subpopulations may be highly productive (e.g., birth substantially exceeds death) for years and serve as sources to other subpopulations, while other subpopulations are sinks (death exceeds birth) and are dependent on immigration from other subpopulations. Populations in each source and sink area can vary considerably in productivity or in numbers over time.

Technology has provided an increase of analytical capabilities in recent years. Computer processing time and storage capabilities that were only recently undertaken with mainframe systems are now routinely available on personal computers. Remote sensing allows the characterization of vast landscape regions at high resolution (e.g., 30 square meters or less) (Wolter et al. 1995; Wolter and White 2002). This information coupled with precise locations on the ground by hand-held geopositioning systems allow rapid analyses of bird survey data within a landscape context (Crozier and Niemi 2003).

To explore bird response at the landscape scale, we used logistic regression analysis at three scales around forest patches within the Chequamegon NF. Percentages of 35 cover type patches plus several summary landscape characteristics (e.g., cover type diversity index and number of cover types) within 200 meter, 500 meter, and 1000 meter radii of 94 forest patches (stands) were used as independent variables to predict the presence of 32 breeding bird species. Presence here was defined as at least one observation of the species during two 10-minute point counts per stand completed in June annually from 1992 to 1996, (Howe et al. 1998; Lind et al. 2003) (Table 3). Overall, logistic regressions for most of 32 species and for each of the buffers were significant. Adjusted R^2 values for significant logistic regressions ranged from 0.06 to 0.65 and similar results were obtained for each buffer. The 500 meter buffer tended to have slightly higher R^2 values than the 200 and 1000 meter buffers.

Response of breeding birds and the predictability of their occupancy of a forest stand varied considerably among species and among different buffers (Table 3). The hermit thrush had relatively low values of adjusted R^2 for each buffer and was best predicted using explanatory variables from the 200 meter buffer, while the pine warbler had the highest adjusted R^2 values and was most predictable using data from the 500 meter buffer. Two of the most common species of these forests, red-eyed vireo and ovenbird, also had relatively high R^2 adjusted values at each buffer. The red-breasted nuthatch was intermediate in predictability among all three buffers. We were unable to determine any common life history characteristics associated with response by the 32 breeding species within this forest at different scales.

TABLE 3–*Mean adjusted R^2 from logistic regression analysis using percentage of patch cover types within 200, 500, and 1000 meters of a surveyed forest patch to predict species occurrence in forest patches (n=94) from 1992-1996 in the Chequamegon National Forest, Wisconsin.*

Species	200 m	500 m	1000 m
Overall (n=32 species)	0.31	0.36	0.31
Minimum	0.06	0.10	0.09
Maximum	0.55	0.65	0.63
Red-breasted Nuthatch	0.25	0.38	0.36
Pine Warbler	0.51	0.65	0.63
Hermit Thrush	0.19	0.10	0.09
Red-eyed Vireo	0.52	0.49	0.57
Ovenbird	0.55	0.55	0.48

Many papers have shown that the landscape context of a forest patch influences the presence of a species within that patch. For example, if a patch is surrounded by suitable habitat for that species, then it is more likely that a species will be found in that patch or be more abundant within that patch compared with a patch surrounded by unsuitable habitat (Hanowski et al. 1997; Pearson and Niemi 2000; Saab 1999). From a metapopulation perspective, if there is a large proportion of suitable habitat in a landscape, and species productivity is relatively high within these patches, then there is a reduced probability for subpopulations to go extinct and colonization of suitable patches within the landscape increases. Hence, the landscape context can have a substantial influence on both suitability and colonization potential for any forest patch independent of its suitability from a microhabitat or patch perspective.

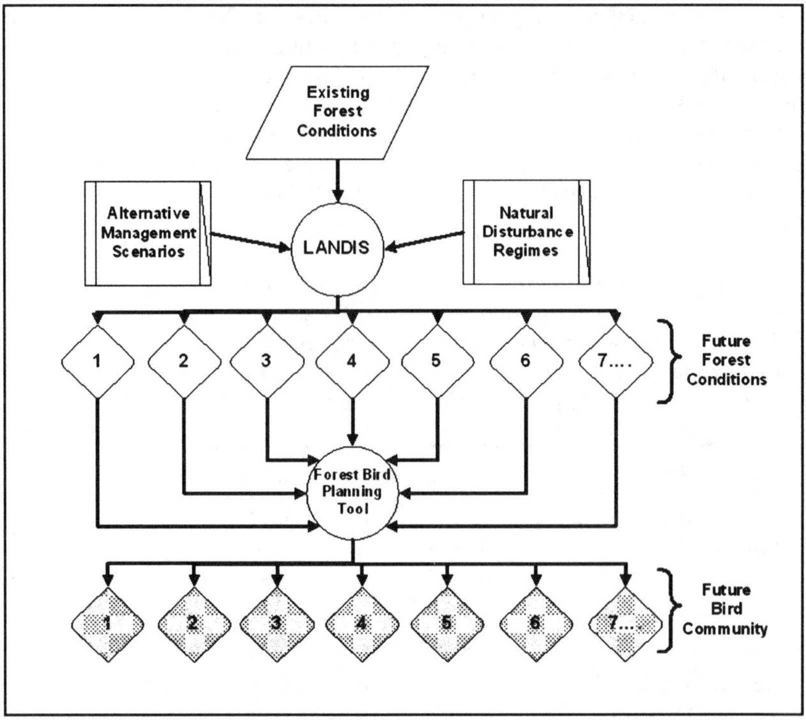

FIGURE 1. *Conceptual diagram linking forest conditions with a landscape disturbance and successional model (LANDIS) and a bird model planning tool to predict future change in bird community biodiversity.*

Scale and Management Integration

The preceding examples illustrate the wide variety of responses of birds to different spatial scales of the environment. To incorporate these scales into management and to maintain bird biodiversity a variety of models that reflect these scales need to be included. We have initiated the development of a bird planning tool that is being integrated with a landscape and disturbance model (LANDIS, Mladenoff et al. 1996, He et al. 1998, He and Mladenoff 1999) (Figure 1). The intent of this effort is to couple state-of-the-art knowledge of forest process dynamics to predict future forest conditions. Predicted forest conditions can then be used to determine how forest birds would respond to a variety of forest management scenarios. Forest landscape disturbances in the western Great Lakes region are in two forms: natural and human-induced changes. Natural disturbances primarily consist of forest fire, wind, and insect outbreaks many of which are partially stochastic. Human-induced changes include logging and urban/industrial expansion. The latter disturbances can be specified in the model and placed within the context of the probability of natural disturbance regimes. Hence, "what if" scenarios can be addressed in the model as well as the consequences of management actions. Conversely, scenarios of management can be explored that are likely to enhance microhabitat, patch, or landscape conditions for specific species (e.g., endangered or threatened).

The conceptual approach to this integration of microhabitat, patch, and landscape scale information on forest birds is initiated by the incorporation of existing forest conditions into a geographic information system (GIS) (Figure 1). Most regions have information on existing forest conditions in the form of patch type, patch age, patch condition or quality (e.g., well-stocked), and the spatial orientation of the patch including size and shape. These data may be modified by additional information that are relevant to selected species of birds such as presence of understory vegetation, availability of dead trees, and coarse woody debris. Because the patch scale information is available in a GIS, a variety of programs are available to calculate the landscape context of the patch. These data could include the proportion of patch types in the landscape surrounding a patch as well as landscape metrics such as edge density, mean patch area, and patch diversity.

These data serve as input to LANDIS which incorporates natural disturbance regimes such as wind and forest fire and specific management scenarios such as silvicultural prescriptions into the modeling of future forest conditions (Mladenoff et al. 1996, He et al. 1998, He and Mladenoff 1999). LANDIS works best at the scale of relatively large landscapes (e.g., 1000s of hectares) and projects future forest conditions at several time intervals (e.g., 10s or 100s of years). The forest bird planning tool subsequently uses these data on patch, microhabitat modifiers, and landscape context as input variables in species-specific prediction models. Forest bird responses are predicted for each of the specified future forest conditions along with estimates of uncertainty derived from the prediction models. At the current time, we have preliminary models developed for over 50 forest bird species that occur in the northern Minnesota and Wisconsin region. Similar analyses and perspectives have been presented previously such as by Scott et al. (2002) and Crozier and Niemi (2003).

Conclusions

Forest birds use a variety of spatial clues to select habitats for nesting, foraging, and cover - their basic requirements for survival. Because of their relatively small size (e.g., most are < 50 g) and their difficulty to observe, the specific habitat requirements of most species are only generally known. Currently we know most about the patch types used by forest birds, but our understanding of landscape context of the patch and the importance of metapopulation distribution of these species is increasing. From empirical observations, we are also aware that certain microhabitat features are essential for the survival of many species. As technology improves, the availability of small transmitters for tracking individual movements of small-bodied species will aid in the identification of these habitat features. A continuing challenge, however, will be a means to manage for these critical microhabitat features in a cost-effective manner. The combination of computer capabilities, geographic information systems, remote sensing technology, and conceptual management models will continue to provide natural resource managers with powerful ways to ask "what if" questions and improve management options to maintain biodiversity values as well as provide the necessary resources demanded by society.

Acknowledgments

Funding for this project, the Minnesota Forest Bird Diversity Initiative, was appropriated by the Minnesota State Legislature from the Environmental Trust Fund as recommended by the Legislative Commission for Minnesota Resources. Partial funding from the US Environmental Protection Agency through cooperative agreement R-82867501 to the University of Minnesota. Additional funding was provided by the Minnesota Department of Natural Resources and the USDA Forest Service, Chequamegon, Chippewa, and Superior National Forests and North Central Research Station. This is contribution number 336 from the Center for Water and the Environment of the Natural Resources Research Institute, University of Minnesota Duluth.

References

Connor, R. N., Miller Jr., O. K., and Adkisson, C. S., 1976, "Woodpecker Dependence on Trees Infected by Fungal Heart Rots," *Wilson Bulletin*, Vol. 88, pp. 575-581.

Crozier, G. E. and Niemi, G. J., 2003, "Using Patch and Landscape Variables to Model Bird Abundance in a Naturally Heterogeneous Landscape," *Canadian Journal of Zoology,* Vol. 81, 441-452.

Gauthreaux, S. A., 1992, "The Use of Weather Radar to Monitor Long-term Patterns of Trans-gulf Migration in Spring," *Ecology and Conservation of Neotropical Migrant Landbirds,* Smithsonian Institution Press, Washington DC, pp. 96-100.

Hanowski, J. M., Niemi, G. J., and Christian, D. C., 1997, "Influence of Within-plantation Heterogeneity and Surrounding Landscape Composition on Avian Communities in Hybrid Poplar Plantations," *Conservation Biology,* Vol. 11, pp. 936-944.

Hanowski, J. M., and Niemi, G. J., 1995, "Experimental Design Considerations for Establishing an Off-road, Habitat Specific Bird Monitoring Program Using Point Counts," *Monitoring Bird Populations by Point Counts,* General Technical Report PSW-GTR-149, Pacific Southwest Research Station USDA Forest Service, Albany, CA, pp. 145-150.

He, H. S., and Mladenoff, D. J., 1999, "Spatially Explicit and Stochastic Simulation of Forest-landscape Fire Disturbance and Succession," *Ecology*, Vol. 80, No. 1, pp. 81-99.

He, H. S., Mladenoff, D. J., Radeloff, V. C., and Crow, T. R., 1998, "Integration of Gis Data and Classified Satellite Imagery for Regional Forest Assessment," *Ecological Applications*, Vol. 8, No. 4, pp. 1072-1083.

Howe, R., Niemi, G. J., Lewis, S. J., and Welsh, D. A., 1998, "A Standard Method for Monitoring Bird Populations in the Great Lakes Region," *Passenger Pigeon*, Vol. 59, pp. 183-194.

Lind, J., Danz, N., Jones, M. T., Hanowski, J. M., and Niemi, G. J., 2003, "Forest Bird Population Trends on the Chequamegon National Forest (1992-2002)," *Passenger Pigeon,* submitted.

McKenney, D. W., and D. B. Lindenmayer, 1994, "An Economic Assessment of a Nest-box Strategy for the Conservation of an Endangered Species," *Canadian Journal of Forest Resources,* Vol. 24, pp. 2012-2019.

Mladenoff, D. J., Host, G. E., Boder, J., and Crow, T. R., 1996, "LANDIS: a Spatial Model of Forest Landscape Disturbance, Succession, and Management," *GIS and Environmental Modeling: Progress and Research Issues.* GIS World Books, Fort Collins, CO, pp. 175-180.

Niemi, G. J., Hanowski, J., Helle, P., Howe, R., Monkkonen, M., Venier, L., Welsh, D., 1998, "Ecological Sustainability of Birds in Boreal Forests," *Conservation Ecology* [on line], Vol., 2, No. 17., Available at http://www.consecol.org/vol2/iss2/art17.

Norris, K., and Pain D.J., 2002, *Conserving Bird Biodiversity, General Principles and Their Application,* Conservation Biology 7, Cambridge University Press, Cambridge, UK.

Pearson, C., and Niemi, G. J., 2000, "Effects of Within-stand Habitat and Landscape Patterns on Avian Distribution and Abundance in Northern Minnesota," *Disturbance in Boreal Forest Ecosystems: Human Impacts and Natural Processes,* USDA North Central Research Station, GTR ND-209.

Pulliam, H. R., 1988, "Sources, Sinks, and Population Regulation, *American Naturalist*, Vol. 132, pp. 652-661.

Robbins, C. S., Dawson, D. K., and Dowell, B. A., 1989a, "Habitat Area Requirements of Breeding Forest Birds of the Middle Atlantic States," *Wildlife Monographs,* No. 103.

Robbins, C. S., Sauer, J. R., Greenberg, R. S., and Droege, S., 1989b, "Population Declines in North American Birds That Migrate to the Neotropics," *Proceedings of the National Academy of Science,* Vol. 86, No. 19, pp. 7658-7662.

Robinson, S. K., Thompson III, F. R., Donovan, T. M., Whitehead, D. R., and Faaborg, J., 1995, "Regional Forest Fragmentation and the Nesting Success of Migratory Birds," *Science*, Vol. 267, pp. 1987-1990.

Saab, V., 1999, "Importance of Spatial Scale to Habitat Use by Breeding Birds in Riparian Forests: A Hierarchical Analysis," *Ecological Applications*, Vol. 9, pp. 135-151.

Scott, J.M., Heglund, P.J., Morrison, M.L., Haufler, J.B., Raphael, M.G., Wall, W.A., and Samson, F.B. (Editors). 2002. *Predicting species occurrences; issues of accuracy and scale,* Island Press, Washington DC.

Terborgh, J., 1989, *Where Have All the Birds Gone?,* Princeton University Press, Princeton, NJ.

Wiens, J, 1989, *The Ecology of Bird Communities,* Cambridge University Press, NY.

Wolter, P. T., Mladenoff, D. J., and Crow, T. R., 1995, "Improved Forest Classification in the Northern Lake States Using Multi-temporal Landsat Imagery," *Photogrammetric Engineering and Remote Sensing,* Vol. 61, pp. 1129-1143.

Wolter, P. T., and White, M. A., 2002, "Recent Forest Cover Type Transitions and Landscape Structural Changes in Northeast Minnesota," *Landscape Ecology,* Vol. 17, pp. 133-155.

Ronald J. McCormick,[1] *Amanda J. Zellmer,*[2] *and Timothy F. H. Allen*[2]

Type, Scale, and Adaptive Narrative: Keeping Models of Salmon, Toxicology and Risk Alive to the World

---

REFERENCE: McCormick, R. J., Zellmer, A. J., and Allen, T. F. H., **"Type, Scale, and Adaptive Narrative: Keeping Models of Salmon, Toxicology and Risk Alive to the World,"** *Landscape Ecology and Wildlife Habitat Evaluation: Critical Information for Ecological Risk Assessment, Land-Use Management Activities, and Biodiversity Enhancement Practices, ASTM STP 1458*, L. A. Kapustka, H. Galbraith, M. Luxon, and G. R. Biddinger, Eds., ASTM International, West Conshohocken, PA, 2004.

ABSTRACT: In modern day toxicology, problems arise in modeling complex ecosystems, such as the Columbia River system. There remain an abundance of models for which responsibility of assumptions is unaccounted and between which cohesion lacks. Models should be evaluated independently, taking into mind issues of scale and type in order to make sure the models actively change in accordance with the adaptive system they try to represent. The authors here suggest using narratives to weave together the inconsistent models. Narratives allow scientists to take responsibility for their assumptions and facilitate improved coherence between models. Problem solving engines, such as soft systems methodology, may then be used to achieve these modeling and adaptive management goals.

KEYWORDS: narrative, model, complex systems, scale, ecosystem, toxicology, salmon, risk assessment, hierarchy theory

Introduction

There was a discussion of allelopathy in an informal evening session of the British Ecological Society Annual Meeting in Northern England in 1982. The question was raised as to what it would take for someone to be convinced that allelopathy was indeed occurring. One commentator wanted a very first order criterion. He wished to see field strength of the allelopathic chemical applied to the target species such that the target suffered measurably more than the allelopath itself. The problem is, for whatever reason, we almost never find it to be so clear cut. Either the chemical at field strengths does not work, or the chemical concentration used in an experimental setting is such that it harms the plant thought to be allelopathic as much as it does the purported target. Geoff Sagar suggested that a better criterion would stem from comparing within a given target species allelopath-exposed and allelopath-naïve populations. His suggestion was to find two populations, one that had been growing with the allelopathic aggressor and one that had not been exposed to the chemical attack. If the naïve population suffers more than the exposed target population, then allelopathy would be presumed to have occurred.

The logic here is that if there was ever a focused advantage to production of an allelopathic chemical, then there would be immediately a selective pressure to overcome the allelopathic effect in the target population. There must be some biological means of

---

[1] Senior Ecologist, Compliance Services International, 1112 Alexander Avenue, Tacoma WA 98421.

[2] Research Aide and Professor, respectively, Botany Department, University of Wisconsin, 430 Lincoln Drive, Madison, WI 53706.

not succumbing to toxic effects, because it is an actuated fact in the allelopath itself. One can reasonably presume that there would be pressure for coming up with some mimic of the allopath's own resistance, and that the target species would find such a strategy sooner rather than later. Plants exude chemicals all the time, but usually there is no particular effect that gives allelopathic advantage. If there was an exudate that was as deleterious to the plant exuding the chemical as it was to potential competitors of other species, there would be no selective pressure to make the exudate more toxic. In fact, the pressure on the plant exuding the chemical would be to make the exudates less toxic all round. There might be some selective pressure to overcome the deleterious effect for all plants regardless of species, but it would not be amplified by a focused competitive pressure from the allelopath, because everyone suffers equally. The narrative of allelopathy would not apply. The irony here is that allelopathy is not demonstrable until it measurably fails to work. The failure to work suggests that there was coherence to allelopathy past, such that it coerced an adaptive response of resistance to allelopathy.

As it is with much of biology, there is never a realistic time zero that is self-evident long enough for the biologist to find it. As soon as an advantage is manifested, there is selective pressure to eliminate it. There may be an identifiable time zero for the release of some anthropogenic chemical, but, even so, natural selection is at work before humans can articulate the deleterious effects.

This absence of an articulated beginning of a toxicological process has consequences for the standard protocol for measuring environmental effects of chemicals. In standard protocols, there is little thought of accommodation to evolutionary response of target populations with regard to unintended spills of industrial chemicals. This is peculiar, because there is always acknowledgement of evolutionary responses to pesticides, and other toxins that are released on purpose. With regard to spills and deleterious effects of other chemical releases, conventionally, measurement is made on some laboratory creature as to fatal levels of assault. Prevailing wisdom says that if the assault is sufficiently weakened, then it will not prove fatal for some large percentage of the target species. That is then deemed to be some sort of safe level allowed in the environment. And yet there are many reports of situations in the field where populations are surviving and doing well, despite being exposed to concentrations at or above the presumed fatal dose.

If field populations are living successfully in what are assigned as being fatal field conditions, clearly something is wrong with the conventional narrative. The problem arises because there is no accounting of evolutionary responses to toxins. We pay attention only when tolerance becomes an issue in itself, as when resistance to some insecticide makes it less effective. Evolutionary response is an inexorable, universal law, whether we focus on it or not. Toxicology must deal with not only the effects of toxins, but also the consequence of ignoring natural selection. Toxicologists cannot focus on everything, so the "focus on it or not" stated above is not a criticism of the field. The improvement should be a tacit realization of constant change in the targets of toxins. The discipline should come to expect and calculate on the assumption of a dynamic world in the narratives that practitioners tell when they build their models.

The essential problem here is that plausible, straightforward narratives for addressing toxin levels are clearly inadequate. In this paper we will attempt to expose the root of the problem and offer a general protocol for creating higher quality narratives. Note that we challenge neither the measurement protocols, nor the standards for meticulous care taken by those measuring toxic concentrations. We admire those with the skill and patience to get the critical numbers. Rather, it is in the setting of the narrative that presents the difficulty in transferring lethal concentrations from lab to field.

A Protocol for Consistent Narrative

Allen et al. (in press) develop a framework that interrelates a series of general systems notions, such that the activities of the observer as scientist or manager are well characterized. When an observer notices an entity, therein arise the criteria for *scale* and *type*, intuitively at first, but soon explicitly. The type is a class that identifies the critical issues as to how the entity is defined. Rosen (2000) refers to equivalence classes, where the point of interest is what is the correspondence across the class between members. The class to which the entity is assigned gives a defining type to the observed entity. In a reciprocal consideration, the entity is checked by the scientist or manager for membership of the class. This check might be as to whether or not the species in question belongs to the class "endangered species." This cycle of definition and verification improves the model that derives from the equivalence class. Choosing the definition of the class offers a general criterion for type, whereas the actual structure of the individual entity meets concretely that general criterion (Figure 1).

Science and adaptive management proceed by achieving a dynamic definition of the equivalence class. Modifying the equivalence class keeps science and management alive to change. Once the entity under observation has been assigned to its type, it can be generalized by challenging the type with some other entity that should fit the class, but may not. For example, having found a species that is susceptible to a toxic material, one could challenge the class by verifying that some related species responds to the poison in the same way. If birds suffer reproductive problems from the presence of organochlorines in the environment, then one might challenge the equivalence class with mammals or fish to see if they too are members of the susceptible class responding with reproductive issues.

The cycle of definition and verification plays upon type generally associated with the class, but it also moves through a concrete example of a type. The same tension between the general and the specific occurs in the issue of scale. In order to see the members of the equivalence class, a protocol for measurement is created, at least implicitly. That is whence comes the general case of scale. Class defined by type is scale-independent. The class *organism* does not denote a collection of objects of a certain size in principle. On the other hand, when one finds a particular organism, it is indeed an object with certain dimensions. The size of the particular item is the local, concrete expression of scale. The issue of scale and type in principle should not be confused with the scale and type of an example.

Related to the difference between scale and type, Ahl and Allen (1996) make much of the distinction between level of observation as opposed to level of organization. Levels of observation are ordered on scaling principles, and have nothing to do with type. The type of thing one sees has nothing much to do with the scaling of the devices for observation or the size of any particular example. When textbooks organize all of biology in a grand hierarchical scheme, they refer to levels of organization, when in fact they mean levels that are ordered by scale, not type (Allen and Hoekstra 1992). Many organisms are much larger than many populations. In fact, mite populations treat their hosts as landscapes.

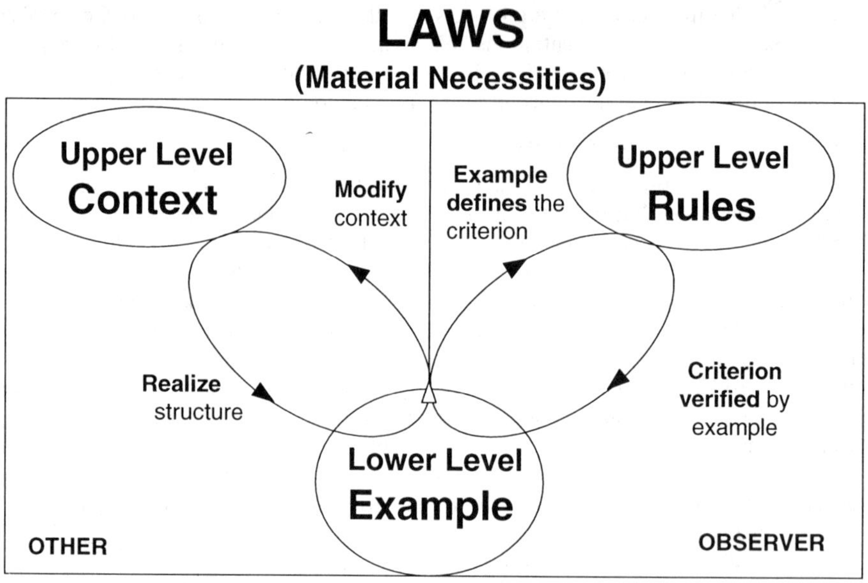

FIG. 1—*The scheme of Allen et al. (in press) for defining observer activity in scientific and management protocols. On the observer side, the concrete observed entity is assigned to an equivalence class. It does not belong to a type in principle; it only belongs by definition. On the other side of the scheme, a loop creates a structure through some realization process. On the return, the structure asserts itself as it modifies the upper level context. The trick is to map the model equivalence class, which is under observer control, to the parallel upper level cause of the pattern, which is not under observer control.*

Levels of organization do not have a scalar component, and are rather a matter of definition.

A level is a class, but one that has asymmetric relationships to a set of other classes. In scale-defined levels the asymmetry is *larger than versus smaller than*, and that gives levels of observation. In levels of organization, the asymmetric relationships are definitional, as when organisms of a given species are aggregated to make a population. Often the term *level* is carelessly introduced as a general indication of holistic intentions, as in the frequent use of *the landscape level* (Allen 1998). Usually, the term *landscape* alone would suffice to denote something defined by proximity, an equivalence class defined by the spatial arrangement of parts. It is a landscape level only when landscapes are defined as somehow above some other ecological class, such as ecosystem. Ecosystems are usually defined as ecological types characterized by relationships of flux and process with regard to matter and energy, a useful class for toxicology. However, ecosystems would belong to a class below landscapes only in certain defined situations. Type generalizes well to the notions of level of organization, as opposed to a scalar level of observation.

Along with the definitional cycle of model-building, we presume there to be an analogous cycle of the material system between the structures being observed and the context for those structures. This cycle, however, stands independent of the observer's decisions regarding scale and type (Figure 1). While we can choose what it is we plan to look at, we cannot decide the actions that the observed entity will perform. The observed structures are the lower level concrete examples and are the same lower level entities we use for building our models. The abstract context is analogous to the upper level, observer-defined equivalence class. Other theorists have alluded to the difference between the upper and lower levels in the material system, but have done so in a more specific manner. For instance, Bailey (1990) uses role and incumbent for his respective upper and lower levels, while Simon (1962) uses relational function and organizational structure, and Rosen (2000) employs essence and realization. Each of these paired terms represents a specific example of abstract context and concrete example, and depending on what the observer wishes to model, one of these specific relationships (or one similar) will be inferred.

A crude characterization of the upper level abstractions could be that they are the observed counterparts to the observer-created models. A realist would say that they are the real material system that is being observed and modeled. More subtly one could say they represent that part of observation above and beyond decisions of type and scale that the model invokes. We know of these upper level contextual considerations by their relationship to the observed structure. The observed structure not only has a reciprocal relationship to the equivalence class through definition and verification, it also is involved in a parallel cycle with the upper level context. The analogous cycle is the self-correcting system of the abstract context and the concrete example. Input to the upper level modifies the context, while output from that upper level consideration is a realization that generates the structure at the lower level. The role embodied in the U.S. Presidency was altered by Richard Nixon and his actions in "Watergate," whereupon the presidency was immediately reoccupied by a newly realized president, President Ford.

Behind every complex physical being is a context for that being. In conventional biological terms, the organism is a realization of its context, that realization being accomplished through DNA by reading and actuating a signal. Again, under the conventional view, an organism sends a signal back to its context through natural selection. The context would be the deme or species that is being modeled. Despite the technical capacity to read whole genomes, we do not and never will have a full characterization of either demes or species, except perhaps in the most extreme artificial experimental settings. The abstract context cannot be defined because it has an aspect that is infinite. While a role is limited, there is an infinity of ways to play that role. We use the separate device of the model or equivalence class to act as a surrogate that we can know.

We do, though, get to define the model and class. In ecology and environmental science, the notion of context applies well beyond a simple genetic system involving DNA and natural selection. For instance, a pesticide is a realization of research and manufacturing processes, with no DNA in sight. Furthermore, the toxin returns a signal to its context through regulation of some undesirable effect, or through profit made with some helpful insecticide. Here the context involves the manufacturer, not some gene pool. It is this greater generality that takes the notion of context beyond some first order

Darwinian model, although the Darwinian version of context makes a particularly clear and intuitive example.

Both of these analogous cycles are limited by the possibilities allowed by the laws. The self-correcting material system is limited by its history. These happenings occur in processes of self-organization. Positive feedbacks drive against negative feedbacks, which emerge as new system structure. There need not be any design to the chronology. By contrast, on the modeling side, design is everything. The laws pertain here as they limit measurement and modeling capability through technology. From within the laws, observer decision limits individual possibilities for modeling. We inherit the word "laws" from Pattee (1978), where he defines them to be universal, but only in the larger context of the discourse. For instance, universals in biology are the significant laws in that field, but would not be universal in physics.

How Narratives Work

Now that we have presented our view of scale and type in observation in relation to upper level abstractions, we are now in a position to see how we use and interpret the equivalence classes or models. Biologists and environmental scientists use models explicitly, but this practice is new, and confined to the latter half of the twentieth century. Having found the benefits of formalization and quantification, practitioners now mistake models as being the point of it all, which they are not. Ultimately, models lead to narratives that link models together in a meaningful way.

Models derive their power from being very local, so that they may be explicitly stated, on the one hand, and modified with precision on the other. Models are embodiments of assumptions, and their usefulness comes from being able to identify assumptions through making them explicit. The model lays out the consequences of going with this versus that set of assumptions. Modeling investigates less the material world, and more the consequences of the assumptions. The cost of being so explicit is that one does not readily envision reasonable assumptions that one could have made but did not. Certainly one can go on to test some other assumptions, but there are limits as to how far the research resources will stretch. On the data side, one must exert great caution in interpretation and prediction when moving outside the range of the data. One can collect and analyze more data, but again the same fiscal limits apply. The confidence one has in the insights derived from models comes at the cost of tight constraints on how much can be achieved. How much of the functioning of the world can one subject to rigorous testing? The answer is, "Not much." And yet, there are far too many situations that must be addressed by science and put under active management for explicit testing to be always part of the scenario. What then are practitioners to do?

Narratives and analogy come to the rescue. There are gaps between models. One can create intermediate models, but then there are new gaps, albeit smaller, between the new models and the old neighbors. There then arises an unwieldy management issue with regard to accommodation between models. One can create larger models containing smaller models nested within them, but within these fully integrated models the sub-models lose their autonomy as they are bound up in a larger meaning. The flexibility of the lower level models is compromised in their forced integration. The gaps between the sub-models disappear in the upper level model, but not in the nature that it models. Both subsuming models into a larger whole, on the one hand, and reduction of larger wholes to sub-models, on the other hand, are limited approaches. In the end one needs some other device beyond forced integration of, or reduction to, lower level models. The solution is narrative.

In a narrative, a series of zero dimensional models are linked through placing their respective meanings in a one dimensional larger whole. The process of building a narrative is not one of integration. Conversely, acknowledging parts of the story is not reduction. Analogy, with its reliance on interpretation, plays a role here and it substitutes for strategies of reduction. Not only is the meaning of each zero dimensional model preserved, meaning itself becomes the currency of the narrative. Meaning is rate-independent, whereas models exhibit rate-dependent dynamics. Meaning in a narrative does not behave, it just is.

There are false narratives, stories that are at odds with accepted facts, but lining up with verified facts does not make a narrative true in a fundamental way. Truth versus falsity is not a zero sum game here. For narratives that are not at odds with the facts, truth and falsity are beside the point. A full, true account of a chronology would have to include an infinity of information. Not only is it impossible to capture absolutely all the details, if one did, it would not be a narrative. This is because there would be no narrator taking responsibility for the meanings that are included, as opposed to other meanings that are for the moment put aside.

Once we enter the realm of narratives, they take on a life of their own. For instance, some narratives include lies. Mimicry is a narrative told to the world by some creature, but it is a lie. That first narrative might be about how good an animal is to eat, and is derived from the homology that, say, an insect has with its close relatives. If one were into eating insects, then apparently cockroaches are delicious. Mimicry invokes co-evolution or convergent evolution, and that invokes evolutionary analogy. Ladybugs are not delicious, in fact they taste very bad. It is in advertising that fact that natural selection has endowed ladybugs with red wing cases with their conspicuous black spots. The ladybug is telling a two dimensional story, the meaning of which is "You can see me, and I am potential food." The second dimension of the narrative is, "I am toxic, or at least I taste bad." Mimicry too invokes a two dimensional story. The cockroaches mimicking lady bugs are laying another dimension of narrative over the one that announces to predators, "Here is another insect that tastes good." The second narrative is, "I am a ladybug, and so taste bad." Notice how meaning here is the currency, and verity is beside the point (Allen et al. in press).

Organisms enter the world with a model of what to expect. That model may not be consciously held, as when the lady bug cannot even see red, because it is outside the range of insect visual wavelengths. Part of the model may be embodied in instinct. Self-knowledge comes not from the organism itself, but from how other organisms with their models of the world read the narrative that is being told to them. When a mimic tells its two dimensional narrative, self-knowledge comes from a further dimension derived from the narrative told by other beings through their interpretation of the mimic. Developing self-knowledge adds yet another dimension of narrative which says, "I believe that you are not a suitable item of prey."

Science is a quest for self-knowledge. We develop models. We investigate these models as we tell the significance of them to the world through experimental protocols. While an experiment is not a lie in the manner of a mimic, it is the result of having interpreted first impressions. There is a sophisticated two dimensional narrative embodied in experimental protocols. The story is something like, "We think the world works this way, but what if it really works that way?" Executing the protocol in an experiment tells a tale to the world, and we obtain knowledge about what we were thinking (self-knowledge) by the world reacting to our narrative in an experimental result. Science is a high dimensional, deeply reflexive narrative.

What Does It Mean for Toxicology?

When salmonids live in water that has observed high levels of potential toxins, the scheme of Allen et al. (in press) shows where to improve the modeling and observation process. It is helpful to run laboratory experiments as to what level of a toxin kills a representative organism. However, problems arise when those values, derived from a few individuals, are sanctified. First, the individuals may not have been tested in sufficient numbers to give reliable estimates of intrinsic variation. Second, they may have been representative of their species in nature at the time they were tested, but the situation has changed. If circumstances are so dire as to warrant expensive experimental protocols, then they are likely to be dire for the wild animals that live in them. Nature does not rest easy in dire straights, it invokes change. Not only has there been enough time for change in the context of the species, one should actively expect change.

There are two fundamental mistakes being made when safe or tolerable levels for toxins are used as benchmarks after their time has passed. First is the failure to come back down again to the observed example from the equivalence class addressed in the original calibration experiments. There are concrete organisms that need to be verified as members of the class. One must ask whether the organisms living for generations in the presence of a toxin are still members of the class that demands regulation of the environment (Figure 2). The science of toxic materials testing cannot be expected to be adequate for more than ephemeral conditions. Toxicology must be a continuously changing field, even in areas where it was once valid and up-to-date, and where the problems appeared solved. While much effort is being made in this direction (Suter et al. 2003), much is still to be done. The second error is in mistaking the map for the territory, the model for that which is modeled (Bateson 1980). While the modeling cycle does share with the cycle of realization and modification the observed structure, the model is only an analogy of the upper level external source of pattern, it is not the same thing. The distinct upper level parts of the modeling cycle relate to the upper level of the other cycle only by inference. There are processes of reinforcement on the other side that are not part of the modeling cycle. Furthermore, the history of the system must be taken into account because the *limitations* due to the laws are in a manner different from the *influence* of laws in the modeling process. In the case of the Columbia River, one must take into account the self-organizing history that drives the system: mining; dams; logging; agriculture; urbanization; and toxic chemicals.

At another level, a problem persists concerning the issue of multiple models attempting to model the same entity, though at different scales. This problem occurs because observer decisions from within the laws of material necessity influence each model in a different way. Indicators from the Columbia River system have been modeled from a number of opposing view points, leaving decision makers at odds with respect to making important decisions. When the Fish and Wildlife Service makes a management decision about the fisheries, they are confronted with opinions based on models made by environmentalists, agriculturalists, commercial harvesters, tribes, chemical companies,

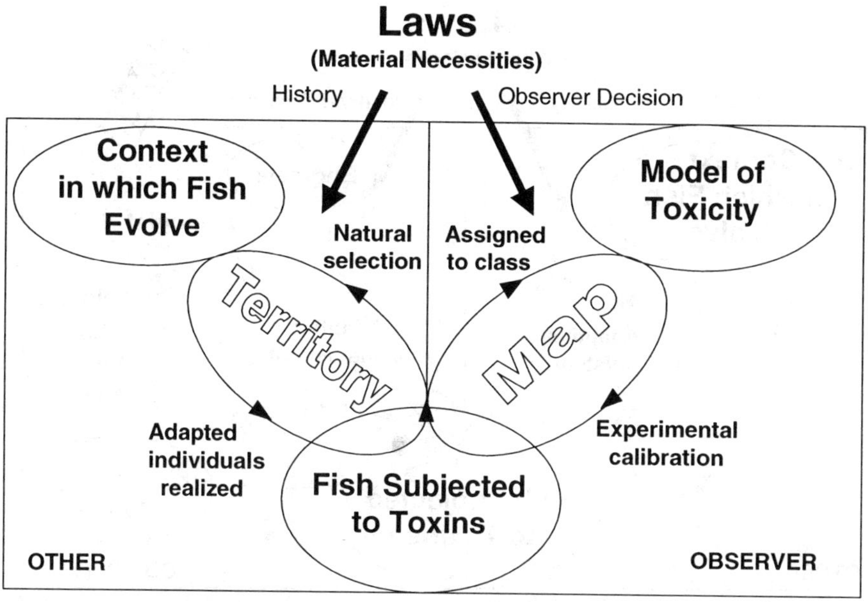

FIG. 2— *The same scheme as in Figure 1, but with indications as to how the modeling cycle must remain alive to changes in the contextual cycle. The second caveat is that the two cycles are separate, and must not be mistaken as the same. The right hand cycle is the map, and it is not the cycle of the territory, the cycle on the left.*

toxicologists, etc. The policy is, therefore, to avoid making new decisions that would bring on new lawsuits from the conflicting parties. The issue then lies in the different models created within the scale-determining biases for each of these players. The toxicologists rely on a model that says this percentage of fish will survive when a toxin in lab experiments is below a certain concentration. In comparison, the ecologists are verifying a model that says the fish populations are sustaining their numbers at higher concentrations. Therefore, what is needed to remove these contradictions is a way to tie the similar, but opposing models together. As a result, the models should become comprehensible to each of the players. That tie is where narratives become particularly useful (Figure 3). The narrative is a way to (1) accommodate to the absence of a time zero point in each of the Columbia River system models, (2) eliminate scale issues, and (3) create a unified story from which each of the parties can begin to understand the similarities in their differing models.

Finally, if the model is to serve management then it needs to be updated so that improved definition and verification can generalize the model. Resting on the original calibration of a toxic effect means that it is not tested in a process of continual improvement. In the management of all complex systems, the problems are never solved for ever (Ackoff 1981; Checkland 1981). The nature of the system being modeled and managed changes through reinforcement from new realizations. Also, the needs of the managers change, such that modeling decisions need to be updated. Societal needs change, and standards for adequacy and control change with social demands. This is the

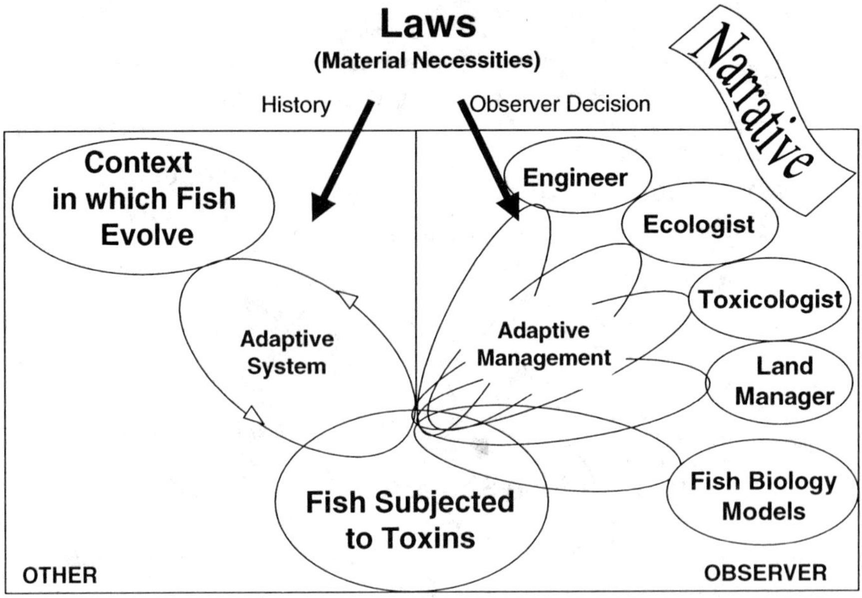

FIG. 3— *Here different players in the science, regulation and management fields each have their own scaled model of the type to which a salmon belongs. Integrated management is achieved by stringing the models together in a comprehensive management plan that is guided by a narrative. The narrative is applied to the context in the left hand cycle, where management action is applied.*

job of the modeling cycle. That cycle becomes maladaptive when some magic number is fixed in the book of regulations.

It is no accident that in *Supply-Side Sustainability*, Allen et al. (2003) recommend managing for the context, so as to offer system parts the material resources they need as well as other services offered by the whole. Management is applied to the context in the left hand cycle. The calculations are made on the right hand modeling side of Figure 3, but action pertains on the side of the material context. As we have insisted above, the modeling cycle must address the whole production system, not just the resource. When the whole is managed from its context, the parts of it that are effectively captured in the modeling exercise can look after themselves. In wolves, salmon, or recreational landscapes, systems held properly in context can function normally. They will offer subsidies to the effort toward sustainability, and that subsidy we can take as a resource. This is the essence of adaptive management.

Beginning Narratives – The CATWOE Process

To this point, we have shown the "why" and the "what" of narratives, but little of the "how." Such ground has been clearly covered by others (e.g., Kay et al. 1999), and a detailed discussion is outside the framework of this paper. However, we do offer an example approach that is easily understood by a variety of stakeholders. Specifically,

Checkland (1981) circumscribes a "soft systems methodology" that aids the development of a narrative about the structures and processes interacting within a given environmental framework. One portion of Checkland's soft systems methodology is the CATWOE process. CATWOE is an acronym, the elements of which define the primary components of a system:

Customer(s)	Who benefits from the system, or, who is affected by the system
Actor(s)	Those operating within the system that can effect transformation
Transformation	A change in state – usually phrased as "from X to Y"
Weltanschauung	Worldview, typically pertaining to why a transformation is desired
Owner(s)	Who can shut down the system, or change the transformation goal
Environment	External constraints on a system, factors outside of the actors control

The process starts with a description of the perceived problem. How the problem is perceived by a group of stakeholders influences the CATWOE elements. Once the CATWOE elements are described, one can write a *root definition* that incorporates the various elements into a narrative, providing problem focus and solution guidance. The description of the problem as perceived and the root definition that derives from the process represent only one of many possible problem/root definition statements that can be made. The interesting aspect of the CATWOE process is that disparate groups of stakeholders can formulate different problem statements and root definitions, thus providing a focused version of their worldview. Stakeholder group members can then read the statements of other groups. We would hope that all participants then will come to a fuller understanding of the biases held and viewpoints taken, not only of other groups, but also their own.

A Single Narrative – The CATWOE of Salmon and Pesticides

Joseph Taylor (1999) uses the title "Taking Responsibility" for the last chapter of his book *Making Salmon.* He notes that for the last 150-years of salmon management in the Pacific Northwest, every special interest group wove a narrative that was "...simple, clear, traditional, and thoroughly flawed." Indeed, salmon continue as pawns in a struggle between disparate groups seeking management power over parts of the context. No one group controls the whole, and none seek such control because they work within a paradigm, and thus can maintain a simple, clear narrative of how salmon work. They point to "the others" as being responsible for the demise of salmon, and do not "take responsibility" for the quality of their narrative. Notably, McCormick has attended salmon harvest allocation meetings and listened in awe as representatives from different ports advocated for their quota and complained of Canadians taking "their fish." In an aside to a new attendee who informed the whole group that he wanted his son to learn the joy of catching a wild fish (as opposed to buying a salmon fillet in the store), a veteran of the process revealed that the knowledgeable groups at the meeting were there each year, in force, and that that was the only way salmon would be allocated to a particular landing.

Even the effects of dam removal on salmon recovery, strongly advocated as an obvious positive in certain circles, are uncertain (Stanley and Doyle 2003). Hatchery reform is generally recognized as necessary, but only with recognition that harvests (tribal, sport and commercial) must continue at acceptable levels. More than 80% of the harvested salmon in the Columbia River come from artificially maintained hatchery stocks (Lackey 2002). Coalitions in California, Oregon and Washington seeking the elimination of toxic substances in the human and natural environment have sued the EPA for failing, under the Endangered Species Act, to protect salmon from pesticides, even though many of the compounds listed have been permitted for use for decades before salmon were ever listed as threatened or endangered. That the EPA actually has reduced or eliminated many uses clearly detrimental to fish and birds is not highlighted in their narrative, presumably because their goal is elimination of toxins, not necessarily restoration of self-sustaining salmonid populations.

If we focus the CATWOE process solely on salmon and water quality, but try to paint a richer, more complex worldview, we could perceive the problem as: inputs from terrestrial runoff to aquatic systems are affecting salmon, water quality and the biotic integrity of our lakes and rivers...inputs affecting salmon and water quality include nutrients and sediments above background levels as well as pesticides and other biotic/abiotic elements at measurable concentrations. From this perceived problem, we might define the CATWOE elements as:

Client: River or Lake Water (receiving water column); Salmon and other aquatic organisms

Actors: Agricultural and horticultural end-product producers; agricultural and horticultural end-product consumers; agrochemical and seed producers; government regulators; salmon and other aquatic organisms; fishers; aquaculturists; hatchery operators; tribal governments; environmentalists; forest products industry; hydroelectric dam operators; urbanites

Transformations: Altered (impaired) waters into unaltered (unimpaired) waters; threatened & endangered salmon populations into self-sustaining populations

World View: That clean water with high biotic integrity and self-sustaining salmon populations represent a desirable environment

Owner of the system: Agricultural and horticultural product consumers; salmon and fish-derived product consumers

Environmental Variables (Other factors): Governing regulations (CWA, ESA, FIFRA) and the registration process; global climate change, long-wave climatic variation, ENSO/PDO, external governing agencies: WTO and other free trade restrictions/requirements; transgenic plants altering nutrient, tilling and pesticide management; transgenic organisms (farmed hybrid giant salmon) altering markets

Limiting the perceived problem to that portion related to pesticides, a *root definition* of the problem might be: agricultural/horticultural product consumers, who, in general, desire clean waters of high biotic integrity for salmonids and humans, want pesticide producers, users and regulators to design, implement and manage a regulatory system that will not degrade waters or adversely affect resident aquatic organism populations, as part of a greater societal goal of restoring those waters and populations to some desired and self-sustaining state, taking full advantage of new models and technologies while also considering potential long-term effects on both waters and fish due to global climate change and external sociopolitical constraints.

The root definitions from a CATWOE of foresters, farmers, and pesticide producers would be much different. To fulfill the demands of adaptive management, we'll need to align those varied definitions via narratives.

Two Additional Narratives – The 4 Hs and the 4 Cs

We have reduced wild salmon stocks to less than 10% of what they were 200 years ago (Lackey 2002) through vast physical alteration of stream habitat as a result of mining, forestry, agriculture and urbanization (Taylor 1999). Additionally, knowing full well that construction of *hydroelectric* dams would block off vast stretches of *habitat*, we sought to compensate *harvest* losses by operating *hatcheries* (the 4 Hs). Those Hs represent the human context that has come to dominate the system, a context within which salmon did not evolve but from which all present day salmon management narratives begin. Salmonids have shown us that they need structurally *complex* streams with *cold, clean* water that *connects* to the ocean (the 4 Cs). That is the context within which West Coast salmonids evolved, and it is the narrative the fish consistently tell. We cannot return to the salmon context of the early 1800s, but we can manipulate the 4 Hs to mimic the 4 Cs, and provide salmon with a context recognizable to their current DNA. Adaptive management is touted as the path toward realigning the narratives implicit in the 4 Hs. We need to become more explicit in what those narratives contain, where the vital interconnections are (e.g., Schindler et al. 2003), and how new data and perspectives will be incorporated. In essence, adaptive management as a whole becomes subsumed under a cohesive set of adaptive narratives (Waltner-Toews et al. in press).

Adaptive Narratives

At least several tens of thousands of years of ecological evolution limit the plasticity of individual and population-based biotic responses, and short-term and long-wave variation in environment set the context for and shape the composition of communities and ecosystems in which salmonids occur. DNA is a very conservative molecule, holding on to seemingly useless information for multiple generations. During the present, relatively stable, interglacial period, salmon have radiated out from their glacial refugia to occupy a landscape that took 10,000 years to evolve. In less than 200 years, humans have completely reconfigured that landscape. In doing so we have come to control many lower level processes, that is, we now set the context for water quality, plant and animal community composition, population size and distribution, and dispersal routes.

Except for random, cataclysmic events (e.g., meteorites leading to the demise of the dinosaurs), evolutionary mechanisms are adapted to operate at much the same rate as climate shifts and geologic processes. In becoming the context, humans have greatly increased the rate at which the context is altered, such that it now changes at a rate much faster than evolutionary processes are geared for. Our management decisions act as evolutionary filters, but at an accelerated rate. Thus, by re-shaping landscape structure to fit our management techniques, humanity has also assumed responsibility for the species that occupy those landscapes. Also, by re-organizing ecosystem pathways we now directly control or greatly affect much of the energy, material and information that flows through the system. Since we actively manage the context, management plans should derive from a viewpoint of the context. Kay et al. (1999) aid in defining context and constraints with a list of questions designed to elicit the operational aspects of complex adaptive systems.

Different world views yield different models, and the argument typically then shifts to "what is the right model?" when a more useful question is "which is the more powerful narrative?" We have shown how narratives allow us to qualitatively interrelate formal models. Formal models help us explore relationships between data, but do not always lead us back to improving and enriching the original narrative. To critically assess risk to a complex system, humans within the system must adaptively weave a continuous narrative about that systems' current functional states, and seek consensus on management goals leading to future desired states.

References

Ackoff, R., 1981, *Creating the Corporate Future*, Wiley, New York.

Ahl, V. and Allen, T. F. H., 1996, *Hierarchy Theory, a Vision, Vocabulary and Epistemology*, Columbia University Press, New York.

Allen, T. F. H., 1998, "The Landscape "Level" is Dead: Persuading the Family to Take It Off the Respirator," In: D. L. Peterson and V. T. Parker, Eds., *Ecological Scale: Theory and Application*, Columbia University Press, New York.

Allen, T. F. H. and Hoekstra, T. W., 1992, *Toward a Unified Ecology,* Columbia University Press, New York.

Allen, T. F. H., Tainter, J. A., and Hoekstra , T. W., 2003, *Supply-Side Sustainability*, Columbia University Press, New York.

Allen, T. F. H., Zellmer, A. J., and Wuennenberg, C. J., (In Press) "The Loss of Narrative," In: K. Cuddington and B. Beisner, Eds., *Paradigms Lost.*

Bailey, K. D., 1990, *Social Entropy Theory*, State University of New York Press, Albany, New York.

Bateson, G., 1980, *Mind and Nature*, Bantam, New York.

Checkland, P., 1981, *Systems Thinking, Systems Practice*, Wiley, New York.

Kay, J. J., Regier, H. A., Boyle, M., and Francis, G., 1999, "An Ecosystem Approach to Sustainability: Addressing the Challenge of Complexity," *Futures*, Vol. 31, No. 7, pp. 721-742.

Lackey, R., 2002, "Salmon Recovery: Learning from Successes and Failures," Northwest Science, Vol. 76, No. 4, pp. 356-360.

Pattee, H. H., 1978, "The Complementarity Principle in Biological and Social Structures," *Journal of Social and Biological Structures*, Vol. 1, pp. 191-200.

Rosen, R., 2000, *Essays on Life Itself*, Columbia University Press, New York.

Simon, H. A., 1962, The Architecture of Complexity, *Transactions of the American Philosophical Society*, Vol. 106, pp. 46-482.

Schindler, D. E., Scheuerell, M. D., Moore, J. W., Gende, S. M., Francis, T. B., and Palen, W. J., 2003, "Pacific Salmon and the Ecology of Coastal Ecosystems," *Frontiers in Ecology and the Environment*, Vol. 1, No. 1, pp. 31-37.

Suter II, G. W., Vermeire, T., Munns Jr., W. R., and Sekizawa, J., 2003, "Framework for the Integration of Health and Ecological Risk Assessment," *Human and Ecological Risk Assessment*, Vol. 9, No. 1, pp. 281-301.

Stanley, E. H. and Doyle, M. W., 2003, "Trading Off: the Ecological Effects of Dam Removal," *Frontiers in Ecology and the Environment*, Vol. 1, No. 1, pp. 15-22.

Taylor III, J. E., 1999, *Making Salmon*, University of Washington Press, Seattle, WA.

Waltner-Toews, D., Kay, J., Murray, T., and Neudoerffer, C., In Press, "Adaptive Methodology for Ecosystem Sustainability and Health (AMESH): An

Introduction," In: G. Midgley and A. E. Ochoa-Arias, Eds., *Community Operational Research: Systems Thinking For Community Development*, Plenum Publications/Kluwer Academic, New York.

Mark C. Andersen[1]

Population Dynamics in Spatially and Temporally Variable Habitats

---

REFERENCE: Andersen, M. C., **"Population Dynamics in Spatially and Temporally Variable Habitats,"** *Landscape Ecology and Wildlife Habitat Evaluation: Critical Information for Ecological Risk Assessment, Land-Use Management Activities, and Biodiversity Enhancement Practices, ASTM STP 1458*, L. A. Kapustka, H. Galbraith, M. Luxon, and G.R. Biddinger, Eds., ASTM International, West Conshohocken, PA, 2004.
ABSTRACT: Populations live in habitats whose quality varies spatially and temporally. Understanding how populations deal with these variable habitats can aid our understanding of theoretical issues, and practical issues of biological invasions and biodiversity conservation. I investigate these issues by superimposing simple models of population growth and dispersal on spatiotemporally fractal landscapes, and examining the properties of the landscapes, and of the populations inhabiting them. The properties of the simulated landscape sequences are comparable to those of real habitats. The simulated populations exhibit a range of dynamic behaviors; these behaviors are strongly influenced by the fractal parameters of the landscapes. The results may help explain several important phenomena seen in reintroductions of threatened and endangered species, introductions of biological control agents, and biological invasions. These phenomena include frequently observed lags between population introduction and initial population growth and spread, and the observed high frequency of failure of introductions.
KEYWORDS: neutral landscape model, spatial population model, coupled map lattice, theoretical ecology, invasive species, endangered species

Introduction

Populations inhabit dynamic landscapes. Thus a more complete understanding of population dynamics must account for the responses of populations to both temporal and spatial variability in habitat quality. However, previous models of spatiotemporal population dynamics have lacked either realism, generality, or both. Here I present results for a class of models that may address these shortcomings of existing models. This class of models basically involves setting up a coupled map lattice model (Kaneko 1985; Sole, Bascompte et al. 1992; Willebordse and Kaneko 1994; Kean 2001) on top of a spatiotemporally fractal neutral landscape model (With 1997; With and King 1997).

A coupled map lattice model is essentially a standard population dynamic model extended to a grid of habitat patches rather than the single patch considered by the simplest models (Kot 2001, chapters 1 through 14), or the system of interconnected

---

[1] Associate Professor, Department of Fishery and Wildlife Sciences, New Mexico State University, Las Cruces, NM, 80003-0003.

patches considered by metapopulation models (Hanski 1999). Coupled map lattice models may be either deterministic or stochastic. Coupled map lattices are discrete-time, discrete-space models, but unlike cellular automata, interacting particle systems, and contact processes (Turchin 1998; Hanski 1999), coupled map lattice models can model continuous variables on these discrete spaces.

Neutral landscape models are used in landscape ecology as a means of generating artificial or simulated landscape patterns with known properties, constraints, and structuring processes. Time sequences of neutral landscapes can be generated with given temporal correlation properties and used to represent variation in habitat quality over time. Here I use such sequences to provide input to a model of population dynamics in spatiotemporally variable habitats.

The specific research question I examine here is "How do spatiotemporal habitat dynamics influence the establishment success of an invasive species or reintroduced species?"

Spectral Synthesis of Spatiotemporally Fractal Landscapes

There are a number of classes of neutral landscape models that have been proposed and used by different researchers. The most useful of these classes of model are amenable to spectral synthesis. Spectral synthesis is the use of spectral basis functions such as wavelet transforms or Fourier transforms to represent neutral landscape models (Keitt 2000). The spectral approach also encompasses a number of different types of scaling relations, although all variants of spectral synthesis rely fundamentally on a scaling relation between amplitudes and frequencies of environmental (spatial or temporal) fluctuations.

One of the most useful neutral landscape models is the fractional Brownian motion or fBm (Keitt 2000). Fractional Brownian motion is controlled by a parameter H called the Hurst exponent; this parameter appears in the scaling relation between the expected variance of increments to the process and their separation distance. The effects of different values of the Hurst exponent are illustrated in Figure 1.

H = 0.01 H = 0.5 H = 0.99

Figure 1 - *Effects of different values of the Hurst exponent H on the spatial properties of a fractional Brownian motion (fBm) process. The effects on temporal properties of an fBm process are analogous to those pictured here; smaller values of H lead to "choppy" sequences with low correlation.*

For the models and analyses reported here, I used Fourier synthesis of spatiotemporal fractional Brownian motion governed by separate Hurst exponents in the time dimension (H_t) and in the two spatial dimensions (H_s). I allowed H_s and H_t to take on any combination of the values 0.01, 0.1, 0.5, 0.9, and 0.99. The spectral synthesis algorithm was implemented in MATLAB ® on a 128 by 128 grid for 128 time steps. The spectral synthesis algorithm employed automatically "wraps" the simulation domain into a toroidal shape; thus edge effects do not arise.

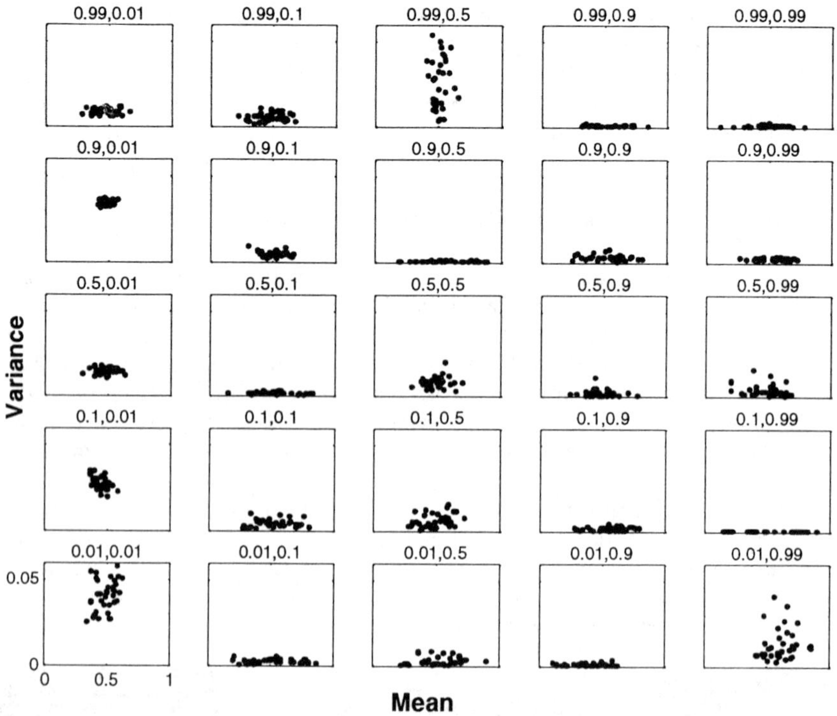

Figure 2 - *Plots of variance (on the y-axis) vs. mean (on the x-axis) of simulated habitat quality at 36 randomly-chosen locations for different values of H_s and H_t from fractional Brownian motion landscapes simulated on a 64×64 grid over 128 time steps. Values plotted are means and variances over all time steps. Values of H_s and H_t are shown (H_s first) above each plot.*

Data analyses reveal some properties of sequences of neutral landscapes generated in this way. Figure 2 shows variances of "habitat quality" (the value of the fBm process scaled to lie between 0 and 1) as a function of the means for all combinations of H_s and H_t values. The plotted values were obtained by randomly selecting the x and y coordinates of 36 "permanent sampling stations" on the grid of simulated locations and taking the means and variances of "habitat quality" at those locations over all time steps in the simulated sequence of landscapes. Note the variety of types of variance-mean relationship that can be generated by this method. Figure 3 shows the same data pooled across all combinations of H_s and H_t values. Note that Davidowitz (2002) presents several graphs of variability in precipitation versus mean precipitation for several habitat types throughout the Southwestern United States. Unlike Figure 3, Davidowitz' graphs typically show a strong exponential decay of variance with mean precipitation. However, it appears from Davidowitz' graphs that this exponential decay only appears when data are pooled across multiple habitat types (e.g., creosote, desert scrub, grassland, and pine forest), with data for individual habitat types somewhat resembling the individual plots in Figure 2.

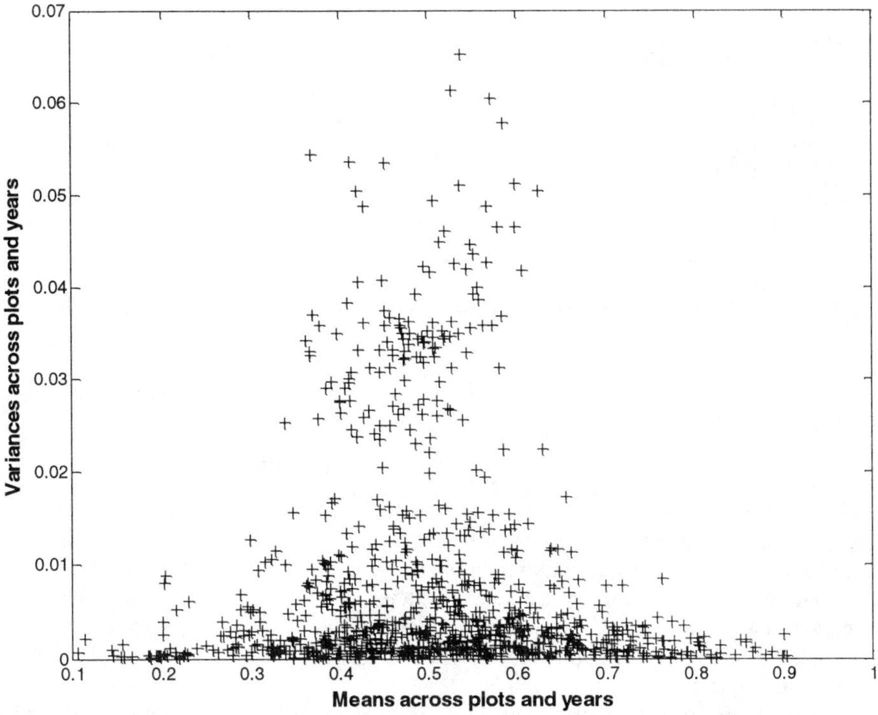

Figure 3 - *Plots of variance vs. mean of simulated habitat quality at 36 randomly-chosen locations for different values of H_s and H_t from fractional Brownian motion landscapes simulated on a 64×64 grid over 128 time steps. Values plotted are means and variances over all time steps; all parameter combinations are pooled.*

The spectral synthesis approach to simulation of neutral landscapes is computationally simple, while encompassing a broad range of scaling relationships. Such models can be used as objects of study in their own right, for comparison with real natural landscapes, as a means of generating replicate landscapes with known properties, or as input to simulation models (Keitt 2000). Spatiotemporal spectral synthesis algorithms can produce realistic landscape sequences with known properties governed by only two parameters.

Coupled Map Lattice Models of Populations in Dynamic Landscapes

The basic structure of a coupled map lattice model is illustrated in Figure 4. The model accounts for emigration and immigration among habitat grid cells, and for birth and death within cells. Such models typically require only a few parameters, yet can model very complex spatiotemporal phenomena. The relatively small number of parameters makes the operation of coupled map lattice models much more transparent than simulation models of similar processes and phenomena.

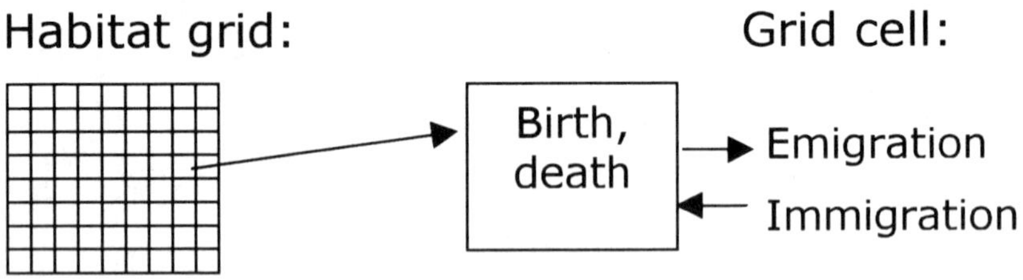

Figure 4 - *Diagram of the basic structure of a coupled map lattice population model.*

For the results I report here, local population dynamics were governed by the Ricker model

$$N_{t+1} = N_t \lambda \exp(-\beta N_t)$$

where λ gives the rate of population growth in the absence of density-dependence and β represents, in a sense, the strength of density-dependence. This model is capable of producing a series of period-doubling bifurcations leading to chaos (Turchin 2003).

Dispersal was governed by one of three possible forms of linkage between neighboring grid or habitat cells. These linkages are the "coupling" referred to in the term "coupled map lattices". The three forms of coupling used for the results reported here are shown in Figure 5.

This model was simulated with 300 different parameter combinations (H_s and H_t =0.01, 0.1, 0.5, 0.9, or 0.99, λ = 2 or 10, a dispersing fraction equal to 0.1 or 0.01, and 4- 8- or 12-cell dispersal). Simulations were always started with only a small patch of grid cells in the center of the grid occupied by the population. Two hundred replicate simulations were run on a 64 x 64 grid for 512 time steps for each parameter combination. The simulations were implemented in MATLAB ®. As with the simulations described above, the spectral synthesis algorithm automatically "wraps" the domain into a toroidal shape. In addition, the dispersal algorithm used also "wraps" the habitat; thus edge effects once again are not an issue in these simulations.

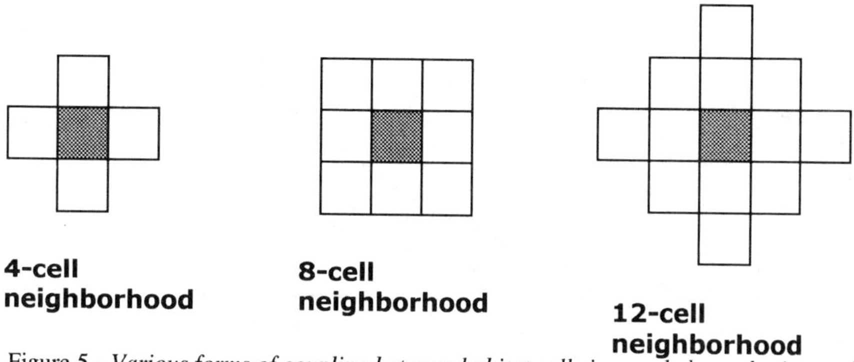

Figure 5 - *Various forms of coupling between habitat cells in coupled map lattice models. These represent different degrees of dispersal ability in the population being modeled.*

Model output for a single parameter combination is shown in Figure 6. Note that most populations that go extinct do so quite early, and that many populations persist at very low levels throughout the duration of the simulation, even if they do not go extinct.

Figure 7 shows the effects of the spatial and temporal Hurst exponents on the fraction of populations going extinct. Note that the fraction of populations going extinct increases with decreasing temporal Hurst exponent, and may increase slightly with increasing spatial Hurst exponent at low values of the spatial Hurst exponent.

Figure 8 shows the effects of the spatial and temporal Hurst exponents on the mean time to extinction. Note that the apparent lack of any strong relationship may be due to poor estimation of mean time to extinction from a relatively small number of extinctions in the 200 replicate simulation runs for each parameter combination.

Conclusions

I have shown that the probability of establishment of a potential invasive species depends on the strength of spatial and especially temporal variability in the environment (Figure 6). The time before a population becomes established (or conversely becomes extinct) is not strongly influenced by the strength of either spatial or temporal variability (Figure 7). The strength and probability structure of environmental fluctuations are known to influence extinction in unstructured populations and in populations with simple spatial structure (Milton and Belair 1980; Ripa and Lundberg 1996; Heino 1998; Morales 1999); my results extend previous results to more complex forms of spatial structure and of environmental variability.

I have also shown that most populations that go extinct (i.e., fail to establish) do so rather early, although populations that do establish may persist at very low numbers for a long time (Figure 5). This phenomenon is well known from empirical studies of invasive species (Rejmanek M 1996; Suarez 2000; Kraft, Sullivan et al. 2002). Although alternative theoretical explanations exist for this phenomenon (Shigesada and Kawasaki 1997), these results provide an alternative hypothesis.

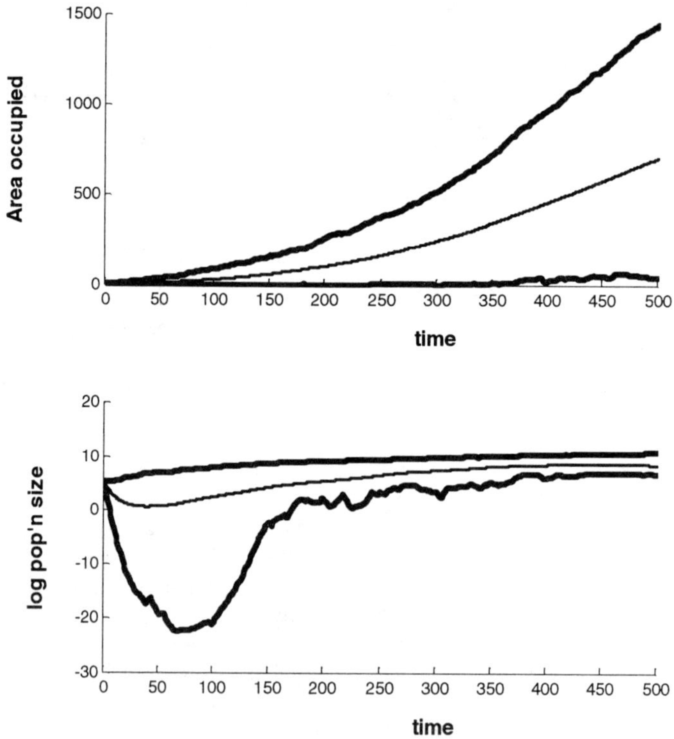

Figure 6 - *Output of the coupled map lattice model described in the text for* $H_s = 0.1$, $H_t = 0.1$, $\lambda = 4$, $d = 0.10$, *and a 12-cell dispersal neighborhood. The upper figure shows area occupied (given a detection threshold of five individuals); the lower figure shows the natural logarithm of population size. Both graphs show the mean over all simulation runs (inner line), and upper and lower simulation envelopes (outer lines).*

Finally, I have shown that fBm sequences have some properties that easily could be, but have not yet been, rigorously compared with data from real environments. Although some notable efforts have been made in this direction (Davidowitz 2002), much remains to be done before we can be confident of the applicability to real environments of results obtained using neutral landscape models such as those described here.

My failure to detect a strong influence of the spatial and temporal Hurst exponents on mean time to extinction may have been due to using a fixed number of replicate simulation runs, rather than running the simulation until a fixed number of extinctions had been observed (a form of acceptance sampling).

Neutral landscape models are an established part of the theoretical ecologist's tool chest. Coupled map lattices, while less well-known, are also potentially important tools for studying situations that previously could only be dealt with using complex over-parameterized simulations. Combining these two tools may provide a way to address some pressing problems in both theoretical and applied ecology.

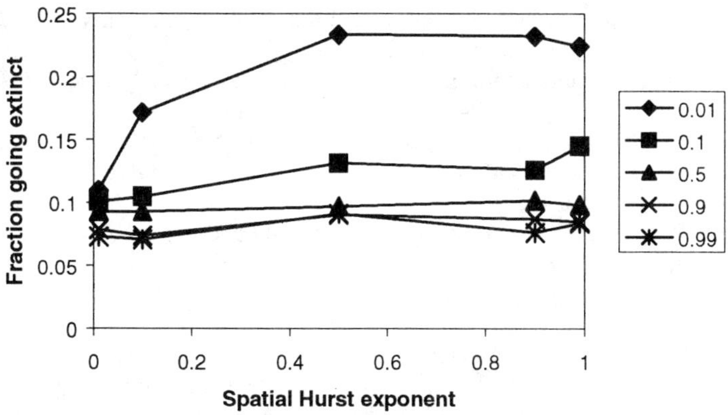

Figure 7 - *Effects of spatial and temporal Hurst exponents on fraction of populations going extinct. Different lines in the figure correspond to different values of the temporal Hurst exponent, as shown in the legend*

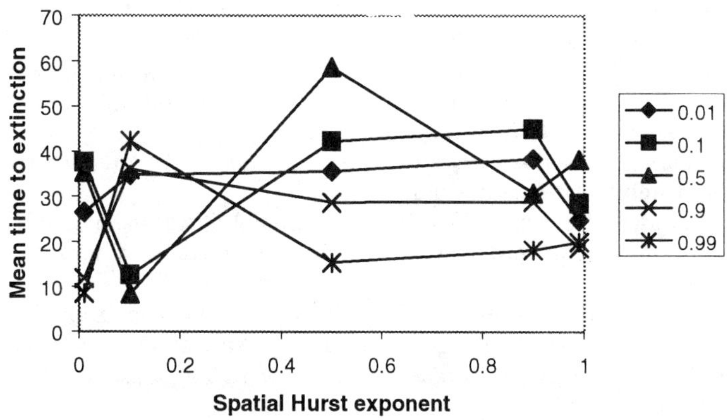

Figure 8 - *Effects of spatial and temporal Hurst exponents on mean time to extinction. Different lines in the figure correspond to different values of the temporal Hurst exponent, as shown in the legend.*

References

Davidowitz, G., 2002, "Does Precipitation Variability Increase from Mesic to Xeric Biomes?" *Global Ecology & Biogeography,* Vol. 11, pp. 143-154.

Hanski, I., 1999, *Metapopulation Ecology,* Oxford University Press, Oxford.

Heino, M., 1998, "Noise Colour, Synchrony and Extinctions in Spatially Structured Populations," *Oikos,* Vol. 83, No. 2, pp. 368-375.

Kaneko, K., 1985, "Spatiotemporal Intermittency in Coupled Map Lattices," *Progress in Theoretical Physics,* Vol. 74, No. 5, pp. 1033-1044.

Kean, J. and N.D. Barlow, 2001, "A Spatial Model for the Successful Biological Control of *Sitona discoideus* by *Microctonus aethiopoides,*" *Journal of Applied Ecology,* Vol. 38, No. 1, pp. 162-169.

Keitt, T. H., 2000, "Spectral Representation of Neutral Landscapes," *Landscape Ecology,* Vol. 15, pp. 479-493.

Kot, M., 2001, *Elements of Mathematical Ecology,* Cambridge University Press, Cambridge.

Kraft, C. E., P. Sullivan, A. Y. Karatayev, L. E. Burlakova, J. C. Nekola, L. E. Johnson, and D. K. Padilla, 2002, "Landscape Patterns of an Aquatic Invader: Assessing Dispersal Extent From Spatial Distributions," *Ecological Applications,* Vol. 12, No. 3, pp. 749-759.

Milton, J. G. and J. Belair, 1980, "Chaos, Noise, and Extinction in Models of Population Growth," *Theoretical Population Biology,* Vol. 37, No. 2, pp. 273-290.

Morales, J. M., 1999, "Viability in a Pink Environment: Why 'White Noise' Models Can be Dangerous," *Ecology Letters,* Vol. 2, pp. 228-232.

Rejmanek M. and R. D. Richardson, 1996, "What Attributes Make Some Plant Species More Invasive?" *Ecology,* Vol. 77, No. 6, pp. 1661-1666.

Ripa, J. and P. Lundberg, 1996, "Noise Colour and the Risk of Population Extinctions," *Proceedings of the Royal Society of London, Series B – Biological Sciences,* Vol. 264, No. 1377, pp. 1751-1753.

Shigesada, N. and K. Kawasaki, 1997, *Biological Invasions: Theory and Practice,* Oxford University Press, Oxford.

Sole, R. V., J. Bascompte, and J. Valls, 1992, "Nonequilibrium Dynamics in Lattice Ecosystems: Chaotic Instability and Dissipative Structures," *Chaos,* Vol. 2, No. 3, pp. 387-395.

Suarez, A. V., D. A. Holoway, and T. J. Case, 2000, "Patterns of Spread in Biological Invasions Dominated by Long-Distance Jump Dispersal: Insights From Argentine Ants," *Proceedings of the National Academy of Sciences USA,* Vol. 98, No. 3, pp. 1095-1100.

Turchin, P., 1998, *Quantitative Analysis of Movement,* Sinauer Associates, Inc., Sunderland, Massachusetts.

Turchin, P., 2003, *Complex Population Dynamics: A Theoretical/Empirical Synthesis,* Princeton University Press, Princeton.

Willebordse, F. H. and K. Kaneko, 1994, "Bifurcations and Spatial Chaos in an Open Flow Model," *Physical Review Letters,* Vol. 73, No. 4, pp. 533-536.

With, K. A., 1997, "The Application of Neutral Landscape Models in Conservation Biology," *Conservation Biology,* Vol. 11, No. 5, pp. 1069-1080.

With, K. A., and A. W. King, 1997, "The Use and Misuse of Neutral Landscape Models in Ecology," *Oikos,* Vol. 79, pp. 219-229.

Laura K. Marsh[1] *and Timothy Haarmann*[2]

Quantitative Habitat Analysis: A New Tool for the Integration of Modeling, Planning, and Management of Natural Resources

---

REFERENCE: Marsh, L. K. and Haarmann, T., **"Quantitative Habitat Analysis: A New Tool for Integration of Modeling, Planning, and Management of Natural Resources,"** *Landscape Ecology and Wildlife Habitat Evaluation: Critical Information for Ecological Risk Assessment, Land-Use Management Activities, and Biodiversity Enhancement Practices, ASTM STP 1458*, L. A. Kapustka, H. Galbraith, M. Luxon, and G. R. Biddinger, Eds., ASTM International, West Conshohocken, PA, 2004.

ABSTRACT: Federal laboratories are often caught between the need to meet mission objectives and the mandate to act as national stewards for natural resources. Following the initiation of the Manhattan Project in 1943, restricted land use at Los Alamos National Laboratory (LANL) retained many undisturbed areas that function as ecosystem sanctuaries for plants and animals throughout the 43 mi^2 site. We developed a Quantitative Habitat Analysis (QHA) that enables both managers and scientists to better meet the goals of ecosystem management and sustainable development for LANL. QHA is a multi-faceted modeling, planning, and management tool with the goal of applying existing models in a new way. QHA provides an objective, standardized, and replicable system for management of wild areas by federal agencies. As part of the development of QHA, we reviewed 42 existing wildlife and habitat models, assessments, or evaluation methods and 12 computer programs. A pilot field study was conducted on 12 plots testing five different methods to determine the most suitable for data collection for the tool. Once methods were selected and models determined, a QHA application tool was created in ArcView to test the pilot data within the tool for user-friendly application. This year, 45 field sites were sampled. QHA currently comprises five main sub-models analyzed within a geographic information system using ArcView: 1) Ecological Land Classification (landscape level effects), 2) Rapid Ecological Assessment (general assessment)/U.S. National Vegetation Classification Element Occurrence (ecosystem "health"), 3) BEHAVE (wildfire and fuels monitoring), 4) Habitat Analysis and Modeling System (wildlife), and 5) ECORSK.6 (bio-contaminants). Key to this QHA tool was the use of "common currencies." The common currencies consist of weighted scores that calculate a "grade" or means of comparison between different programs and scoring methods. Development and calibration of the QHA continues.

KEYWORDS: Quantitative Habitat Analysis, ArcView, Los Alamos National Laboratory, natural resources management, computer application, resource management tool

---

[1] Staff Scientist, Ecology Group, Los Alamos National Laboratory, P.O. Box 1663, MS M887, Los Alamos, NM 87545.

[2] Biology Team Leader, Ecology Group, Los Alamos National Laboratory, P.O. Box 1663, MS M887, Los Alamos, NM 87545.

"What we observe is not nature itself,
but nature exposed to our method of questioning."
--Heisenberg (1958)

Introduction

Modeling has become an important tool for ecologists and for the study of ecological systems. Many models exist to answer a multitude of questions with regard to management and status of habitats (Verner et al. 1986, Liu and Taylor 2002, Marsh and Haarmann 2000a, b, c). Part of what ecologists do is revise hypotheses and collect new data about the environment and how organisms interact with it, thus the model and the view of nature represented often undergo many changes from the initial conception to what is deemed the final product (Jackson et al. 2000). Quantitative models translate ecological hypotheses into predictions that can be evaluated in light of existing or new data (Jackson et al. 2000). That said, it is important to remember models are only tools not reality. Based on this concept, we recognized a need to develop a multi-faceted, user-friendly tool that we named Quantitative Habitat Analysis (QHA) for use at Los Alamos National Laboratory (LANL). QHA is a tool for assessing habitat, contaminants, wildlife, and wildfire in a way that allows the user to quickly and simply identify areas that are appropriate for preserving over areas suitable for projects (i.e., new buildings) that disturb the environment.

Following the establishment of the Manhattan Project in 1943, restricted land use at LANL created a unique opportunity to play a significant role in the conservation and sustainability of important aspects of biodiversity and other natural resources in the area. While some of this role has been obligated by compliance with federal regulatory mechanisms such as the Clean Water Act, the Endangered Species Act, and the National Environmental Policy Act, much also comes from the Department of Energy's (DOE's) stated goal of "using thoughtful planning to sustain the natural systems for which we are stewards." In 1976, the lands that comprise LANL were federally designated as a National Environmental Research Park. In addition, a significant portion of the site has been previously designated and preserved as environmental research/buffer area under LANL's Site Development Plans. These land-use practices have resulted in the preservation of undisturbed and sensitive natural resources that function as sanctuaries for biota throughout the LANL region. Over the past ten years, LANL has moved from a strictly project-specific, compliance-oriented strategy for environmental protection to a more landscape- and stewardship-focused approach.

DOE policy is to manage all of its land and facilities as valuable national resources. Long-term sustainable development goals are needed to focus this effort. There is a need for a comprehensive system of habitat evaluations. Various methods have been used to determine the vulnerability of a habitat or species within a region. Most of these are qualitative. Those quantitative tools each operate with different definitions, calculations, and values for natural resources from one organization to the next. Thus, each organization may provide different outcomes and recommendations based primarily on subjective classifications. QHA provides an objective, standardized, replicable, and accessible system for accurately determining the direction of stewardship is necessary, not only for continued management of wild areas by federal agencies, but for all professionals working in the field. QHA enables both managers and scientists to better meet the goals of ecosystem management and sustainable development.

Development of QHA

QHA was developed as a tool for quantifying different aspects of compliance over a large landscape. These features are not only topological, they also cover species, behaviors, and attributes of animals and plants, climate, fire, and contaminant distribution patterns unique to the region.

Background

The initiation of QHA was to develop a method for "ranking" fragmented habitat in tropical rain forest (Marsh 1999; Marsh et al. 2003). QHA for LANL was developed as an outgrowth of those efforts. We first determined the volume of information available on habitat models and assessments. All possible sources of information were searched for habitat models. This included literature searches, particularly through BIOSIS and other libraries, and web-searches for federal (e.g., Department of Defense, DOE, Environmental Protection Agency, U.S. Forest Service, U.S. Department of Agriculture, US Geological Survey, National Park Service, U.S. Air Force, U.S. Fish and Wildlife Service, and other national labs) and non-profit/public organizations who use habitat models (e.g., The Nature Conservancy, Conservation International, National Center for Ecological Analysis and Synthesis, and Conservation Breeding Specialist Group). To do these searches, the following keywords were used: habitat analysis, habitat assessment, habitat evaluation, quantitative biological assessment, quantitative habitat assessment, biological assessment, biological classification, ecological classification, ecological assessment, ecological evaluation, biodiversity classification, biodiversity analysis, and biodiversity assessment (c.f., Marsh and Haarmann 2000a, b).

Existing programs and models compiled in the first phase of this project were reviewed for use by LANL. Those selected represent all of the major categories necessary for management and decision-making. We reviewed 42 models and assessment or evaluation methods and 12 computer programs. The categories created for QHA during this phase include, but are not limited to, the following: Ecosystem Management, Ecosystem Health, Wildlife Properties, Threatened and Endangered (T and E) Species, Wildfire Properties, and Ecorisk. These categories are collectively called, "QHA categories."

Next, we determined the format of the QHA process. It is easiest to understand QHA by thinking of it as the "umbrella" under which selected programs to work (Fig. 1). Each program will ultimately work both independent of and intertwined within the QHA process. For example, we link all of the selected programs and methods to a QHA category (see "Selected Programs and Methods"). These links between the categories are what we call "common currency," or integrating language and ranking system of QHA (Marsh and Haarmann 2000a, b). Programs and methods can be interchanged, replaced, or added with more or fewer categories as determined by the user's needs (see "Common Currency").

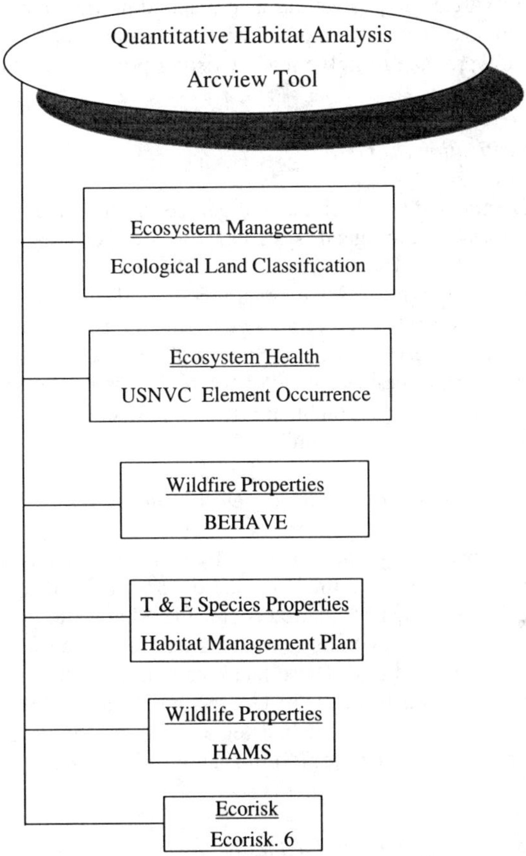

FIG. 1—*The QHA model, where QHA itself is an overarching concept encompassing many datasets, programs, and methodologies. The lines connecting each subsection are the common currencies, where the connection of all programs ultimately lead to one score. Ecological Land Classification is a GIS based system, US National Vegetation Council (USNVC) Element Occurrence is a methodology with a ranking system, BEHAVE is a name for a fire modeling computer program, Threatened and Endangered (T & E) species Habitat Management Plan is a computer map- based dataset in ArchView, Habitat Analysis and Modeling System (HAMS) is a computer based program, and Ecorisk.6 is a computer based program.*

Selected Programs and Methods

The programs and data collection methods were tested in the lab and in the field before final selection. Cirteria for selection was user accessibility and an output we could use for the common currencies for evaluating scores across programs. We wanted to have an analysis for each QHA category as well as a corresponding field method for supplying the data. The entire tool is a series of command windows within the ArcView application. Within this format we were able to link already existing programs and data to the QHA categories.

Ecological Land Classification (ELC)

ELC is a scientific endeavor that attempts to organize, stratify, and evaluate ecological systems (and complexes of ecological systems) for the purposes of land resource management (Sims et al. 1996). The end-products of ELC are context and some spatially-based rule sets for the design, practice, evaluation, and iterative improvement of integrated resource management. ELC is a prerequisite for the evaluation of different trade-off decisions regarding resource issues (Sims et al. 1996). The use of ELC in QHA is to provide an overall landscape-level view of the region as it contributes to the other components of the tool through a geographic information system (GIS). ELC supports the QHA category of "Ecosystem Management."

U.S. National Vegetation Classification Element Occurrence (EO)

This is a method for classifying whole habitats by using a simple ranking system. An "element" is defined as a plant or animal population or stands of a plant community where only one occurrence of such population is considered at a site (Groves and Valutis 1999). The EO ranking of a plant community within a site focuses on three sets of factors: condition, landscape, and size. These factors are used extensively by The Nature Conservancy, and are described in detail in Groves and Valutis 1999. The element is given ranks between 1.0 and 4.0 for each category and subset, where 1.0 is highly degraded and 4.0 is nearly "pristine." This information supports the QHA category of "Ecosystem Health."

BEHAVE: Fire Behavior Prediction and Fuel Modeling

This program gathers available fire models into a system that is driven by direct user input (Andrews 1986, Andrews and Chase 1989). BEHAVE produces tables of fire behavior given user defined environmental conditions. It provides methods for projecting the behavior of active fires, for prescribed fire planning, for fuel assessment, and for many fire management applications. This program is used for QHA category "Wildfire Properties."

Habitat Analysis and Modeling System (HAMS)

This program is designed to combine graphical, analytical, and modeling capabilities into Windows-based applications (Cooperative Wildlife Research Lab 1996). It primarily analyzes habitat suitability of original or modified images using pattern recognition (PATREC). PATREC is a method that evaluates of habitat suitability based on probability that a particular habitat condition is consistent with a set of observed landscape conditions. This program seems to be more suitable than the traditional Habitat

Suitability Index (HIS) for users unfamiliar with the HSI methodology. By using HAMS, data can be collected that satisfies this model in a variety of ways depending on the desired outcome. This program is used for the QHA category "Wildlife Properties."

Habitat Management Plan (HMP)

The HMP was developed for use specifically for compliance and management issues surrounding Threatened and Endangered (T and E) species at LANL (LANL 1998). Within the construct of the HMP are definitions developed for Areas of Environmental Interest (AEIs) that define core and buffer zones for corresponding species habitat use. This tool is a map-based program that allows the user to determine the location of a site in question as it corresponds to a T and E species zones. This program is used for QHA category "T and E Properties."

ECORSK.6

ECORSK.6 is a custom FORTRAN model that was developed as a tool specifically for conducting ecological risk assessment at LANL (Gonzales et al. 2002). The program integrates GIS data on environmental contamination and animal distribution with many other types of information, such as contaminant toxicity, so that animal exposures to contaminants can be estimated and compared to no adverse effect levels or animal safe limits. Integration of datasets results in the production of hazard indices which, when compared to risk evaluation criteria, estimate the potential for impacts to an organism from consumption of contaminants in food and soil. This program is used for the QHA category "Ecorisk."

Common Currency

The concept behind common currency is simple. We developed a scoring system that links different program outputs in a way that produces a single rank or series of ranks for an individual habitat or study site. Each program generates a score or value as a product of the program or model's calculations. In each program or in the case of EO habitat ranking by the researcher, generates a score based on the data collected for the specific program.

For example, ECORSK.6 generates values between or greater than 0.01 and 100.0 based on the chemical analytes found within a sample (typically soil). Their toxicity levels are scored and these scores are then used for ranking within QHA. To create a score for QHA, these values were given "grades" and a corresponding score. A score of ≤0.3 was an "A," 0.31 to 1.0 a "B," 1.1 to 10.0 a "C," through an "F" grade of >100.0 (Gonzales et al. 1997). An "A" rank transcribed to a value of 4.0, "B" a 3.0, "C" a 2.0, and so on. Some of the categories are determined both by a program and by a ranking system. For QHA category "Wildfire Properties," we use BEHAVE, a program that gives data output on fuel models for predicting fire behavior. We use this model along with the common currency ranks for wildfire hazards developed by R. Balice (in Marsh and Haarmann 2000c). The currency again sets an A through F scale based on additional fire behavior not covered in BEHAVE such as inter-crown distance, canopy cover, and tree density.

Once we have assigned a score to each output for each category, the user (LANL ecologists) is asked to weight the importance on a scale from 1 to 10 based on the question asked of the data, where 1 is lowest and 10 is highest. For instance, if the question is "Is site 1 a less degraded habitat for use by Mexican spotted owls (an endangered species) to nest in than site 2?" More weight may be given to Ecosystem Health, T and E Properties, and Eco-Risk categories than Wildlife Properties, Wildfire Properties, or Ecosystem Management categories. Once the categories have been weighed, they can be summarized for a project or question. However, this step is not recommended as it is better to leave scores separate for interpretation.

Field Methods

Once we had determined a set of categories for QHA, we proceeded with a pilot study. The pilot study was conducted primarily to 1) compare field data collection methods, 2) ultimately select a standard field method, 3) gain preliminary comparisons between the sites, and 4) begin analysis by creating an application interface through ArcView. Species richness is important for evaluating ecosystem health and habitat availability for wildlife and T and E species and contributes to ecosystem management decisions. Thus, in each habitat we decided to place three 20 by 50 m plots to capture the most diversity by plot shape (Stohlgren et al. 1995). Plots were placed randomly within the selected habitats. Plot layout was constrained by canyon or drainage edge, burned areas, roads, or change in habitat type (e.g., transition zones). For the pilot study, we determined that it would be beneficial to compare two different habitat types to see if the QHA tool would provide ranks appropriately based on the data collected. We chose ponderosa pine forest (PIPO) and piñon-juniper woodland (PJ) as the focal habitats. Plots were in pairs with a "control" (or "desired future state") and "experimental" (degraded or untreated) plots. We assumed that all study areas were homogeneous within themselves. That way a single "type" of habitat would then be comparable. In total we sampled 12 plots, two sites in PIPO and two sites in PJ.

Within each 20 by 50 m plot we compared several vegetation data collection methods to determine which one would provide the most data to satisfy as many QHA categoris as possible in the shortest amount of field time: Gentry Method for stems greater than 2 cm diameter at breast height (Gentry 1986), Dallmeier Method used to charaterize vegetation type and estimate plant diversity and abundance (Dallmeier 1992), Modified Whittacker Method use to sample all species regardless of size for presence and cover (Stohlgren et al. 1995), and Vegetation and Fuels Method used specifically at LANL for sampling soils, vegetation, and fuel loads (Balice et al. 1999, 2000). We determined that the Modified Whittacker method and the Gentry Method suited our QHA data fields in a way that covered all categories necessary for the QHA tool (Marsh and Haarmann 2000c).

During field season 2002, we collected additional QHA plot data to further test the QHA tool. We refined the data collection categories by adding more fuels and wildlife information. We collected data in 45 plots, 16 different sites and in four types of habitat and conditions (PIPO, PJ, transitional, and disturbance/tree thinning areas in PJ). These data will be used to further add to data fields within the QHA tool.

Using QHA

The QHA tool was designed to be a user-friendly yet robust system for answering questions regarding resource management with respect to institutional goals (Marsh et al. 2001). We intentionally created a tool that allows the flexibility of adding components, such as statistical packages for data analysis, as needed. To date we have populated the fields on the initial run of the QHA tool with data from the pilot study. During 2003, we anticipate working on adding analysis programs and the 2002 field datasets.

On our QHA ArcView tool, we have the following menu items: Digital Orthophoto, Ecosystem Health, T and E Properties, Wildlife Properties, EcoRisk Properties, QHA Ratings, Raw Data, Window, and Help (Fig. 2). We describe each section and its use in detail below. We have not included in the menu bar a selection for Ecosystem Management but will add this item in 2003. If more information is needed about a method, a box pops up with details from a QHA report on the methods used to collect data, information about the program used and the methods for scoring in that category. Each of the categories in the menu has an option for the user to get more information.

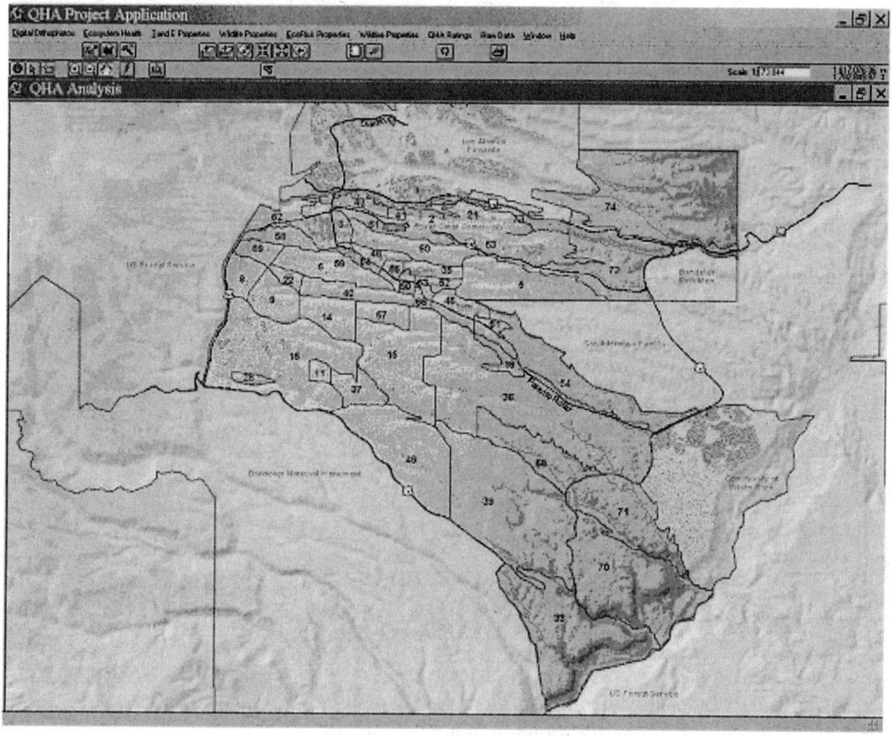

FIG. 2—*QHA ArcView Tool start page. Notice the map of LANL boundaries in the center and the menu bar at the top.*

Digital Orthophoto

This is a satellite image of the exact site where data are collected. Maps are called up both as photos or as topo-images. Plot location and site location are overlaid on these maps, along with any other notable features as desired, such as land cover maps, roads, or buildings.

Ecosystem Health

When this heading is selected, a drop down menu appears that allows the user to select from different items, including Condition Factor, Landscape Context Factor, Size Factor (c.f. Groves and Valutis 1999 for description of factors), List Cover Types (based on a LANL land cover map), Plant Species List, Ecosystem Health Grade (generated from the QHA common currencies), and More Information on Ecosystem Health. If one of the factors is selected, a score box appears with all of the EO rankings (Fig. 3).

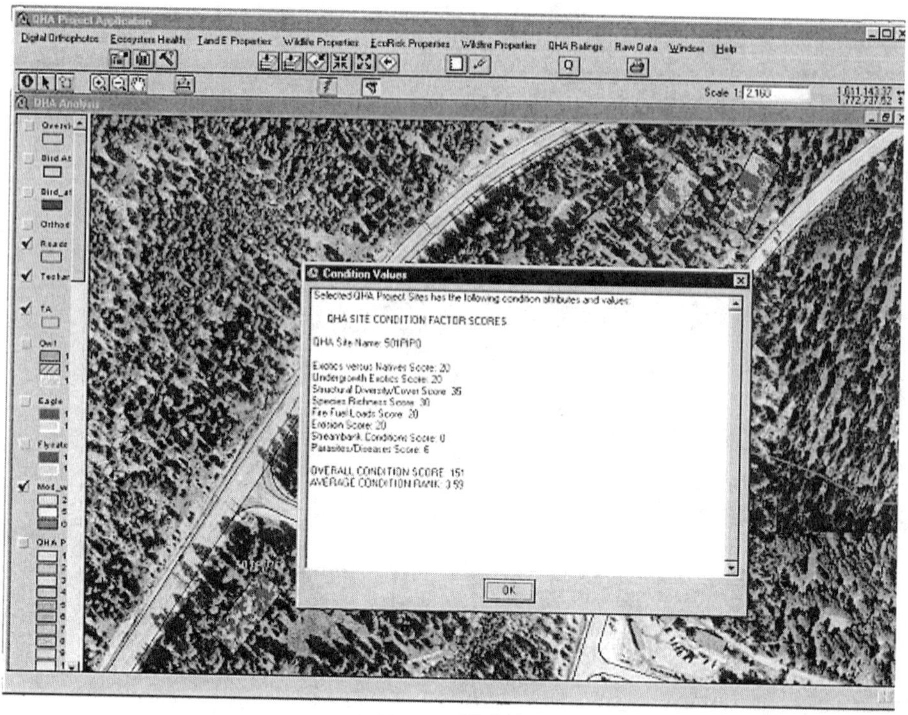

FIG. 3—*QHA work screen showing digital orthophoto with plots overlaid and drop down menu showing items for selection.*

T and E Properties

The data available in this section are primarily from the HMP defined AEIs. Here we include selections of Potential Species (that might be found in the study plot), List of AEIs in the area, Closest Distance to AEIs (from the plot to the nearest AEI), T and E Grade, and More Information.

Wildlife Properties

There are many selections under the menu item for Wildlife Properties. Options include Arthropod (Species) List, Arthropod Abundance, Bat List, Bat Abundance, Bird List, Bird Frequency, Small Mammal List, Small Mammal Detection (includes sign of animals in study area), Large Mammal List, Large Mammal Detection, Wildlife Property Grade, and Information. We will be adding a link for HAMS and data analysis links to that system.

Ecorisk Properties

In this category we include List Individual HQs (hazard quotient for specific analytes), List Overall HQ (average for an area), Ecorisk Score, and Information. In this section we are also adding a link to the ECORSK.6 program for data to be analyzed directly.

Wildfire Properties

This category has several item selections: Intercrown Distance, Tree Density 1 (overstory trees), Tree Density 2 (understory trees), Canopy Cover, Ladder Fuels, Down Fuels, Diseased Trees, Wildfire Grade, and Information. We will map a link to the BEHAVE program in this category for direct fuels analysis.

QHA Rankings

Under this menu item, the user can directly find scores for each of the QHA categories. All of the categories include the score within each drop down menu, but in this section, all of them are listed and can be selected and compared. A final composite grade from all of the QHA ranks can also be calculated. For a composite score to be calculated, the user is prompted to supply weighting factors to sort out the importance of each category before the final composite score is calculated.

Raw Data

For the user who is interested in looking at the original data in its unprocessed form, a menu selection for this option is included. The drop down items currently only include the vegetation data and are VegPlots 0.5 by 2 m, VegPlots 2 by 5 m, VegPlots 20 by 50 m, and Tree Data. We will be adding raw links to all of the categories' datasets.

Discussion and Future Development

QHA is a tool in progress. In addition to the programs that fit specific categories, we intend to map a statistics program, S-Plus, as an additional menu item for calculations within the raw datasets. Since the tool is flexible in respect to data analysis and interpretation, it will become useful to many resource managers for many different kinds of questions. For instance, while the original use for QHA is for ecologists at LANL to better inform Lab project managers to better understand the quality of the habitat proposed for destruction, it is also setup to work directly on projects within ecology. Once datasets represent larger areas of LANL land, there will be greater potential for asking lab-wide question, such as whether the feeding behavior of large ungulates (elk, deer) are within habitats that are consistently of a high grade or not, based on the QHA criteria.

There are limitations to a tool such as this. We are still working to discover all of the potential mistakes the grading system creates. Does it really reflect "reality" of the habitat in question? Like many tools, models give us a measure for understanding and comparing complex systems. We are using this tool to span many categories that are in themselves separate. The common currency concept is new and will need a great deal of testing before we are comfortable with using the tool for a wider audience. It is limited currently by few datasets available for each program to analyze, and we are still working to better quantitative interpretation among the QHA categories.

Management decisions are based carefully on output from the QHA ranking system. The tool is too new to rely on it as a sole basis for all decisions, so at present it is used to augment decisions along with basic compliance assessments.

Acknowledgments

Many thanks to all of the staff and students of the Ecology Group, LANL, who have helped us over the years in the various aspects of this project, including Gil Gonzales, Todd Haagenstad, Kathy Bennett, Carey Bare, Rhonda Robinson, Branson Pearson, Chuck Hathcock, Ginny Seamster, Tina Sommer. We also thank the Biological Resources Management Plan Special Studies for funding the initial stages of this project.

References

Andrews, P. L., 1986, "BEHAVE: Fire Behavior Prediction and Fuel Modeling System—BURN Subsystem, Part 1," *General Technical Report INT-194*, USDA Forest Service, Intermountain Research Station, Ogden, Utah.

Andrews, P. L. and Chase, C. H., 1989, "BEHAVE: Fire Behavior Prediction and Fuel Modeling System—BURN Subsystem, Part 2," *General Technical Report INT-194*, USDA Forest Service, Intermountain Research Station, Ogden, Utah.

Balice, R. G., Oswald, B. P., and Martin, C., 1999, *Fuels Inventories in the Los Alamos National Laboratory Region*, 1997, LA-13572-MS, Los Alamos National Laboratory, Los Alamos, New Mexico.

Balice, R. G., Miller, J. D., Oswald, B. P., Edminster, C., and Yool, S. R., 2000, *Forest Surveys and Wildfire Assessment in the Los Alamos Region; 1998-1999*, LA-13714-MS, Los Alamos National Laboratory, Los Alamos, New Mexico.

Cooperative Wildlife Research Lab, 1996, "HAMS: Habitat Analysis and Modeling System," *Wildlife Society Bulletin*, Vol. 24, No. 2, pp. 340–341.

Dallmeier, F., 1992, "Long-term Monitoring of Biological Diversity in Tropical Forest Areas: Methods for Establishment and Inventory of Permanent Plots," *MAB Digest*, Vol. 11, Paris, UNESCO.

Gentry, A. H., 1986, "Species Richness and Floristic Composition of Choco Region Plant Communities," *Caldasia*, Vol. 15, pp. 71–91.

Gonzales, G. J., Gallegos, A. F., Foxx, T. E., Fresquez, P. R., Pratt, L. E., Mullen, M. A., and Gomez, P. E., 1997, *Preliminary Risk Assessment of the Bald Eagle at the Los Alamos National Laboratory*, LA-UR-97-4732, Los Alamos National Laboratory, Los Alamos, New Mexico.

Gonzales, G. J., Gallegos, A. F., Ryti, R. T., Newell, P. G., Bare, C. M., Bennett, K. D., Biggs, J. R., Kock, S. W., Mullen, M. A., Myers, O. B., Soholt, L. F., and Volke, R. W., 2002, *Tier 2 Ecological Risk Assessment of Los Alamos National Laboratory Institutional Issues on the Pajarito Plateau Using ECORSK.6*, LA-13956, Los Alamos National Laboratory, Los Alamos, New Mexico.

Groves, C. and Valutis, L., 1999, *Guidelines for representing ecological communities in ecoregional conservation plans.* The Nature Conservancy, Arlington, Virginia, pp. 1–73.

Heisenberg, W., 1958, *Physics and Philosophy: The revolution in modern science*. New York: Harper.

Jackson L. J., Trebitz, A. S., Cottingham, K. L., 2000, "An Introduction to the Practice of Ecological Modeling," *BioScience* Vol. 50, No. 8, pp. 694–706.

LANL, 1998, *Threatened and Endangered Species Habitat Monitoring Plan Overview*, LA-LP-98-112, Los Alamos National Laboratory, Los Alamos, New Mexico.

Liu, J., Taylor, W. W., 2002, Cambridge Studies in Landscape Ecology: Integrating Landscape Ecology into Natural Resource Management, Cambridge University Press, Cambridge, UK.

Marsh, L. K., 1999, Ecological Effect of the Black Howler Monkey (*Alouatta pigra*) on Fragmented Forests in the Community Baboon Sanctuary, Belize, PhD Dissertation, Washington University, St. Louis, Missouri.

Marsh, L. K. and Haarmann, T. K., 2000a, *Quantitative Habitat Analysis: A progress report on Phase I of the Work Plan*, LA-UR-00-6018, Los Alamos National Laboratory report, Los Alamos, New Mexico.

Marsh, L. K. and Haarmann, T. K., 2000b, *Quantitative Habitat Analysis: A progress report on Phase II of the Work Plan*, LA-UR-00-6017, Los Alamos National Laboratory, Los Alamos, New Mexico.

Marsh, L. K. and Haarmann, T. K., 2000c, *Quantitative Habitat Analysis: Final Report for FY00*, LA-UR-00-6016, Los Alamos National Laboratory, Los Alamos, NM.

Marsh, L. K., Haarmann, T. K., and Bennett, K., 2001, *Quantitative Habitat Analysis ArcView Application Interface: Final report FY01*, LA-UR-01-6694, Los Alamos National Laboratory, Los Alamos, New Mexico.

Marsh, L. K., Chapman, C. A., Norconk, M. A., Ferrari, S. F., Gilbert, K. A., Bicca-Marques, J. C., and Wallis, J., 2003, "Fragmentation: Specter of the Future or the Spirit of Conservation?" In: Marsh, L. K. (Ed.), *Primates in Fragments: Ecology and Conservation*. Kluwer Scientific/Plenum Press, New York, pp. 381–395.

Sims, R. A., Corns, I. G. W., and Klinka, K., 1996, "Introduction," In: Simms, R. A., Corns, I. G. W., Klinka, K. (Eds.), *Global to Local: Ecological Land Classification*, Kluwer Academic, Netherlands, pp. 1–10.

Stohlgren, T. J., Falkner, M. B., and Schell L. D., 1995, "A Modified-Whittaker Nested Vegetation Sampling Method," *Vegetatio,* Vol. 117, pp. 113–121.

Verner, J., Morrison, M. L., and Ralph, C. J., 1986, Wildlife 2000: Modeling Habitat Relationships of Terrestrial Vertebrates, University of Wisconsin Press, Madison, Wisconsin.

Lawrence A. Kapustka,[1] *Hector Galbraith,*[2] *Matt Luxon,*[3] *Joan M. Yocum,*[4] *William J. Adams*[5]

Predicting Biodiversity Potential Using a Modified Layers of Habitat Model

REFERENCE: Kapustka, L. A., Galbraith, H., Luxon, M., Yocum, J. M., Adams, W. J., **"Predicting Biodiversity Potential Using a Modified Layers of Habitat Model,"** *Landscape Ecology and Wildlife Habitat Evaluation: Critical Information for Ecological Risk Assessment, Land-Use Management Activities, and Biodiversity Enhancement Practices, ASTM STP 1458*, L. A. Kapustka, H. Galbraith, M. Luxon, G. R. Biddinger, Eds., ASTM International, West Conshohocken, PA 2004.
ABSTRACT: Ecological resources of the 360-km^2 Kennecott Utah Copper Corporation property in the Oquirrh Mountains has been studied since the 1970s. Detailed descriptions of the vegetation were undertaken in 1994 as part of a site-wide Ecological Risk Assessment (EcoRA). A vegetation map was produced from infrared aerial photographs and floristics lists were compiled for a site-wide ecological risk assessment. Vegetation cover was characterized using a relevé sampling procedure. Data from georeferenced relevé sampling points were used to propagate vertical structure characteristics across polygons. We adapted the Layers of Habitat HSI Model and calculated an index value for each polygon. Submontane shrub and the submontane shrub/desert shrub cover types comprised 32% and 12% respectively of the site-wide index; contributing more to the site-wide index than their area would suggest. Conversely, the grassland and salt desert shrub cover types contributed less to the site-wide index than their area would suggest. Index values were grouped into pentiles and the polygons were displayed in GIS format to illustrate the spatial distribution of structural diversity. The Layers of Habitat model is a surrogate for biodiversity (richness) potential. The spatial representation of structural diversity is being used to guide management practices to enhance site-wide biodiversity.

KEYWORDS: Biodiversity, landscape characterization, vertical structure

[1] President and Senior Ecotoxicologist, ecological planning and toxicology, inc, 5010 SW Hout street, Corvallis, OR 97330-9540

[2] Galbraith Environmental Sciences, 289 Wiswall Hill Rd., Newfane, VT 05345

[3] Ecotoxicologist, Windward Environmental LLC, 200 West Mercer Street, Suite 401, Seattle, WA 98119

[4] Environmental Scientist, , ecological planning and toxicology, inc, 5010 SW Hout street, Corvallis, OR 97330-9540

[5] Principle Environmental Scientist, Rio Tinto Health Safety and Environment, 5295 South 300 West, Suite 300, Murray, UT 84107

Introduction

Biodiversity is a description of the collection of species found in an area. In its simplest form, this is a mere count of all species, referred to in technical literature as "species richness." In most situations, however, additional value judgments are applied in the evaluation of biodiversity. For example, exotics or "weedy" species (e.g., starling) are valued less than endemic (native) ones (e.g., mountain bluebird or lazuli bunting). Other considerations such as resident versus transient (migratory) or abundance (single sighting versus common) add to our technical understanding and appreciation of biodiversity beyond the simple numbers. The degree to which abundance of individuals is spread across many species is referred to as a measure of "evenness." Taken together, "richness" and "evenness" define many indices of biodiversity. Rare species, ones that are present due to uncommon habitat features of a region, also become highly valued as contributors to the biodiversity profile, even if the requisite habitat supports only a few species. One example of this for Kennecott Utah Copper Corporation (KUCC) property is the northern goshawk, which requires conifer forests that tend to have relatively low species richness, compared with other forest types.

Patterns of biodiversity vary across large areas. Generalized global trends indicate that maximum biodiversity occurs in equatorial rain forests with a gradient of decreasing numbers of species toward the polar regions (Whittaker 1970). Structural features of the landscape described in terms of vertical and horizontal heterogeneity are related to biodiversity. Increased heterogeneity results in greater potential for different species to share an area, thus elevating the level of biodiversity.

Ecologists also have described relationships between biodiversity and ecological services. Generally, higher biodiversity corresponds to higher levels of primary and secondary productivity of the landscape, improved efficiency of energy use within the ecological system, improved efficiency in nutrient cycling processes, improved efficiency in water use ratios, and improved capacity to withstand internal and external stresses imposed on the ecological system. Collectively, these features enable higher sustained levels of use for human purposes, including harvest, lowered maintenance costs to manage water flows (e.g., flood prevention), and many intangibles such as enhanced aesthetics, photographic opportunities, botanizing, birding, and the like.

KUCC and its parent company Rio Tinto have adopted policies and begun to incorporate management objectives that promote landuse management practices to enhance biodiversity. The KUCC holdings in the Northern Oquirrh Mountains and wetlands along the south shore of the Great Salt Lake have the potential to support impressive levels of biodiversity. Considerable study of ecological resources of the property has occurred since the 1970s, initially to document effects mining operations had on vegetation and wildlife, and subsequently to characterize the recovery of these resources. Detailed descriptions of the vegetation were undertaken in 1994 as part of a site-wide Ecological Risk Assessment (EcoRA; **ep** and **t** 1994a, b, 1995a, b, c, 1996). Those and other historical data (Blanchard 1973, BioWest 1991, SWCA 1996) as well as newer compilations (**ep** and **t** 1997, TNC 1997) have been used to document the biodiversity observed or suspected to occur on the site. Moreover, these data have been used to assess the site-wide potential biodiversity using a modification of a Layers of Habitat Model (Short 1984). In our analysis, we draw attention to particular sources of uncertainty both in the documented and in the forecasted biodiversity (primarily richness). Finally, we present suggestions regarding means to reduce uncertainties and to institute a monitoring program to guide future environmental management decisions.

Materials and Methods

Previous work on the site characterized the biological features of the landscape in relation to environmental impact assessments, ecological risk assessments, and to assist decision-makers in evaluating various land management options for such matters as erosion-control, flood abatement, and the like.

Site Description

A major component of structural diversity relates directly to the composition and character of the physical environment. Clearly, interactions of climate over long periods and weather extremes over shorter periods help to define the type of vegetation occupying a site. The great variation in elevation, slope, aspect, and parent material across the KUCC property contributes significantly to the observed and potential biodiversity.

Physical Setting

The Oquirrh Mountains are located in central Utah at the southeast end of the Great Salt Lake, just west of the Wasatch Range and Salt Lake City (40° 28' to 40° 45' N latitude; 112° 05' to 112° 15' W longitude). They lie in a north-south orientation between the Jordan and Cedar Valleys to the east and the Tooele and Rush Valleys to the west. Approximately 360 km^2 (140 sq mi) of this area is owned by KUCC in support of their mining operations.

Precipitation varies greatly in the Oquirrh range, from approximately 40 cm annually at the Garfield monitor, to 60 cm at Bingham, to an estimated 100 cm at higher elevations (approximately 15 to 40 inches annually). Conspicuous geological folding is evident throughout the range, generally in a northwest-southeast orientation, with fault block uplifts and tears visible in many locations. The crest of the Oquirrh Mountains is generally about 2500 meters (8000 ft) in elevation, with Nelson Peak, Clipper Peak, and West Mountain exceeding 2,800 meters (9000 ft). Elevations in the Salt Lake Valley east of the Oquirrhs range from about 1600 meters (,100 ft) in the south (east of Copperton) to about 1300 m (4200 ft) at the Great Salt Lake (Figure 1). Sampling in 1994 was designed to distribute points across elevation zones (Figure 2).

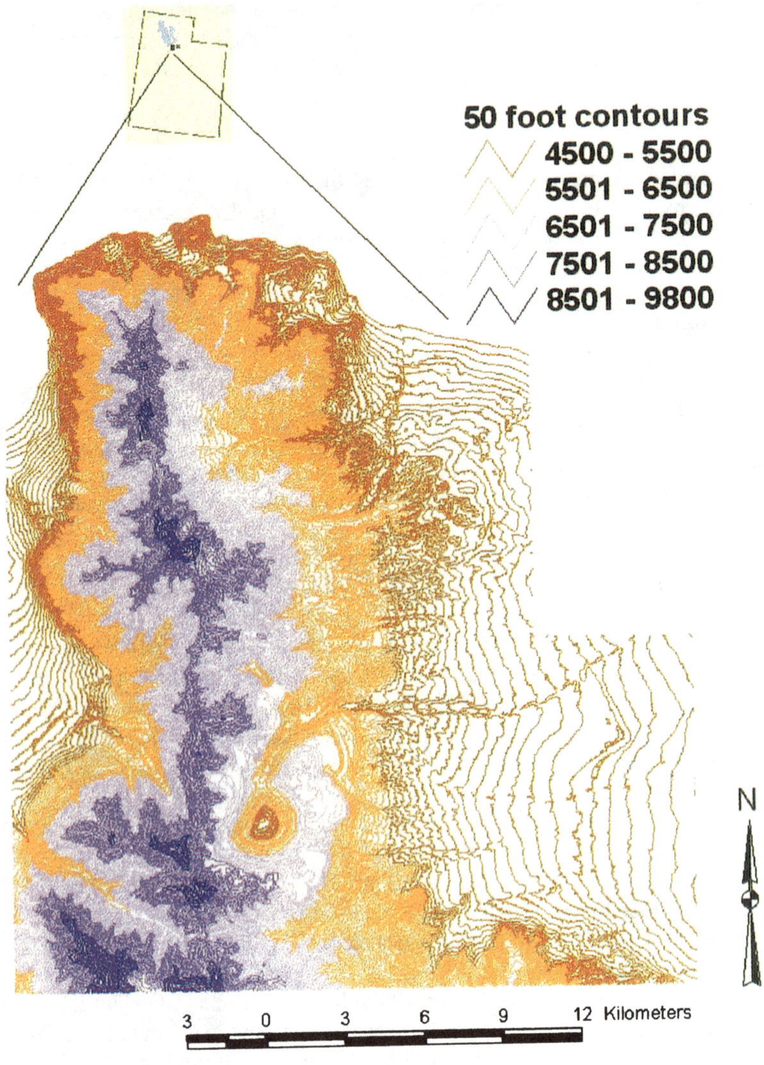

Figure 1. *Elevation contours of the Oquirrh Mountains.*

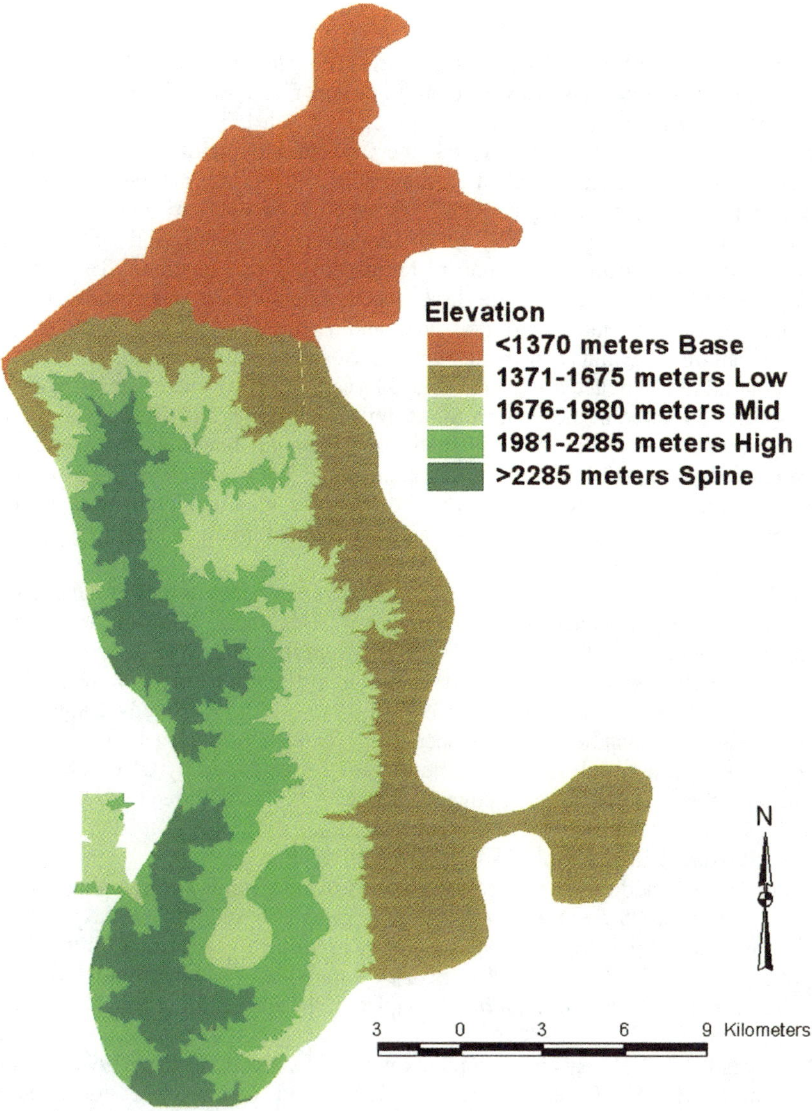

Figure 2. *Elevation zones of the Oquirrh Mountains used to distribute sampling locations across potentially important zones in the 1994 survey.*

Vegetation

Central Utah is part of the Great Basin Desert. Mountain ranges within this area are biological islands and differ greatly in their flora and fauna from those on the valley floors. Fresh and saline wetlands adjacent to the Great Salt Lake have additional biotic differences. The elevation gradient within the Oquirrh Mountains has resulted in the formation of several different plant communities, including natural communities associated with valley, foothill [1600 to 2000 m; 5250 to 6500 ft] and mid-montane [2000 to 3000 m; 6500 to 9800 ft)] zones (Arnow *et al.* 1980), and anthropogenic communities resulting from agriculture, disturbance, and revegetation. They include coniferous forest, juniper, sub-montane shrub, aspen, aspen-conifer, desert shrub, riparian forests, salt desert shrub, marshland, grassland, and a number of introduced cover types such as Russian olive savannas, town-site woodlands, and agricultural fields.

Several studies conducted over the past 30 years have included lists of plant and wildlife species observed in the Oquirrh Mountains. The studies had different objectives. Also, they were conducted on different spatial and temporal scales. Consequently, the data are not fully compatible in terms of locating the points of observation. Few of the studies were intended to catalogue species occurrence in a comprehensive manner. Despite these limitations, the data provide considerable information regarding levels of biodiversity.

Floristics lists were compiled by canyon and elevation zones as part of the site-wide EcoRA field work conducted in 1994 (**ep** and **t** 1995c) and 1995 (**ep** and **t** 1997). Taxonomy followed that of Welsh *et al* (1993). Supplemental work added additional species to the KUCC species list (TNC 1997), though the observations were not made with the same level of precision as in the 1994 and 1995 efforts. The study design for the EcoRA focused on documentation of species by canyon and elevation zone. The stratified random design did not conform to the cover types used by Blanchard (1973). Consequently, the quantitative plant cover data cannot be linked directly to the wildlife observations.

The 1994 data tallied 356 plant species from 56 families (

Table *1*). Two additional species were observed in a limited re-sampling of low-elevations areas in 1995 associated with trapping grids. In addition, of the 99 species identified by Paul Rokich (**ep** and **t** 1996) as those used in re-vegetation efforts, 39 taxa were not represented in the compiled floristics for the site. Though cross-tabulation of species listed in TNC (1997) has not been completed, the floristics list will likely exceed 410 taxa.

More than 7,000 observations of plants were recorded during the 1994 sampling effort. Nearly 60% of the species were identified as mid- or late successional species, though the areas most affected by previous smelter emissions (e.g., Kessler Canyon, Black Rock Canyon) or physical disturbance (e.g., Butterfield Canyon) had the highest proportion of early-successional species (Table 2). The proportion of late-successional species increased with increasing elevation (Table 3).

Table 1. *Top ten plant families in terms of species richness from 1994 relevé data.*

Family		Number of species
Compositae	sunflower	68
Gramineae	grass	47
Leguminoseae	legume	26
Chenopodiaceae	goosefoot	18
Rosaceae	rose	18
Scrophulariaceae	figwort	15
Cruciferae	mustard	14
Polygonaceae	buckwheat	10
Onagraceae	evening primrose	9
Liliaceae	lily	8
Remaining 46 families		111
	Total	356

Table 2. *Distribution of plants (frequency of occurrence) by successional position across sampling areas.*

Sample Unit	Early-successional	Mid-successional	Late-successional	Unknown
Black Rock Canyon	58.1%	23.9%	16.8%	1.2%
Butterfield Canyon	54.3%	39.6%	5.5%	0.6%
Coon Canyon	44.3%	31.6%	23.4%	0.7%
Harkers Canyon	32.2%	35.6%	30.6%	1.6%
Kessler Canyon	67.8%	25.5%	5.8%	0.9%
Little Valley	59.0%	24.6%	14.6%	1.7%
Pine Canyon	60.5%	27.8%	10.4%	1.3%
Spine	7.1%	56.8%	35.5%	0.7%
Tailings Berm – new	55.7%	38.2%	2.3%	3.8%
Tailings Berm – old	65.5%	25.6%	5.0%	3.9%
Wetlands	31.7%	49.8%	18.1%	0.5%
Grand Total	41.9%	36.2%	20.6%	1.2%

The distribution of native versus introduced species followed a pattern similar to successional position (Table 4-5). Approximately 75% of the taxa were native, a proportion similar to that of other Utah mountain ranges (Kass 1988). Patterns of native and introduced species across watersheds or canyons followed historical disturbance levels, with those areas in the lowest elevation and closest to the smelter (e.g., Black

Rock or Kessler Canyons) or recreational use (Butterfield Canyon) having the highest proportion of introduced species (Table 5)

Table 3. *Distribution of plants (frequency of occurrence) by successional position across elevation zones.*

Elevation Zone	Early-successional	Mid-successional	Late-successional	Unknown
(1) Base-elevation	53.9%	35.1%	8.7%	2.3%
(2) Low-elevation	69.9%	20.7%	8.7%	0.7%
(3) Mid-elevation	53.5%	30.1%	15.7%	0.7%
(4) High-elevation	30.2%	37.3%	29.8%	2.7%
(5) Spine	7.1%	56.8%	35.5%	0.7%
Grand Total	41.9%	36.2%	20.6%	1.2%

Table 4. *Distribution of native versus introduced plants (frequency of occurrence) across elevation zones.*

Elevation Zone	Introduced	Native	Unknown
(1) Base-elevation	47.5%	50.2%	2.3%
(2) Low-elevation	36.6%	62.8%	0.7%
(3) Mid-elevation	28.5%	70.9%	0.7%
(4) High-elevation	15.7%	81.6%	2.7%
(5) Spine	4.9%	94.4%	0.7%
Grand Total	24.4%	74.4%	1.2%

Table 5. Distribution of native versus introduced plants (frequency of occurrence) across sampling areas.

Sample Unit	Introduced	Native	Unknown
Black Rock Canyon	30.9%	67.9%	1.2%
Butterfield Canyon	45.7%	53.7%	0.6%
Coon Canyon	22.2%	77.1%	0.7%
Harkers Canyon	13.4%	85.0%	1.6%
Kessler Canyon	38.3%	60.8%	0.9%
Little Valley	29.6%	68.7%	1.7%
Pine Canyon	28.9%	69.9%	1.3%
Spine	4.9%	94.4%	0.7%
Tailings Berm -- new	61.8%	34.4%	3.8%
Tailings Berm -- old	55.9%	40.2%	3.9%
Wetlands	27.6%	71.9%	0.5%
Grand Total	24.4%	74.4%	1.2%

A Geographic Information System (GIS) base map of the area was prepared at a scale of 1:24000, equivalent to USGS 7.5 minute topographic scale with a 100 m grid overlay to facilitate random sampling. In June of 1994, fixed wing infrared aerial infrared photographs were taken (1" = 2000', equivalent to 1:24000), and orthophoto maps were generated at the same scale. Vegetative cover types were mapped into 21 vegetative community types (mapping criteria are presented in Appendix A) and 3 other environment types, including "Developed," "Barren," and "Water" categories (Figure 3). Polygons were identified initially as "breaks" in color and visual texture of photo images discernable using stereo pairs 1" = 2,000' scale and refined after making field observations. The smallest polygons generated were ~1 ha where there is sharp contrast in vegetation, and 5-10 ha where types are more similar. The vegetation types were further refined by Relevé data [see ASTM Standard guide for sampling terrestrial and wetlands (E 1963-98) for description of the relevé method that was standardized in 1998], photograph, and reconnaissance information from the eight vegetation types originally identified. Six cover types, led by submontane shrub and grassland, occurred at 2000 ha (Figure 4).

In May 2002, both the maps and the photos were compared to selected vegetation patches in various locations to confirm that the maps prepared from 1994 photos were still valid. Because succession occurs so slowly in the Basin and Range area, there were only a few places where noticeable changes had occurred in vegetation cover. The "No Data" cover type on the 1994 vegetation map is a sparsely vegetated slag-scree area in Pine Canyon. "Marshlands" are erroneously labeled as there was a shrub component detected in the relevé sampling. In several other places, individual trees or shrubs, which were visible on the IR photos were located in the field. Nearly all appeared to be portrayed accurately on the maps, vis-à-vis 2002 conditions.

The major exceptions appeared to be in Kessler Canyon, where grasses and shrubs appear to have colonized areas that were barren in 1994. The qualitative increase in the expanse of grassland both on the floor of the valley and up the slopes (including in erosion gullies) would not likely result in substantively different portrayal of the overall vegetation as mapped in 1994. Changes in landuse in the industrial areas also resulted in significant changes since 1994. The wetlands area from the slag pond to Kessler Springs and northward to Interstate-80 was modified extensively. The area should now be mapped as part of the active industrial area (collectively "Developed" lands on the 1994 map). Secondly, much of the old tailings pond surface has been seeded to grass. This area should be re-assigned as "Recently Re-vegetated."

Reconnaissance of submontane shrub communities in the southern portion of Coon Canyon in late September 2002 identified substantial overgrazing. The shrub and herbaceous layers in several portions of the mountain maple and gambel oak communities were essentially absent over considerable areas. Though it is not possible to know what these specific areas looked like earlier, it is likely that the extent of grazing impacts fluctuate among years depending on drought conditions and grazing intensity. In 1995, small mammal populations were noted to be low in the Soldier's Bench area (**ep** and **t** 1996) due to habitat loss attributed to grazing.

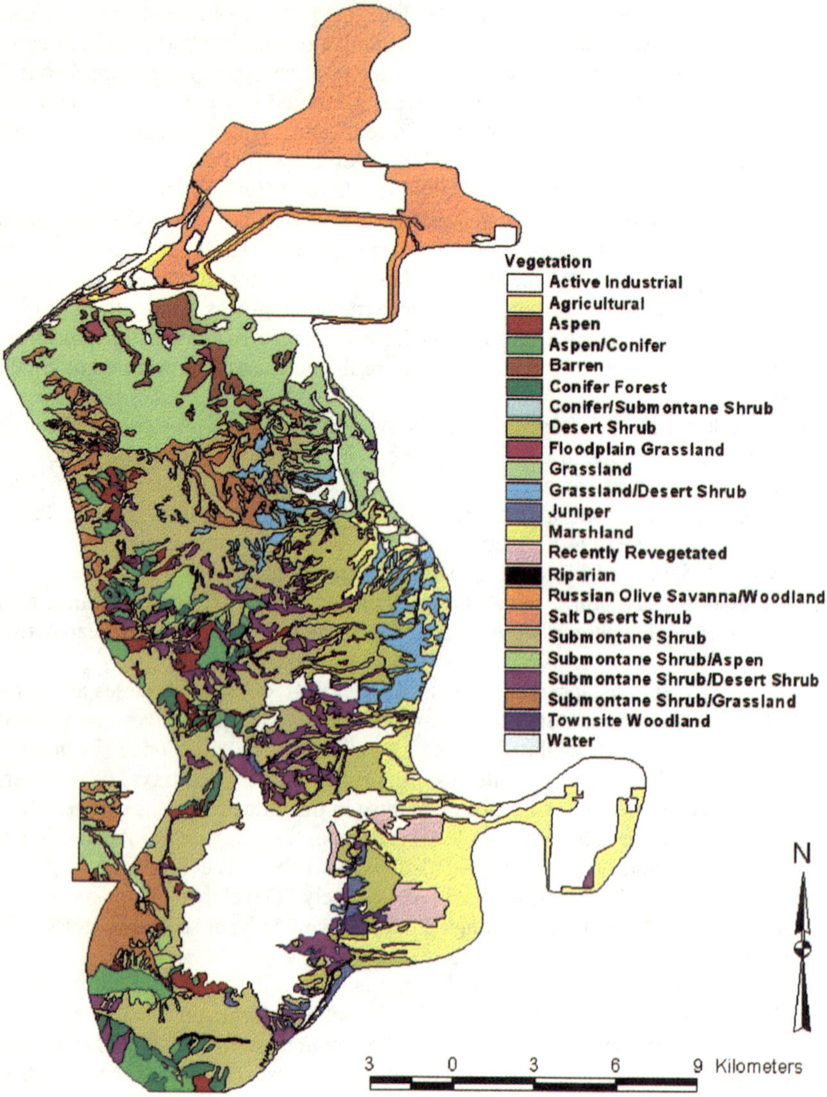

Figure 3. *Vegetation map produced from 1994 IR photos.*

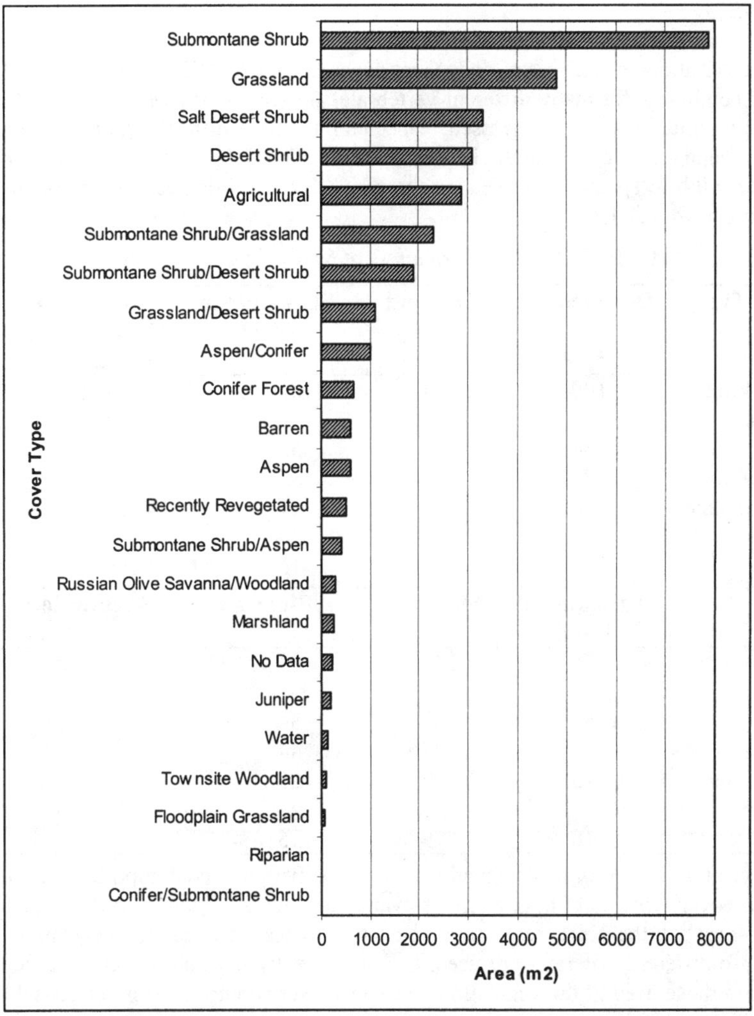

Figure 4. *Amount of cover types mapped from 1994 IR photos.*

Wildlife

The variation in land form and cover types across the >360 km^2 area of KUCC provides conditions for many different vertebrate species. As successional development of plant communities has progressed, especially in the Northern Oquirrh Mountains, wildlife abundance and richness has increased. There have been upwards of 300 vertebrate wildlife species observed on the KUCC, with birds accounting for >60% of species richness (Table 6).

Table 6. *Taxonomic richness of KUCC by cover types.*

Cover Type[1]	Desert Shrub	Juniper	Submontane Shrub	Aspen-Conifer Forest	Subalpine Herbland
All Wildlife Species	109	114	105	96	18
Birds	56	55	55	57	9
Mammals	33	38	33	28	7
Herpetofauna	20	21	17	11	2
Number of Orders	16	17	16	15	8

	Agricultural Land	Marshland	Riparian	Area-wide Totals
All Wildlife Species	74	144	103	274
Birds	57	108	34	177
Mammals	2	27	44	72
Herpetofauna	15	9	25	25
Number of Orders	16	19	15	24

[1] The original cover type designations prior to mapping contemporary vegetation. These cover types may have more relevance in understanding wildlife use patterns than some of the finer resolution vegetation types used in the vegetation map. Because different criteria were used, it is not directly possible to relate these cover types to those used in the vegetation map. Data were compiled from **ep** and **t** 1995 b, c, 1996, and 1997.

Layers of Habitat Model

The Layers of Habitat (LoH) model is intended to provide an approximate indicator of potential wildlife diversity, especially for birds (Short 1984). It provides a relatively simple approach that can be used to quantify potential biodiversity at a site and that can assist in the delineation of "diversity contours." Beyond this, the results of this model also may be valuable in the planning of diversity enhancement or maintenance.

The basic assumption of the model is that with increasing vertical and horizontal diversity, there will be a greater number of "niche opportunities" and, hence, more species likely to occupy the site. One consideration in devising the tiered approach is that the progressive improvements in resolution may lower the uncertainty inherent in the index. Thus a model based solely on herbaceous, shrub, and tree layers is likely to underestimate the structural diversity, and therefore the potential wildlife diversity, compared to one that incorporates life forms (e.g., graminoid, forb) components and sub-layers (e.g., low-growing shrubs versus tall shrubs). One example relevant to the Basin and Range Province is that the chipping sparrow uses tall shrubs for territorial displays and calling perches, whereas the lark sparrow uses low-growing shrubs for its singing perches.

In addition to the structural features of the LoH (regardless of the level of resolution), there are several wildlife species that key on the presence of specific plant species (e.g., sage sparrow). The transition from the somewhat abstract LoH model to a more detailed consideration of dominant plant taxa is made using a guild analysis. Fundamentally, this involves an evaluation of specific wildlife species, which are expected to be present based on plant community composition. By using a hybrid of the LoH and the guild approach, one can test the reliability of the different tiers of the LoH model for a specific locality.

The mapping criteria were used to assign the expected layers for each cover type. For example, one might assume that an area mapped as "barren" would have no vegetative layers. However, the mapping criterion for the cover type was "<15% herbaceous cover." According to the Layers of Habitat, a layer "exists" if there is at least 5% cover. Relevé sampling associated with this cover type confirmed >5% herbaceous canopy cover.

The fieldwork in 1994 used orienteering techniques to locate randomly selected sampling points and to transfer these onto the GIS base map. At that time, GPS units proved to be too inaccurate in the mountainous terrain to be useful. Consequently, the precise location of sampling points for relevé data could not be ensured and in some cases may have been located on the border or near a border of a mapped polygon. In selecting data for use in characterizing vegetation types, in this post facto manner, we chose to eliminate those points that were mapped as being within 50 m of the polygon borders (corresponding to the mapping resolution). This resulted in a substantially reduced data set that could be used to characterize the various cover types (Table 7). However, for the general purpose of using these data to provide a qualitative check on the presence of particular cover types, we believe these data are useful.

Calculation of the LoH Index was adapted from the *Habitat Suitability Index Models: Arizona Guild and Layers of Habitat Model* (Short 1984), but was expanded to consider area applications. The Arizona model compares each location to the presumptive optimum structural diversity embodied in a reference riparian zone. Here we used the reference condition as an internal part of the project area.

Table 7. *Distribution of relevé sampling locations by cover type.*

Cover Type	**Number of Sample Locations**	**Number of Sample Locations clearly within a Polygon**
Agricultural	3	1
Aspen	1	1
Aspen/Conifer	2	1
Barren	5	2
Conifer Forest	1	1
Conifer/Submontane Shrub	0	0
Desert Shrub	7	3
Developed	3	0
Floodplain Grassland	2	1
Grassland	42	36
Grassland/Desert Shrub	2	1
Juniper	0	0
Marshland	3	1
No Data	2	1
Recently Re-vegetated	1	0
Riparian	3	0
Russian Olive Savanna/Woodland	10	4
Salt Desert Shrub	2	1
Submontane Shrub	21	14
Submontane Shrub/Aspen	1	1
Submontane Shrub/Desert Shrub	9	1
Submontane Shrub/Grassland	16	4
Totals	136	74

Steps to calculating the Layers of Habitat Index

The following sequential procedure was used in calculating the LoH index:

1. Delineate polygons of the project area by cover type.
2. Assign each cover type a set of recognizable layers, beginning with the ground layer and moving upward through the potential herbaceous, shrub, tree canopy, and tree bole layers.[6] The overall maximum number of layers is five for a mature forest.
3. Designate the maximum (optimal) composition (cover) for each layer. The default value is 100 units (%) for each layer. However, if documentation is readily available to set a lower value as the cap, then the lower value should be selected. [Once these designations are made (i.e., maximum value for each layer), this becomes the

[6] N.B. For some systems as in the Arizona Riparian model, the ground layer is assessed in terms of the presence of "penetrable" soil, such that soil invertebrates and burrowing animals are capable of inhabiting the area. For other systems, such as the Basin and Range Province, presence of soil, cobble, scree, or rock outcrops may be important features that promote use by a wide range of wildlife (e.g., canyon or rocks wrens).

denominator for the LoH Index. For example, if there are only four layers of the possible five, each with a cover of 100%, then the numerator becomes 100+100+100+100/2500

4. Examine polygon cover types and assign a score for each layer for which the canopy cover of that layer is ≥5%. Thus for cover type within the site with four layers, the score might be 100,100, 100, 100. In this example the numerator would be 4 x (400) = 1,600. Thus for this polygon, the LoH Index = 1,600/2,500 = 0.640. Similarly, a polygon with values of 100,100, 0, 0 would have an Index = 200/2,500 = 0.080.
5. After completing this for each polygon, the site-wide LoH Index value may be expressed as an area-weighted sum.

Results and Discussion

Several of the cover types achieved their maximum index value, (i.e., an area with potentially three layers having three layers (based on the "presence ≥5%" criterion). When normalized to the maximum potential index value (here, 5 layers), the site-wide "Area-weighted LoH Index," was determined to be 0.462. The cover type contributing the most to the site-wide index was the submontane shrub area (Figure 5).

Comparing the LoH Index contribution relative to the area of that cover type indicates that eight types contribute more than might be expected according to their area (led by the submontane shrub community) whereas seven cover types contribute less than might be expected according to their area (with the grassland and salt desert shrub communities being the poorest (Figure 6). Maps of the LoH values calculated from the presence of particular layers (Figure 7) provide a rapid visualization of spatial relationships for wildlife biodiversity.

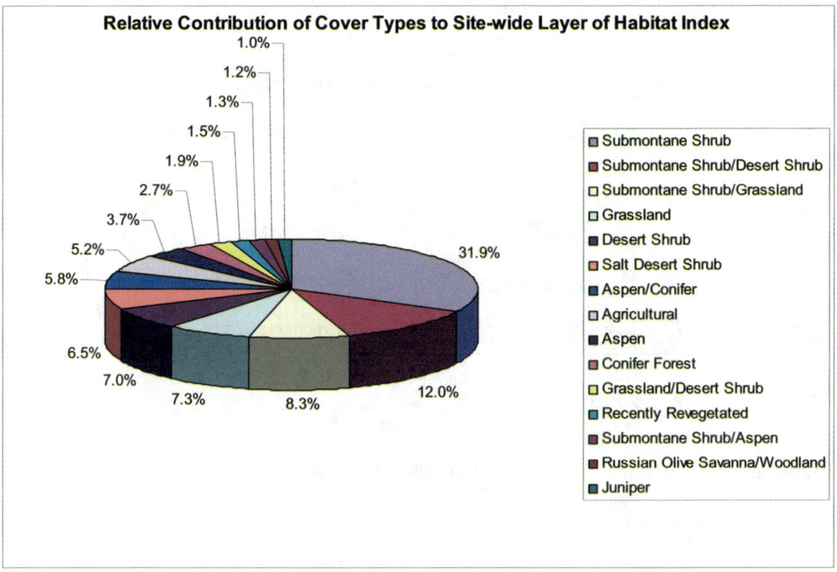

Figure 5. *Relative contribution of cover types to site-wide LoH Index value.*

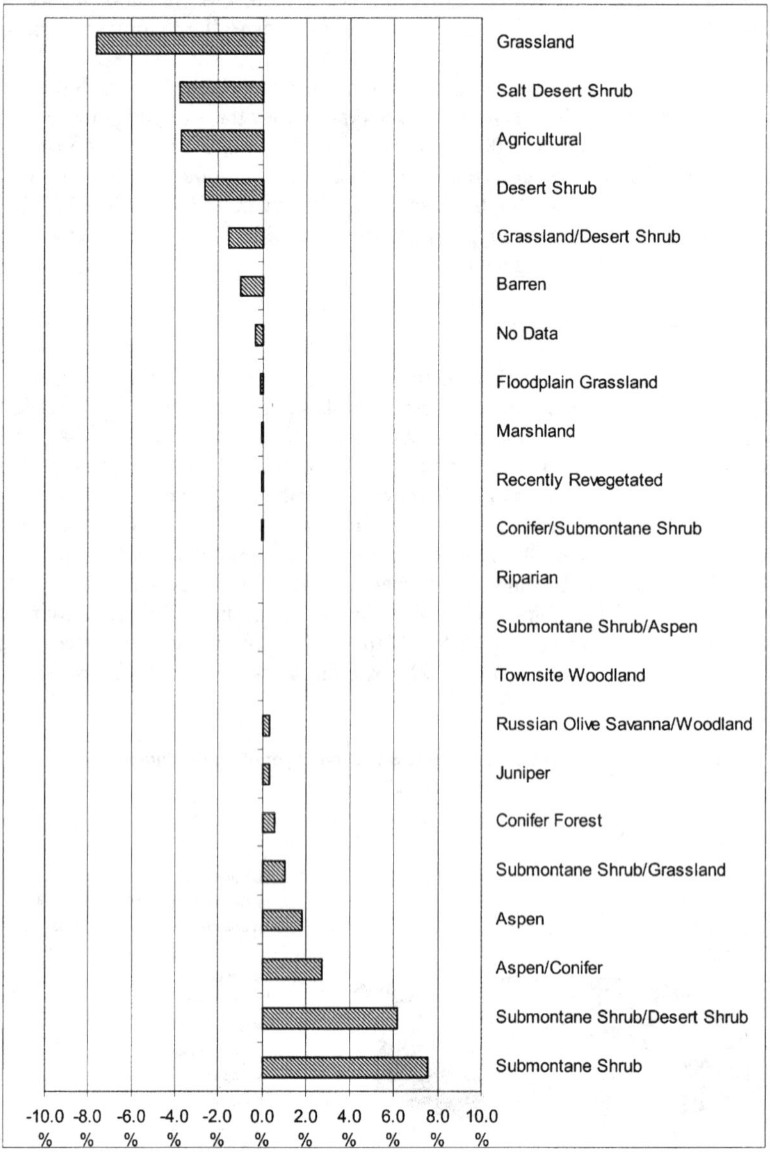

Figure 6. *Proportional contribution of cover types to site-wide LoH Index value.*

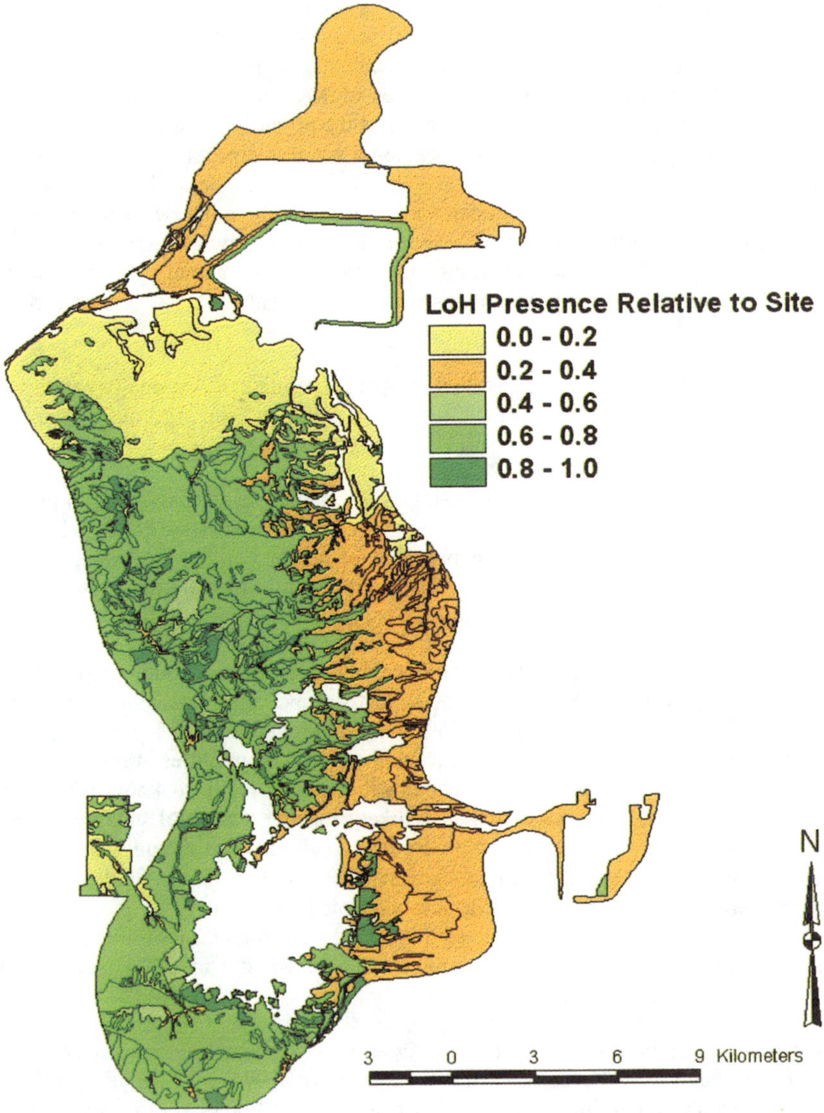

Figure 7. *Spatial distribution of LoH Index values classified into pentiles.*

The LoH Index value is based on conditions as depicted in 1994. Landuse changes, such as transition of the old tailings pond from an active industrial area to a recently revegetation area, have increased the potential useable habitat area. Expansion of grassed areas in the valley and along the erosion gullies of Kessler Canyon and Black Rock Canyon have also improved structural complexity of the property since 1994.

Several important caveats should be considered when interpreting the results of this analysis:

- ✓ Modifications of the LoH Model were based on broadly defined ecological principles. The original model was devised for selected areas in Arizona and have not been applied to other ecoregions. Consequently, the actual optimum vertical structure for the Oquirrh Mountains may be different than we assumed in the model.

- ✓ The LoH model has been field-tested in riparian habitats in Arizona (where it was found to perform well (Short 1984)), but it has not been rigorously tested or calibrated in the habitats characteristic of the Oquirrhs.

- ✓ Among the modifications in habitat layers, ground cover was assumed to present in all cover types. For some areas, this may be soil that is sufficiently loose to permit burrowing, in others the presence of scree, boulders, or rock outcrop provide cover for various species. The optimum mix of different ground cover types in terms of maximizing wildlife diversity, is unknown.

- ✓ Reconnaissance of submontane shrub communities in the southern portion of Coon Canyon in late September 2002 identified potential problems in the LoH Index because of overgrazing. The shrub and herbaceous layers in several portions of the mountain maple and gambel oak communities were essentially absent over considerable areas. Though it is not possible to know what these specific areas looked like earlier, it is likely that the extent of grazing impacts fluctuate among years depending on drought conditions and grazing intensity. In 1995, small mammal populations were noted to be low in the Soldier's Bench area (**ep** and **t**, 1996) due to habitat loss attributed to grazing. To the extent that this situation is the norm, one could infer that diminution of the shrub and herbaceous layer due to grazing, lowers the potential and the actual wildlife diversity and use of the area.

- ✓ Parameterization of the different cover types was based on 1994 data. Limitations of the vegetation maps and relevé data from 1994 may have significant consequences regarding the LoH Index calculations. Potential effects are greater when using relevé percentage cover values instead of the"$\geq$5% cover" test in assigning values to the LoH numerator.

Nevertheless, even at this generalized level of analysis, (i.e., the mere presence of layers) the LoH model appears to provide a means of presenting a structured qualitative estimate of the spatial representation of biodiversity potential. Additional refinements of the model could add greater realism to the process. For example, refinements of the LoH

Model to incorporate incremental increases in structural complexity (i.e., not merely "≥5% cover") may improve the value of the model predictions.

Recommendations

The current level of analyses provides a useful method to portray biodiversity potential of an area. The information can be used to conduct "what if" scenarios to explore likely impacts to site-wide biodiversity and can be used to guide landuse management decisions to promote biodiversity. Even so, there are several important considerations for future development of this approach. In particular, efforts to link predicted biodiversity levels with site observations could establish a level of verification to the models. Such data could determine whether additional refinements to the model were needed for it to apply broadly to other sites. Also, such data could indicate whether inclusion of actual percentage cover rather than just presence of each life-form would be beneficial. For example, in the basic approach, the model does not distinguish between one site having an herbaceous canopy occupying 6% of the area from one having 6% cover. All of the steps above are followed, except that in Step 4, instead of applying a simple test (if X <5%, then 0, else 100), the test is more robust (if X<5%, then 0, else X). Scores computed in this fashion will always be less than with the basic model, but the relative spread among cover types becomes more apparent.

If such refinements were shown to improve the estimates in a substantial degree, then additional segregation of cover data could be pursued. For example, within the herbaceous layer, fern/fern allies, forbs, and graminoids would be distinguished. Within the shrub and tree layers, evergreen versus deciduous would be distinguished. And finally, sub-layers could be divided further to incorporate information on canopy height recognizing that a grassland may be comprised of low growing plants (e.g., Muhlenbergia) and thus should support fewer wildlife species than a grassland comprised of tall grasses (e.g., various wheatgrasses). Further, the herbaceous layer could be segregated into graminoid and forb components; the shrub layer may be divided into low-profile shrubs versus tall shrubs; and the tree layer may be segregated into deciduous versus coniferous components, as well as short versus tall trees.

References

Arnow, L., Albee, B. and , A. 1980. 'Flora of the Central Wasatch Front, Utah. Second edition, revised.' University of Utah Printing Service, Salt Lake City, UT.

BioWest. *Jurisdictional wetlands on selected Kennecott Utah Copper Properties.* Prepared for Kennecott Utah Copper Corporation, Salt Lake City, UT, 1991.

Blanchard, J. F. 1973. *An ecologic and systematic study of the birds of the Oquirrh Mountains, UT.* Master of Science Thesis, Department of Biology, University of Utah, Salt Lake City, UT.

ep and **t**. 1994a. *Work plan for the ecological risk assessment of Kennecott Utah Copper, Salt Lake City.* Prepared for Kennecott Utah Copper Corporation, Salt Lake City, UT.

ep and **t**. 1994b. *Sampling and analysis plan for the 1994 screening level field studies for the ecological risk assessment of Kennecott Utah Copper, Salt Lake City.* Prepared for Kennecott Utah Copper Corporation, Salt Lake City, UT.

ep and **t**. 1995a. *Problem formulation report: Screening level for the ecological risk assessment of Kennecott Utah Copper, Salt Lake City.* Prepared for Kennecott Utah Copper Corporation, Salt Lake City, UT.

ep and **t**. 1995b. *Analysis report: Screening level for the ecological risk assessment of Kennecott Utah Copper, Salt Lake City.* Prepared for Kennecott Utah Copper Corporation, Salt Lake City, UT.

ep and **t**. 1995c . *Risk characterization report: Screening level for the ecological risk assessment of Kennecott Utah Copper, Salt Lake City, UT.* Prepared for Kennecott Utah Copper Corporation, Salt Lake City, UT.

ep and **t**. 1996. *Final Report: Ecological Risk Assessment, Northern Oquirrh Mountains.* Prepared for Kennecott Utah Copper Corporation, Salt Lake City, UT.

ep and **t**. 1997. *Final Report: Ecological risk assessment, South Shore Wetlands.* Prepared for Kennecott Utah Copper Corporation, Magna, UT.

ep and **t**. 1995b *Analysis report: Screening level for the ecological risk assessment of Kennecott Utah Copper, Salt Lake City.* Prepared for Kennecott Utah Copper Corporation, Salt Lake City, UT.

Kass. R. J. 1988. 'A checklist of the vascular plants of the House Range, Utah'. *Great Basin Naturalist* Vol. 48, pp. 102-116.

Short, H. L. 1984. *Habitat Suitability Index Models: the Arizona Guild and Layers of Habitat Model.* FWS/OBS-82/10.70, US Fish and Wildlife Service, Fort Collins, CO.

SWCA. 1996. *Baseline Monitoring for the Kennecott Tailings Modernization Project Mitigation Site, 1995. Volume I – Results.* Prepared for Kennecott Utah Copper Corporation, Magna, UT.

TNC. 1997. *Northern Oquirrh Mountains Bio-inventory.* The Nature Conservancy, Utah Field Office, Salt Lake City, UT; [prepared by Dr. Ron Kass]. Prepared for Kennecott Utah Copper Corporation, Magna, UT.

Welsh, S. L., Atwood, N. D., Goodrich, S. and Higgins, L. C. (Eds.). 1993. *A Utah Flora. Second Edition, Revised.* Brigham Young University, Provo, UT.

Whittaker, R.H. 1970. *Communities and Ecosystems.* Macmillan Company, New York.

Appendix A. Vegetation cover type mapping definitions.

- ✓ **Conifer Forest**: >50% cover by white or Douglas-fir.
- ✓ **Pinyon-Juniper**: >10% cover by juniper[7].
- ✓ **Submontane Shrub**: >70% cover by mesic mountain shrubs, with < 10% aspen or conifer. Curlleaf mountain mahogany stands are included in this mapping unit; on air photos they appear similar to pinyon-juniper but occur at a higher elevation.
- ✓ **Aspen**: >50% cover of aspen.
- ✓ **Aspen/Conifer**: >50% cover combined by aspen and white fir or Douglas fir, without clear dominance by either aspen or conifer.
- ✓ **Desert Shrub**: Vegetation dominated by big sagebrush or rabbitbrush, with <5% submontane shrub, <10% junipers, and <25% grassland inclusions.
- ✓ **Riparian**: Wetland vegetation, willows, cottonwoods, and tall oak and maple in stream bottoms. The limited areas mapped were determined primarily from field observations, and are not readily distinguishable on the aerial photography.
- ✓ **Salt Desert Shrub**: A mosaic of low elevation salt influenced communities associated with the Great Salt Lake, including shadscale, greasewood, pickleweed, saltgrass, and playa. These types occur intermingled in a complex mosaic and were mapped together.
- ✓ **Marshland**: Permanently or semi-permanently inundated areas dominated primarily by common reed. Includes wet meadows with Baltic rush and saltgrass.
- ✓ **Grassland**: Areas dominated by grasses or forbs (> 15% herb cover, < 5% shrub cover). These areas occur primarily where desert shrub grassland has been removed by fire, and in re-vegetating areas of the North Oquirrh Mountains.
- ✓ **Recently Re-vegetated Areas**: Areas which are dominated by early succession herbaceous vegetation, apparently planted within the last three years.
- ✓ **Russian Olive Savanna/Woodland**: Areas on Tailings Berm dominated by Russian olive and tamarisk (>5% cover of trees), with associated herbaceous and shrub species. Includes narrow strips of common reed and cattail marsh in some areas.
- ✓ **Town-site Woodland**: Areas of former habitation with mixed deciduous trees, and primarily a grassland understory. Tree canopy is approximately 10-30% cover.

---

[7] Though pinyon pine does not occur in the Oquirrh Mountains this vegetation type is referred to as Pinyon Pine by plant geographers.

- ✓ **Agriculture**: Irrigated or dryland cropland.
- ✓ **Developed**: Areas occupied by buildings, pavement, and horticultural landscaping and areas of gravel, bare dirt, and weedy vegetation near these, where they appear to occur because of active or recent mechanical clearing, vehicle movement or similar human activities.
- ✓ **Barren**: Areas with <15% vegetative cover, not associated with recent mechanical clearing or human disturbance.
- ✓ **Water**: Areas covered by open water without emergent wetland vegetation, including portions of the Great Salt Lake and some adjacent ponds.
- ✓ **Conifer/Submontane Shrub Mosaic**: 10-50% conifer cover, mixed primarily with submontane shrub.
- ✓ **Aspen/Submontane Shrub Mosaic**: 10-50% aspen cover, mixed primarily with submontane shrub.
- ✓ **Submontane Shrub/Grassland Mosaic**: 5-70% submontane shrub cover, with herbaceous vegetation (grasses and forbs) in the openings.
- ✓ **Grassland-Desert Shrub Mosaic**: Areas of mixed, intergrading or poorly distinguishable grassland and desert shrub vegetation. Grassland covers approximately 25-95% of this type.
- ✓ **Floodplain Grassland**: Areas of tall dense (> 80% cover) grasses and forbs within the beds of three flood control reservoirs.
- ✓ **Submontane Shrub/Desert Shrub**: 5-70% submontane shrub cover, with desert shrub (mainly sagebrush-dominated) in the openings.

Thomas R. Ryon,[1] *Mike J. Bonar,*[2] *Kirsta L. Sherff-Norris,*[3] *and Robert A. Schorr*[4]

Habitat Ranking System for the Threatened Preble's Meadow Jumping Mouse (*Zapus hudsonius preblei*) in Eastern Colorado

---

Reference: Ryon, T. R., Bonar, M. J., Scherff-Norris, K. L., and Schorr, R. A., **"Habitat Ranking System for the Threatened Preble's Meadow Jumping Mouse (*Zapus hudsonius preblei*) in Eastern Colorado,"** *Landscape Ecology and Wildlife Habitat Evaluation: Critical Information for Ecological Risk Assessment, Land-Use Management Activities, and Biodiversity Enhancement Practices, ASTM STP 1458*, L. A. Kapustka, H. Galbraith, M. Luxon, and G. R. Biddinger, Eds., ASTM International, West Conshohocken, PA, 2004.

ABSTRACT: An objective, quantitative method was devised to delineate and rank habitat of the federally listed Preble's meadow jumping mouse in the Monument Creek watershed in eastern (El Paso County) Colorado. Methods involved review of field studies and devising a set of instructions to develop a habitat map based on field study results. Information was included from field research on individual mouse movement and home range, as well as the habitat continuum that includes vegetation characteristics at sites that supported high densities of Preble's mice as well as those that had low densities of mice. Preble's mice are considered to use lush riparian vegetation as well as dense upland vegetation adjacent to riparian areas. Radio telemetry data were reviewed to establish a relationship between riparian edge and the extent (i.e., linear distance from riparian edge) to which Preble's mice use upland areas. Riparian vegetation mapping, aerial photos, and hydrology were used as a basis for mapping habitat. The ultimate purpose of the Habitat Ranking System is to provide a foundation for sound conservation of habitat in El Paso County.

KEYWORDS: Preble's mouse, habitat, watershed, riparian, Colorado

---

[1] Senior Wildlife Biologist, MFG/TetraTech, Inc., 3801 Automation Way, Suite 100, Fort Collins, CO 80525-3434.

[2] Natural Resource Manager, Environmental Services Department, El Paso County, 2002 Creek Crossing, Colorado Springs, CO 80906-1225.

[3] Wildlife Biologist, Colorado Springs Utilities, 121 South Tejon, Fourth Floor, Colorado Springs, CO 80947-0940.

[4] Wildlife Biologist, Colorado Natural Heritage Program, Colorado State University, College of Natural Resources, 254 General Services Building, Fort Collins, CO 80523.

Introduction

A methodology was needed to identify the locations and extent of habitat for the federally threatened Preble's meadow jumping mouse (*Zapus hudsonius preblei*) within the Monument Creek Watershed in El Paso County, Colorado near Colorado Springs (Figure 1). This methodology was required because El Paso County is developing a Regional Habitat Conservation Plan (RHCP) for the Preble's mouse and a scientifically sound habitat map was needed to base conservation strategies for the RHCP. This plan encompassed the Preble's mouse, which was listed as threatened by the U.S. Fish and Wildlife Service (USFWS) in May 1998.

However, it was realized that it would be time consuming and costly to conduct habitat mapping over the large geographic area under the jurisdiction of El Paso County or the City of Colorado Springs. Hence, a habitat ranking system (HRS) was developed to categorize and inventory habitat for the Preble's mouse within the Monument Creek Watershed and other lands within the planning area of the RHCP. The upper portion of Monument Creek was chosen as a Study Area for the HRS.

The HRS allows a specific area to be numerically evaluated for its value as potential habitat based on similarities in vegetation and topography to habitats where Preble's mice have been captured. The HRS reduces the need for expensive and time-consuming fieldwork, and allows identification of the location and extent of *potential* habitat for Preble's mice in as-yet unstudied stream drainages.

The intent of this paper is to present a methodology for ranking habitat in the Monument Creek Watershed. First, we present the purpose and objectives of the HRS and give definitions of terms used. Then we present the existing spatial information used to develop maps and present existing biological information that is used in the HRS. Finally, we develop criteria for ranking Preble's mouse habitat and present the resulting categories within the HRS. A flow chart and tabulation of ranking values are used to illustrate the HRS.

Purpose and Objectives of the Habitat Ranking System

The purpose of the HRS is to specifically identify potential Preble's mouse habitat in the Monument Creek Watershed and predict the relative value of this habitat. This purpose is achieved through the following objectives:

- Illustrate extent of potential habitat in Monument Creek Study Area
- Categorize Preble's mouse potential habitat as varying degrees of quality
- Identify linkages among potential habitat areas
- Identify where potential habitat does not exist
- Use existing information to develop ranking criteria and associated potential habitat maps

Definitions

Several terms are defined here for their exclusive meaning to the HRS.

Study Area – the upper portion of the Monument Creek Watershed. It includes land that is owned by either El Paso County or the City of Colorado Springs below 7600-foot

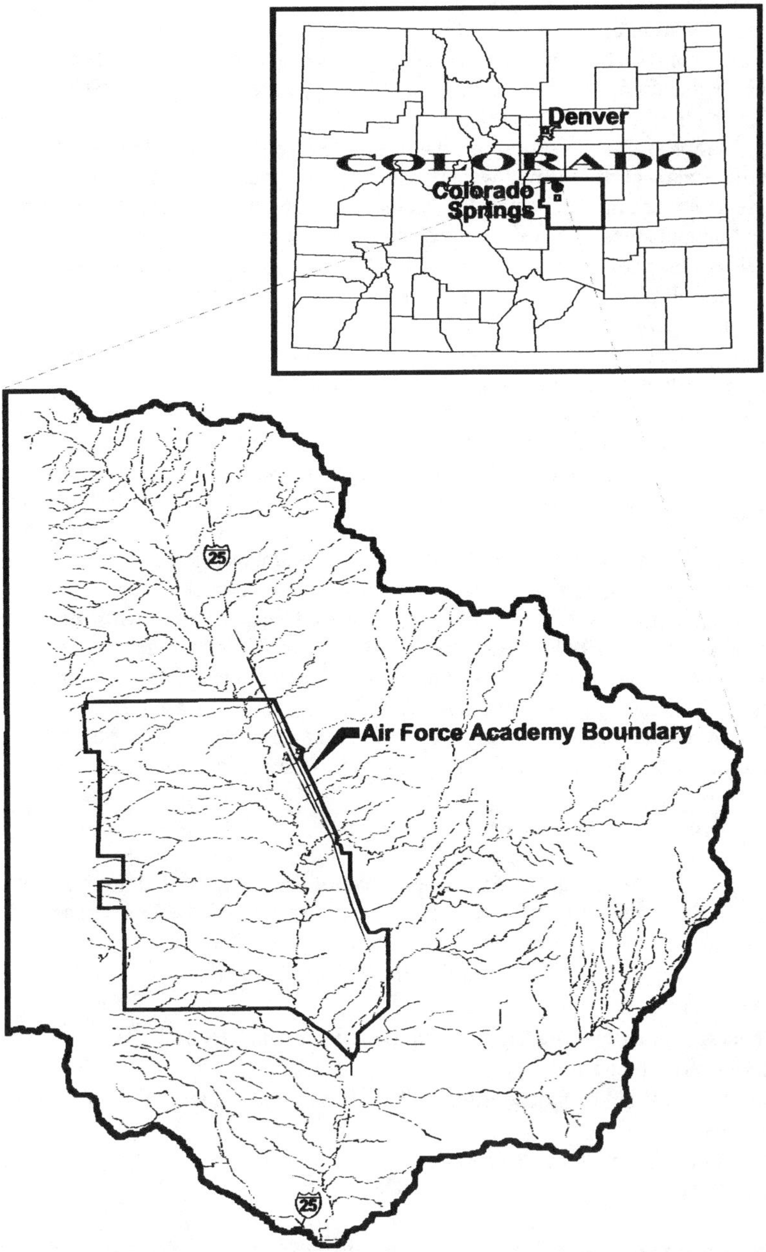

Figure 1 - *Location Map of the Monument Creek Watershed Study Area*

(2316 meters) elevation. No federal lands are included in the Study Area. Figure 1 presents a location map.

Habitat Unit – a habitat unit is an area inclusive of a riparian vegetation patch, either shrub or herbaceous, that extends 1/4 mile (0.4 km) upstream and downstream from the riparian patch.

Riparian vegetation – groups of plants adjacent to, and affected by, surface or ground water of permanent or seasonal water bodies such as rivers, streams, "on channel" ponds and reservoirs, or drainage ways. These areas have distinctly different vegetation than adjacent upland areas or have plant species similar to surrounding areas that exhibit a more vigorous or robust growth form (CDOW 2001b, adapted from USFWS 1997).

Riparian shrub – low-growing streamside woody vegetation such as willow (*Salix* spp.) or alder (*Alnus tenufolia*).

Riparian herbaceous – non-woody riparian vegetation, including rushes, sedges, grasses, and other wetland plants.

Patch – a distinct variation in habitat that differs in identity, size, or quality from surrounding habitats (modified from Art 1993).

Preble's mouse habitat – well-developed plains riparian vegetation along perennial and ephemeral (seasonal) streams. A water source is usually nearby. Preble's mice seem not to depend upon specific vegetation species, but there is evidence that they prefer moist, lowland habitats with considerable cover (Bakeman 1997a; Whitaker 1972). The general preferred vegetation structure consists of dense herbaceous and woody vegetation, including a variety of grasses, forbs, and thick shrubs (Armstrong et al. 1997; Bakeman and Deans 1997b; Meaney et al. 1997). These thick shrubs are typically willows (*Salix* sp.). Successful trapping sites typically consist of a multistoried cover, but the cover species vary greatly (Bakeman and Deans 1997b; Meaney et al. 1997).

Upland habitat/vegetation – any vegetation outside of the riparian vegetation, including grasslands, shrubs and trees of any height and thickness.

Methodology of the Habitat Ranking System

Biologists reviewed and analyzed available information and data from a range of sources and consulted with recognized experts in the field of Preble's mouse biology. Information and data were reviewed from the following sources:

- Existing scientific literature (published and "gray")
- Recent survey data
- Vegetation, topographic, and development disturbance maps
- Aerial and satellite imagery
- Data on the Preble's mouse from the U.S. Air Force Academy (USAFA) in Colorado Springs
- Consultations with local experts
- Site visits

Reports, research results, and agency documents were reviewed to assess the current state of knowledge on habitat for the Preble's mouse in El Paso County. Local experts were interviewed as well to gain further understanding of methods and results of habitat surveys, population estimates, and radio telemetry studies. These efforts resulted in an extensive compilation and evaluation of the latest results and expert opinions.

Existing Spatial Data

Vegetation mapping in riparian areas had been previously completed. The Colorado Division of Wildlife (CDOW) created a Riparian Vegetation Map System (CDOW 2001a) that includes vegetation and non-vegetation types, such as willow shrubs and open water. The Riparian Vegetation Map was created in 1988 from aerial photography (CDOW 2001b). Riparian habitat mapping for the HRS is based on this system, with some minor modifications. Because no mapping is available for vegetation outside of riparian areas, a 1998 aerial photo was used to identify upland vegetation.

Hydrologic Features

Hydrologic features include stream coverages and the boundary of the Monument Creek Watershed. The City of Colorado Springs Planning Department provided complete stream coverage by amending a map from CDOW (Friesen[5]). The CDOW coverage is from the 1:24,000 scale digitized creek coverage of Monument Creek. County stream maps were used for portions of the Study Area that were not included in the CDOW coverage. The boundary of the Monument Creek Watershed developed by CDOW was used to delineate the outer bounds of the Study Area.

Existing Biology Data

Biological criteria were developed from existing information on the Preble's mouse behavior and habitat use to categorize and rank potential habitat within the Study Area. Based on the literature review and consultation with local experts familiar with the biology of the Preble's mouse, the following information was deemed useful in developing biological criteria for a HRS for El Paso County.

White and Shenk (2000) found that shrub and tree cover corresponded well with large population densities of Preble's mice. Areas of less shrub cover supported lower densities. They summarized densities of the Preble's mouse over two years from nine study sites in Colorado and compared these densities to the vegetation cover found at the nine sites.

The habitat found on USAFA support some of the highest population densities of Preble's mice compared with other sites in Colorado (Schorr 2001; White and Shenk 2000). Shrubs dominate the riparian areas at study sites on USAFA. Based on habitat mapping provided by CDOW, Bakeman (Ensight 1999, 2000) demonstrated how sections composed mostly of shrub habitat along Dirty Woman Creek within the same watershed held more mice (81 mice/mile; 50.3 mice/ km) than areas composed of herbaceous vegetation (43 mice/mile; 26.9 mice/ km). We examined these different sites as

---

[5] Friesen, P. 2001. Personal communication [March 8 and May Email to T. Ryon, Greystone Environmental Consultants, Greenwood Village, Colorado. *RE: Electronic Stream Coverages*]. GIS Specialist, City of Colorado Springs Planning Department, Colorado Springs, Colorado.

examples of habitat that would support high (USAFA) and low (Dirty Woman Creek) population densities.

Preble's mice can live in streamside vegetation with very little shrub cover, although these areas in El Paso County may hold mice only in low densities. Recent studies in El Paso and Boulder counties have found Preble's mice in riparian areas that contain very little shrub habitat. In nearly all cases, there is still some shrub component to the mouse habitat, but forbs (in this case, thick herbaceous [non-woody] wetland plant species) can and do replace shrubs in terms of the cover they provide (Meaney[6]).

Biologists from the Colorado Natural Heritage Program (CNHP) have been studying the Preble's mouse at the USAFA since 1995 (Corn et al. 1995, Schorr 2001). Radio telemetry has been used to identify active season home ranges as well as movement patterns. Home ranges were estimated from a set of telemetry points collected over 30 days in summer. The home range of a Preble's mouse varies from 0.27 acres (0.11 ha; Schorr 2001) to 3.2 acres (1.3 ha), with a mean home range of 1.3 acres (0.53 ha; Schorr[7]).

Preble's mice are highly mobile creatures. They travel almost exclusively along streamside corridors. A Preble's mouse at Rocky Flats, Colorado, was observed traveling 1.0 mile (1.6 km) in one night (Kaiser-Hill 1999). Schorr[7] has also observed this long-distance travel. However, these observations appear not to be typical. In fact, in three studies (Ensight 1999; Kaiser-Hill 1999; Schorr 2001), all but a few observations of mouse movement along the riparian corridor are less than 1/4 mile (0.4 km). A distance of 1/4 mile would include the majority of nightly distances traveled along streamside corridors.

General Criteria Development for Ranking Habitat

Based on the information gathered and the existing data on Preble's mouse habitat, we concluded that a HRS could be developed by:

- Consulting a vegetation map of riparian areas, combined with information on vegetation in upland areas.
- Determining the extent to which Preble's mice use upland areas.
- Establishing a relationship between the type and extent of riparian vegetation and the population density of the Preble's mouse.
- Incorporating the mobility of the Preble's mouse into the ranking of habitat.

---

[6] Meaney, C.A. 2001. Personal communication [March 21 Notes in Comments of Habitat Ranking System Review to T. Ryon, Greystone Environmental Consultants, Greenwood Village, Colorado. *RE: Rank Forbs as Preble's Mouse Habitat and Barriers to Movement at I-25 Creek Crossings*]. Consulting Mammalogist.

[7] Schorr, R.A. 2001. Personal communication [Feb 21 telephone conversation with T. Ryon, Greystone Environmental Consultants, Greenwood Village, Colorado. *RE: Home ranges of Preble's mouse at the U.S. Air Force Academy Grounds*]. Zoologist, Colorado Natural Heritage Program, Fort Collins, Colorado.

Using the bounds of the Study Area, we evaluated the extent that the CDOW riparian vegetation map could be used. Nearly all of the mapping units were useful, but we removed two features before habitat was ranked. These features were open water and grassy stream channels. All open water features were removed before habitat was ranked, as these do not represent habitat for the Preble's mouse. We removed grassy stream channel features listed as upland grasses by CDOW if they were within the riparian zone and only if they were in headwater areas of the Monument Creek Watershed. These types were removed because we concluded that the portions of creek channels that support only grassy stream channels within the headwaters are not considered suitable habitat for the Preble's mouse. These headwater areas do not contain any shrubs or wetland types and are too dry, holding water only during storms; therefore, they should not be considered habitat. Conversely, grassy creek segments among shrub patches found outside of the headwaters remained as part of the riparian vegetation map.

Designations of Upland Areas

No upland vegetation mapping is available for the Monument Creek Watershed. Instead, we used the 1998 aerial photograph and developed a coverage of vegetated and disturbed areas by digitizing human-altered locations generally within 300 feet (91.4 m) from the edge of riparian vegetation. These included areas of commercial and residential real estate development and associated landscapes. In commercial developments, we digitized buildings, parking lots, and landscaped-grounds surrounding buildings. Areas as small as one acre were included as disturbed areas. In residential developments, we digitized entire subdivisions when the lot size was less than five acres (2.02 ha). A single building associated with a residence was not considered a significant disturbance. Additionally, agricultural land uses were not designated as disturbed areas. In the Monument Creek Watershed, agricultural activities are mostly restricted to livestock operations. Specifically, grazed uplands or irrigated hay meadows were not designated as disturbed areas because land management practices may allow such areas to be Preble's mouse habitat. Agricultural land management practices were unknown to us, so we did not rank these areas as disturbed areas in order to conservatively include potential habitat areas. All remaining areas that are not disturbed were considered vegetated and could be included as part of upland vegetation mapping.

It was necessary to know the extent that Preble's mice use upland areas to further refine the bounds of areas that were considered for ranking. As a result, we examined radio telemetry data from five sites in Colorado. Data from radio telemetry studies in El Paso (Schorr 1998, 2001), Douglas (Shenk[8]; Shenk 2000, 2002; Shenk and Sivert 1998) and Jefferson counties (Kaiser-Hill 1999, 2000) describe observations of mice from the edge of riparian vegetation as far away as 600 feet (183 m). However, the average value (69 feet [21 m] from riparian vegetation) suggests a skewed data distribution, with most upland movement contained within the first 100 feet (30.5 m) from the edge of the riparian zone.

---

[8] Shenk, T.M. 2000. Personal communication [Oct 24 handout from meeting with T. Ryon, Greystone Environmental Consultants, Greenwood Village, Colorado. *RE: Colorado Division of Wildlife Research Findings for the Preble's Mouse presented at Preble's Mouse Research Group Meeting*]. Research Biologist, Colorado Division of Wildlife, Fort Collins, Colorado.

Radio telemetry data points represent multiple movements of many individual mice over the active season (June to September). After we had identified the points that were found outside of riparian vegetation (4557 upland points), we measured the distance from these upland points to the nearest edge of riparian vegetation. The edge of riparian vegetation was created by using the CDOW map. The 95th percentile of the combined (pooled) data sets is at 182 feet (55.5 m) from the edge of the riparian vegetation. This area encompasses 95 percent of observed upland movements of all observations in Colorado and would protect virtually all studied individuals. Therefore, it was assumed that potential upland habitat for Preble's mice in Colorado extends 182 feet from the edge of riparian vegetation.

Relationship between Riparian vegetation and the Preble's Mouse

The habitat ranking system uses population densities for the mouse with associated vegetation patterns and average home range combined with a surrogate size for the shrub patch to rank riparian vegetation. Two sites in the Monument Creek Watershed, the USAFA and Dirty Woman Creek, represent the range of population densities reported to date. Researchers at USAFA report relatively large densities, and researchers at Dirty Woman Creek report small densities, especially in riparian herbaceous vegetation that lacks shrubs.

We reviewed study reports from USAFA and Dirty Woman Creek, visited both of these sites to observe vegetation patterns, and reviewed how riparian vegetation was displayed in the CDOW riparian vegetation map. We found that USAFA sites were composed mostly of riparian shrubs. We further found that the shrub patches tended to be continuously distributed along Monument Creek (up to 1/4 mile long) and are relatively large in aerial extent. This pattern of riparian vegetation, large and continuous patches of riparian shrubs, coincides with areas of high population density. Therefore, we consider these patterns of riparian vegetation to represent riparian habitat of high value.

The portions of Dirty Woman Creek where low densities of Preble's mice were found generally lacked shrubs and instead bore long patches of herbaceous riparian vegetation. They also likely hold water for at lease a portion of the year. In fact, portions of Dirty Woman Creek contained sections of herbaceous riparian vegetation up to 750 feet (228.6 m) long. This pattern was seen repeatedly where the low-density estimates were derived. This pattern of riparian vegetation, consisting of long stretches of herbaceous riparian vegetation of approximately 750 feet, is considered to represent riparian habitat of lower value. These areas still contain habitat, but they likely yield lower densities of mice when compared with habitat similar to USAFA.

A review of habitat at USAFA and Dirty Woman Creek yields patterns of riparian vegetation that represent high-value and low-value habitat. However, many areas with smaller shrub patches that are not continuous and are intermixed with riparian herbaceous vegetation fall between the patterns of high-value and low-value habitat. A distinction is needed to differentiate between high- and medium-value habitat. However, no sites are known from Monument Creek that represent medium population density.

Instead of finding an alternative site outside of the Study Area, we explored how values for the home range from USAFA could be used to distinguish between high- and medium-value riparian habitat and between medium- and low-value habitats. Home range values for Preble's mice have been reported only from two sites in Colorado (Kaiser-Hill 1999; Schorr 2001). Home ranges vary considerably among individuals from the same site and often overlap (Kaiser-Hill 1999). The typical home ranges reported from USAFA and the Rocky Flats site near Denver, Colorado, are composed of a shrub component along with other vegetation types (Schorr[9]; Ryon[10]).

We know these home ranges can support at least one Preble's mouse, by definition. If increasing densities are related to increasing shrub (and tree) vegetation and the average home range (1.3 acres) from USAFA represents a generalized value from an overall site composed of large population densities, then the average home range from USAFA likely represents high-quality habitat for at least one Preble's mouse.

Shrub patches are known to be a key indicator of habitat in El Paso County. If the average home range (1.3 acres) with a smaller shrub patch can support at least one mouse, then a shrub patch as large or larger than the average home range, along with surrounding vegetation (see section on Habitat Unit), could support many mice. Therefore, average and minimum home range values can be used as surrogate threshold values of shrub patch size if surrounding vegetation is included. The shrub patches indicate areas where many mice could be supported in high-value habitat. Shrub patches below the average value would be considered of medium value, and shrub patches below the minimum home range value (0.27 acres) would represent areas of mixed shrub and herbaceous riparian vegetation. In this case, the extent of herbaceous vegetation would have to be considered. If the riparian vegetation is similar to the patterns found in low-value habitat, it can be ranked as low value.

Finally, some areas may not represent habitat for the Preble's mouse at all. These riparian areas include dry grassy stream channels in headwater areas, unvegetated sections of stream channels, or surrounding floodplains.

Minimum Habitat Unit

The habitat units are sections of stream with riparian vegetation that are used as the basis for ranking vegetation. A habitat unit is created by first identifying a riparian shrub patch larger than the average home range of a Preble's mouse. This shrub patch becomes the focal point for the habitat unit, and an area is identified that extends 1/4-mile upstream and downstream from the shrub patch. Thus, a unit is created that is more than one-half mile long inclusive of the shrub patch length. Each ranking unit is initially based on either riparian shrub patches or areas of riparian herbaceous vegetation as mapped in the CDOW riparian vegetation map. It is also based on the daily average movements of Preble's mice (for example, a quarter-mile up- and down-stream from a shrub patch). In this manner, key indicators of habitat such as shrubs or herbaceous

---

[9] Schorr, R.A. 2001. Personal communication [Feb 21 telephone conversation with T. Ryon, Greystone Environmental Consultants, Greenwood Village, Colorado. *RE: Home ranges of Preble's mouse at the Air Force Academy Grounds*]. Zoologist, Colorado Natural Heritage Program, Fort Collins, Colorado.

[10] Ryon, T.R. 2003. Personal communication [Feb 8 telephone conversation with M. Bonar, El Paso County Environmental Services, Colorado Springs, Colorado. *RE: Home ranges of Preble's mouse at the Air Force Academy Grounds*]. Zoologist, Colorado Natural Heritage Program, Fort Collins, Colorado.

vegetation are captured within a reach of stream that corresponds to the distance typically traveled in a day. The habitat unit forms the basic (minimum) ranking unit within the HRS and may capture a mosaic of riparian vegetation, but is based on a shrub patch or a patch of herbaceous vegetation. Within a habitat unit, other shrub patches may be found, as may any other type of riparian vegetation mapped by CDOW.

Identifying the minimum habitat ranking units begins a stepwise process of categorizing vegetation, creating units, and ranking the units. Several biological criteria are used to accomplish this process. First, the average and minimum home ranges, reported from the USAFA, are used as surrogate values to categorize stream reaches into habitat units. Shrub patches are incorporated into habitat units based on size (area). Finally, riparian areas where no shrubs grow but support herbaceous vegetation are assigned to a habitat unit last.

This process was carried out until all riparian areas within the Study Area are assigned to a habitat unit. Once all possible habitat units are created, each unit is ranked based on the type and extent of riparian vegetation. This portion of the process will be explained in detail in the Riparian Vegetation Ranking Criteria section, below. Upland areas that surround each habitat unit are considered next. Adjacent to each habitat unit, a section of upland that extends perpendicular to the stream out to 182 feet from the upland edge of riparian vegetation is considered. Two upland areas (one on each side of the riparian habitat unit) are considered according to criteria for upland habitat. In general, upland areas are ranked depending on the presence of vegetation, followed by the proximity of human disturbance. This portion of the process also will be explained in detail in the Upland Vegetation Ranking Criteria section. This discussion provides a general description of the stepwise process of creating and ranking habitat units. More details on how units are ranked follow in the next section.

Specific Ranking Criteria

Based on the criteria presented in the previous section, we established five categories to reflect the relative value of areas within the Study Area to support and sustain the Preble's mouse. The HRS also incorporates habitat linkage corridors that act as connectors between habitat areas. Movement within habitat areas and between upland and riparian habitat components are accommodated in the size of individual habitat units. Therefore, the HRS is made up of the following five categories:

- High-Value Habitat
- Medium-Value Habitat
- Low-Value Habitat
- Linkage Corridors
- No Habitat

These habitats are found within the proper elevation range (7600 feet and lower) for Preble's mice and are within the boundaries of the Study Area. Each category was considered in detail based on the scientific information presented and cited in this document. Table 1 is provided to illustrate how the value of each component is calculated. Values are assigned for riparian vegetation and adjacent upland areas in the following manner.

Riparian Vegetation Ranking Criteria

Ranking criteria for riparian habitat involves identifying certain vegetation types of specific extent based on:

- Identifying patterns at sites that support high or low population densities of Preble's mice,
- Using surrogate home range values to identify riparian shrub patch sizes that are equivalent to high or medium value, and
- Including typical daily movement distances (1/4 mile) to incorporate vegetation beyond a shrub patch into a minimum habitat unit size.

High-Value Riparian Vegetation – Habitat units with large riparian shrub patches (where the total area of riparian shrub is 1.3 acres or larger – the average home range for the Preble's mouse) were assigned the highest ranking for riparian vegetation. These shrub patches are continuously distributed along the stream up to and beyond 1/4 mile. Because Preble's mice are highly mobile, the habitat unit was extended beyond each shrub patch for 1/4 mile upstream and downstream. This extension is prudent in light of the results of telemetry and trapping studies that indicated that Preble's mice typically travel within 1/4 mile along the stream in a single night. In most cases, this extension of the habitat unit includes other non-shrub vegetation types that Preble's mice would use for foraging areas within the riparian zone. Patches that are within 1/4 mile from each other or closer would, therefore, be combined into one large unit of the same ranking. These areas are ranked high because they could support high densities of Preble's mice and resemble sites along Monument Creek on the USAFA. No human disturbances, such as large dams or structures that would pose a barrier to movement, are allowed in this category of habitat units.

Medium-Value Riparian Vegetation – Medium-value habitat in the HRS is defined as units where the riparian shrub patches are more moderately sized and are not as contiguous as were found in high-value riparian habitat. These shrub patches may be as large as 1.2 acres, but are at least 0.27 acres. The minimum home range reported for a Preble's mouse is 0.27 acres, and a smaller habitat patch is likely not inhabitable. Again, the habitat unit was extended beyond each shrub patch for 1/4 mile upstream and downstream. As in high-value riparian habitat, no human disturbances that would pose a barrier to movement are allowed.

Low-Value Riparian Vegetation – Habitat units of low-value riparian habitat include small shrub patches (less than 0.27 acres) that exhibit extensive (more than 750 feet in length) non-woody wetland areas. These areas are likely to support low densities of mice and may be found in the smaller tributaries or the upper reaches of creeks. Some human disturbance may be present in these habitat units.

Non-Habitat in Riparian Areas – These areas are not categorized as habitat units and are likely small herbaceous wetland areas or unvegetated areas that are isolated from riparian shrub areas. They often indicate where human disturbance is apparent. Some areas are barriers to movement of the Preble's mouse and act as complete barriers or as filters (Ensight 2001). All stream reaches that were not ranked as high-, medium-, or low-value habitat were assigned the lowest ranking and are designated non-habitat. These areas are generally unvegetated reaches that are occasionally vegetated. Often, these areas are the result of disturbance from recent real estate development or may be formed from stream down-cutting that removes riparian vegetation. A few areas designated as non-habitat riparian areas are large flood control impoundments or diversion structures that pose a barrier to movement along the stream channel. These barriers include large rip-rap sections of stream channel, concrete-lined channels, and large reservoirs. Any large areas of open water are considered non-habitat.

Ranking Criteria for Adjacent Upland Areas

Upland habitat is an important component of habitat for the Preble's mouse. The lack of vegetation mapping in the Monument Creek Watershed limited the options available to rank habitats in upland areas. This simplified and conservative process identifies areas that support vegetation by noting areas that are currently undisturbed by human activities. In addition, uplands adjacent to riparian areas are considered more valuable if the uplands on both sides of the stream support vegetation. If only one side of the stream has upland vegetation, then a lesser value is assigned. Finally, a zero value is assigned in the HRS if no upland habitat is available on either side of the riparian zone.

High-Value Upland Areas – Upland areas on both sides of the riparian habitat unit that are vegetated, regardless of vegetation type, out to 182 feet received the highest ranking for upland areas. No human disturbance is allowed in High-Value Upland areas. Note that areas that are currently grazed, but still vegetated, may be included. Hay meadows may also be included as habitat under this designation.

Medium-Value Upland Areas – Uplands where only one side of the riparian habitat unit is vegetated extending to 182 feet wide were placed in the medium value category. The remaining side is vegetated, but also is disturbed and therefore does not extend to 182 feet. Alternatively, uplands on the remaining side may be completely disturbed and not vegetated at all. These may be upland sites that have a residential subdivision close to upland habitat but not immediately adjacent to it. Note that areas that are currently grazed, but still vegetated, may be included. Hay meadows may also be included as habitat under this designation.

Low-Value Upland Areas – Upland areas where neither side is vegetated out to 182 feet were placed in the low value category. These areas have some vegetation present, but disturbance has occurred on both sides of the creek. Some upland habitat is still found, so a low value is designated.

Non-Upland Habitat – These are uplands where development or disturbance is found immediately adjacent to the edge of riparian vegetation. No vegetation is found surrounding the riparian habitat unit. These include locations where commercial development is next to stream-side vegetation or where landscaped grounds are mowed up to the edge of riparian vegetation. These are areas recognized as having no value as upland habitat.

Combining Riparian and Upland Rankings

As presented in the HRS (Table 1), riparian habitat units and adjacent uplands are given a numerical ranking based on the vegetation quality in terms of Preble's mouse habitat. Converting vegetation quality to a numerical ranking provides a simplified method for comparing vegetation quality in terms of Preble's mouse habitat and provides a way to compare various habitat areas. In the final step of ranking, riparian ranks and upland ranks are combined. This results in four categories of habitat ranking:

- High Value Habitat
- Medium Value Habitat
- Low Value Habitat
- Linkage Corridors
- No Value (non-habitat)

Within these categories, one or more cases are possible. Table 1 presents all the possible outcomes of combining the various riparian habitat units with those of upland areas. We assigned a greater value to habitat areas that have riparian and upland habitat components, based on the definition of Preble's mouse habitat. The combined riparian and upland score is multiplied by three, and the areas that rank as High, Medium, or Low Value Habitat have their scores multiplied. This creates a value difference that is warranted based on habitat requirements and separates areas that do not have both habitat components, namely Linkage Corridors and non-habitat areas.

Linkage Corridors have not been previously addressed because a case for Linkage Corridors can be made only by combining riparian and upland rankings. When the upland area is highly disturbed (ranked as zero) and riparian vegetation still provides cover as indicated by a ranking of High, Medium, or Low Riparian Value, then the habitat unit is delineated as a Linkage Corridor. These areas must be equal to or less than one mile, corresponding to the maximum reported length that a Preble's mouse can travel in a short period (24 hours or less). These areas link habitat units and therefore are important for conservation efforts.

Discussion

The HRS is a scientifically based habitat prediction model directly applicable to the Monument Creek Watershed (Figure 1). Areas with large shrub patches and vegetated adjacent uplands are properly identified as high value habitat. These areas should be considered as potentially supporting high densities of mice such as those found at the USAFA. Areas of low value habitat have few shrub patches and more herbaceous vegetative cover with adjacent upland vegetation. These low value habitats potentially support Preble's mice in low densities when compared to the USAFA.

The HRS also identifies linkage corridors valuable to maintaining connectivity of habitats and population viability. Linkage corridors provide vegetative cover in riparian areas to allow Preble's mice to move among preferred habitat even where human activities may have limited the uplands. Linkage corridors are also important in incorporating seasonal and long-term Preble's mouse movement that is vital to their natural history. Preble's mice need to travel to find new food sources and proper food sources as their physiological needs shift during their active season, especially in preparation for hibernation. Additionally, Preble's mice need to find hibernation sites and may travel considerable distances along riparian corridors in selecting a site (Kaiser-Hill 1999, 2000; Shenk and Sivert 1998).

The HRS also identifies habitat gaps. In riparian areas, potential linkage corridors could be enhanced via restoration efforts to reconnect disjunct habitats. Whether upland or riparian, habitat gaps represent areas for future mitigation to offset future development impacts or sites where development can occur with little impact to Preble's mouse habitat.

Mapping upland vegetation, at least to the level of grasslands, shrublands, and unvegetated areas, could refine upland habitat mapping. Preble's mice may prefer upland areas that have both grasslands and shrubs intermixed (Clippinger 2002). This would aid in better defining upland habitat and may help to eliminate grasslands from consideration that would not provide foraging areas or hibernation sites. This may further restrict the amount of upland habitat designated in this model.

Another useful input to upland habitat mapping would be to incorporate land management. Areas that are grazed or otherwise harvested annually such as hayfields could be given a lower rank versus ungrazed uplands. It would be helpful in conservation efforts if there were a means to recognize over-grazed uplands as having some habitat potential. Such areas can become habitat with changes in land management practices and could have implications to mitigation efforts.

The HRS will be used as a means to rank habitat for the Preble's mouse throughout the El Paso County planning area inclusive of the Monument Creek Watershed (Figure 1). With the habitat rankings as a foundation, conservation strategies for the El Paso County RHCP can be successfully identified. The HRS uses the current mapping information, such as the CDOW riparian mapping and aerial photos, and therefore represents a snapshot in time in terms of habitat quality. However, the resulting habitat map provides a baseline for future comparisons and the HRS provides the means to rank habitats in the Monument Creek Watershed in the future.

The HRS methodology could be adapted to other watersheds outside of El Paso County. Similar field data about Preble's mice specific to the watershed of interest would be needed. Alternatively it may be easier to simplify the ranking categories if vital data have not been researched in the area of interest. For example, the HRS incorporates field measurements for Preble's mice from the Monument Creek Watershed that may not be appropriate for other watersheds. For example, home range values from USAFA were used as surrogate threshold values to distinguish between high-value and medium-value riparian vegetation. Because only one other site in Colorado has home range estimates, it may be appropriate to combine high-value and medium-value categories and eliminate the need for home range values. Such a consolidation would simplify the ranking of riparian vegetation to shrub, non-shrub, and non-vegetation categories. This may be a means to more generally apply the HRS to other watersheds in Colorado. It is apparent, however, that adequate extent of riparian shrubs and adjacent uplands are key components to Preble's mouse habitat.

Acknowledgements

Thank you to Tanya Shenk of the Colorado Division of Wildlife for sharing radio telemetry data from her three study sites in Douglas County. Thanks to Dan Heidenreich for his GIS abilities in creating maps. Thanks to Michael Jones of Greystone, Lisa Ross of the City of Colorado Springs and Art Willden at MFG, Inc. for their support and review of this manuscript. Thanks to Mark Bakeman and Carron Meaney for sharing information and their insights to the Preble's mouse.

References

Armstrong, D. M., Miller, C., Sanders, M., and Marguiles, M., 1997, "Habitat of the Meadow Jumping Mouse (Zapus hudsonius preblei) on Boulder City and Boulder Open Space Lands," *Report on Habitat Findings on the Preble's Meadow Jumping Mouse,* M. Bakeman, Ed., pp. 1-18.

Art, H. W., Ed., 1993, *The Dictionary of Ecology and Environmental Science,* Henry Holt and Company, Inc., New York.

Bakeman, M. E., 1997a, "Conclusions on Habitat of the Preble's Mouse," *Report on Habitat Findings on the Preble's Meadow Jumping Mouse,* M. Bakeman, Ed., p. 40.

Bakeman, M. E. and Deans, A., 1997b, "Habitat of the Preble's Meadow Jumping Mouse at Rocky Flats, Colorado," *Report on Habitat Findings on the Preble's Meadow Jumping Mouse,* M. Bakeman, Ed., 91 pp. 19-35.

Clippinger, N. W., 2002, "Biogeography, Community Ecology, and Habitat of Preble's Meadow Jumping Mouse (Zapus hudsonius preblei) in Colorado," Doctoral Thesis, Department of Environmental, Population and Organismic Biology, University of Colorado, Boulder, 131 pp. + appendices.

Colorado Division of Wildlife (CDOW), 2001a, "Riparian Vegetation Mapping". *[Web Page]. Located at: http://ndis.nrel.colostate.edu/ndis/ftp/riparian.htm,* Accessed: February 8, 2001.

Colorado Division of Wildlife (CDOW), 2001b, "Riparian and Wetland Mapping: Wetland versus Riparian," *[Web Page]. Located at: thtp://ndis.nrel.colostate.edu/ndis/riparian/ripwetdef.htm,* Accessed: February 7, 2001.

Corn, J. G., Pague, C. A., Ellingson, A. R., Sherman, M., Zwiejacz, T., Kittel, G., and Fleming, C., 1995, "Final Report on the Geographic Extent of the Preble's Meadow Jumping Mouse Population on the United States Air Force Academy," Presented to the U.S. Air Force Academy, 44 pp.

Ensight Technical Services, 1999, "Report on Preble's Meadow Jumping Mouse Movement Assessment at Dirty Woman and Monument Creeks, El Paso County, Colorado," Prepared for Sugnet & Associates, Durango, Colorado, 24 pp. + Figures and Tables.

Ensight Technical Services, 2000, "Preble's Meadow Jumping Mouse 1999 Data Summary for Dirty Woman Creek, Town of Monument, El Paso County, Colorado", Submitted to Wilson and Company, Colorado Springs, Colorado, 10 pp. + Figures and Tables.

Ensight Technical Services, 2001, "Preble's Meadow Jumping Mouse Issues Within the I-25 Project Area in Northern El Paso County, Colorado," Submitted to Wilson and Company, Colorado Springs, Colorado. 29 pp. + figures, tables, and appendices.

Kaiser-Hill Rocky Flats, 1999, "Preble's Meadow Jumping Mouse Study at Rocky Flats Environmental Technology Site, Appendix B, 1998," *Annual Wildlife Report. Kaiser-Hill Company, Rocky Flats Environmental Technology Site, Golden, CO.*

Kaiser-Hill Rocky Flats, 2000, "Preble's Meadow Jumping Mouse Study at Rocky Flats Environmental Technology Site, Appendix B, 1999," *Annual Wildlife Report. Kaiser-Hill Company, Rocky Flats Environmental Technology Site, Golden, CO.*

Meaney, C. A., Clippinger, N. W., Deans, A., and O'Shea-Stone, M., 1997, "Habitat Preferences of Preble's Meadow Jumping Mice in Colorado." *Report on Habitat Findings on the Preble's Meadow Jumping Mouse,* M. Bakeman, Ed., pp. 36-57.

Schorr, R. A., 2001, "Meadow Jumping Mice (Zapus hudsonius preblei) on the U.S. Air Force Academy, El Paso County, Colorado," Unpublished Report to the Natural Resources Branch, U.S. Air Force Academy, 53 pp.

Shenk, T. M, 2002, Unpublished radio telemetry data, Maytag, Pinecliff Ranch, Woodhouse Ranch Properties, 1998-2000.

Shenk, T. M. and Sivert, M. M., 1998, "Movement Patterns of Preble's Meadow Jumping Mouse (Zapus hudsonius preblei) as they vary across time and space," Unpublished Report, Colorado Division of Wildlife, 27 pp. + Figures.

U.S. Fish and Wildlife Service (USFWS), 997, "A System for Mapping Riparian Areas in the Western United States," *U.S. Fish and Wildlife Service National Wetlands Inventory*, December.

White, G. C., and Shenk, T. M., 2000, "Relationship of Preble's Meadow Jumping Mouse Densities to Vegetation Cover," Draft Document.

Whitaker, J. O., Jr., 1972, "Zapus hudsonius," *Mammalian Species*, 11:1-7.

TABLE 1 - *Flow chart results of all possible habitat suitability outcomes.*

Habitat Suitability	Riparian Conditions	Upland Conditions*	Total	Upland Multiplier	Score
High Value Habitat (Score 24)					
Case 1	Riparian vegetation is composed primarily of large shrub patches equal to or larger than 1.3 acres and may be continuous along the stream channel for up to or beyond ¼ mile. (4)	The upland vegetation extends to 182 feet from the edge of riparian vegetation on both sides of the riparian area. (4)	8	3	24
Medium Value Habitat (Score 18-23)					
Case 1	Riparian vegetation is composed primarily of large shrub patches equal to or larger than 1.3 acres and may be continuous along the stream channel for up to or beyond ¼ mile. (4)	Adjacent uplands are vegetated on both sides of the riparian area. The upland vegetation extends to 182 feet on one sides of the riparian area. The remaining side is vegetated but not to 182 feet. (3)	7	3	21
Case 2	Riparian vegetation is composed primarily of shrub patches greater than 0.27 to 1.3 acres. Site is dominated by riparian shrub, but may also be interspersed with sections of non-woody wetlands. (3)	The upland vegetation extends to 182 feet from the edge of riparian vegetation on both sides of the riparian area. (4)	7	3	21
Case 3	Riparian vegetation is composed primarily of shrub patches greater than 0.27 to 1.3 acres. Site is dominated by riparian shrub, but may also be interspersed with sections of non-woody wetlands. (3)	Adjacent uplands are vegetated on both sides of the riparian area. The upland vegetation extends to 182 feet on one sides of the riparian area. The remaining side is vegetated but not to 182 feet. (3)	6	3	18
Case 4	Riparian vegetation is composed primarily of small shrub patches, but less than 0.27 acres. Site is not dominated by shrubs, but is a combination of riparian shrubs and sections of large, non-woody wetlands that are over 750-ft in length. (2)	The upland vegetation extends to 182 feet from the edge of riparian vegetation on both sides of the riparian area. (4)	6	3	18

*When upland and riparian habitat components are present, the sum of their combined ranks are multiplied by 3.

TABLE 1 - *Continued.*

Habitat Suitability	Riparian Conditions	Upland Conditions*	Total	Upland Multiplier	Score
Low Value Habitat (Score 6-17)					
Case 1	Riparian vegetation is composed primarily of shrub patches greater than 0.27 to 1.3 acres. Site is dominated by riparian shrub, but may also be interspersed with sections of non-woody wetlands. (3)	Adjacent uplands are vegetated on both sides of the riparian area and the extent of upland vegetation is less than 182 feet from the edge of riparian vegetation on both sides of the riparian area; or the upland vegetation extends to 182 feet on one side and the other side is highly disturbed. (2)	5	3	15
Case 2	Riparian vegetation is composed primarily of small shrub patches, but less than 0.27 acres. Site is not dominated by shrubs, but is a combination of riparian shrubs and sections of large, non-woody wetlands that are over 750-ft in length. (2)	Adjacent uplands are vegetated on both sides of the riparian area. The upland vegetation extends to 182 feet on one sides of the riparian area. The remaining side is vegetated but not to 182 feet. (3)	5	3	15
Case 3	Riparian vegetation is composed primarily of small shrub patches, but less than 0.27 acres. Site is not dominated by shrubs, but is a combination of riparian shrubs and sections of large, non-woody wetlands that are over 750-ft in length. (2)	Adjacent uplands are vegetated on both sides of the riparian area and the extent of upland vegetation is less than 182 feet from the edge of riparian vegetation on both sides of the riparian area; or the upland vegetation extends to 182 feet on one side and the other side is highly disturbed. (2)	4	3	12
Case 4	Riparian vegetation is composed primarily of small shrub patches, but less than 0.27 acres. Site is not dominated by shrubs, but is a combination of riparian shrubs and sections of large, non-woody wetlands that are over 750-ft in length. (2)	Adjacent uplands are vegetated only on one side of the riparian area. The extent of upland vegetation is less than 182 feet from the edge of riparian vegetation on the side that is vegetated. (1)	3	3	9
Case 5	Herbaceous continuous over 750-ft in length - very few shrubs (1)	The upland vegetation extends to 182 feet from the edge of riparian vegetation on both sides of the riparian area. (4)	5	3	15

*When upland and riparian habitat components are present, the sum of their combined ranks are multiplied by 3.

TABLE 1 - *Continued.*

Habitat Suitability	Riparian Conditions	Upland Conditions*	Total	Upland Multiplier	Score
Low Value Habitat, continued					
Case 6	Herbaceous continuous over 750-ft in length - very few shrubs (1)	Adjacent uplands are vegetated on both sides of the riparian area. The upland vegetation extends to 182 feet on one sides of the riparian area. The remaining side is vegetated but not to 182 feet. (3)	4	3	12
Case 7	Herbaceous continuous over 750-ft in length - very few shrubs (1)	Adjacent uplands are vegetated on both sides of the riparian area and the extent of upland vegetation is less than 182 feet from the edge of riparian vegetation on both sides of the riparian area; or the upland vegetation extends to 182 feet on one side and the other side is highly disturbed. (2)	3	3	9
Case 8	Herbaceous continuous over 750-ft in length - very few shrubs (1)	Adjacent uplands are vegetated only on one side of the riparian area. The extent of upland vegetation is less than 182 feet from the edge of riparian vegetation on the side that is vegetated. (1)	2	3	6
Linkage Corridors (Score 3-4)					
Case 1	Riparian vegetation is composed primarily of large shrub patches equal to or larger than 1.3 acres. (4)	Uplands adjacent to riparian area are distrubed by real estate development or otherwise not supporting grasses. (0)	4	1	4
Case 2	Riparian vegetation is not composed primarily of large shrub patches equal to or larger than 1.3 acres, but shrubs are present. (3)	Uplands adjacent to riparian area are distrubed by real estate development or otherwise not supporting grasses. (0)	3	1	3
Non-Habitat (Score <3)					
No suitablility	Riparian vegetation is composed primarily of small shrub patches, but less than 0.25 acres. Site is not dominated by shrubs, and may have very few shrubs at all. May have large sections of non-woody wetlands that are over 750-ft in length. (2, 1)	Uplands adjacent to riparian area are distrubed by real estate development or otherwise not supporting grasses. (0)	2	1	2
No suitablility	No riparian or wetland vegetation in or near creek (0)	Inconsequential since no riparian habitat present	0		0

*When upland and riparian habitat components are present, the sum of their combined ranks are multiplied by 3.

Hector Galbraith,[1] *Jeff Price,*[2] *Mark Dixon,*[3] *and Julie Stromberg*[3]

Development of HSI Models to Evaluate Risks to Riparian Wildlife Habitat from Climate Change and Urban Sprawl

REFERENCE: Galbraith, H., Price, J., Dixon, M, Stromberg, J., **"Development of HSI models to evaluate risks to riparian wildlife habitat from climate change and urban sprawl,"** *Landscape Ecology and Wildlife Habitat Evaluation: Critical Information for Ecological Risk Assessment, Land-Use Management Activities, and Biodiversity Enhancement Practices, ASTM STP 1458*, L. A. Kapustka, H. Galbraith, M. Luxon, G. R. Biddinger, Eds., ASTM International, West Conshohocken, PA, 2004.

ABSTRACT: Hitherto, HSI models have largely been utilized to quantify the quality of existing habitat for wildlife species, without reference to how that habitat may have been altered in the past or how it might be altered in the future. In this study, we are using HSI models as part of an integrated modeling approach to estimate the risk of habitat quality gain or loss for a variety of indicator species due to future climate change and aquifer management decisions at the San Pedro Riparian National Conservation Area (SPRNCA). Current anthropogenic stressors, including agricultural and municipal water use, are having adverse impacts on the extent and quality of riparian habitat in the SPRNCA. Future climate change, through its potential effects on hydrology and water demand by local communities, may exacerbate these effects. Because of these current and potential future changes, vertebrates that depend on riparian habitats for their breeding, wintering or migration sites are at risk. Combining climate, hydrology, and vegetation modeling with HSI models allows us to predict the effects these risks.

KEYWORDS: Biodiversity, Riparian habitat, Hydrology, Water use, Climate change, HSI models

[1] Galbraith Environmental Sciences LLC, 289 Wiswall Hill Rd., Newfane, VT 05345.
[2] American Bird Conservancy, Boulder, CO.
[3] Department of Plant Biology, Arizona State University, Tempe, AZ.

Introduction

The upper San Pedro Riparian National Conservation Area (SPRNCA) ecosystem in southeastern Arizona and northern Sonora is of critical importance in maintaining regional biodiversity at the ecotone between the Sonoran and Chihuahuan deserts and the Plains grassland. It contains one of the richest assemblages of species and supports one of the most important western migratory bird habitats in North America (Arias Rojo et al. 1999). For its size, the SPRNCA has one of the highest avian diversities found anywhere in the U.S. Almost 390 bird species have been recorded there, of which 250 are migrants that winter in Central or South America and depend on the San Pedro as a staging post on their journeys to and from their breeding areas in the U.S. and Canada. Between one and four million songbirds use the SPRNCA as a migratory corridor each year.

Riparian ecosystems are fragile, especially those found in arid climates. Water is the lifeblood of these communities and the abundance, diversity and health of these ecosystems are strongly influenced by the hydrologic regime, particularly depth to groundwater and the amount, timing and pattern of surface flow. A water table within a few feet of the ground surface is an essential prerequisite for the growth and survival for riparian tree species and other vegetation, while frequent and strong surface flows are essential for the recruitment of tree species such as cottonwoods and willows (Stromberg et al., 1996; Stromberg, 1998; Auble et al., 1994).

The hydrologic regime in the SPRNCA depends on both the local climate (evaporation rates, rainfall) and the state of the groundwater system. Groundwater pumping to provide water for agricultural, industrial, and municipal uses impacts the state of the groundwater system. Models developed by the Arizona Department of Water Resources and others *(in* Arias Rojo et al. 1999) indicate that groundwater pumping in the nearby city of Sierra Vista area has already impacted baseflow in the river. The population of Sierra Vista has grown at an annual rate of 2.4% over the last 20 years and is projected to continue at that rate for the foreseeable future. This has already resulted in an increase in the depth to groundwater in the shallow aquifer. Groundwater modeling indicates that as the population continues to grow, the baseflow of the river may be further affected (Vionnet et al., 1992). This, in turn, could lead to reductions in the extent of the riparian vegetation, invasion by xerophytic species such as mesquite and non-natives (e.g., salt cedar) and to a reduction in faunal biodiversity in the area (Stromberg *et al.*, 1996).

Previous hydrologic models, while taking into account human population growth in the area around the Upper San Pedro River, have not taken into account potential changes that could accompany climate change resulting from an increase in greenhouse gases. Models prepared for the western megaregion of the U.S. National Assessment show potential average temperature increases of around 1.8° C by 2030 and of between 4° and 5.3° C by the year 2095 (VEMAP 2000). Even with precipitation increases, the overall amount of soil moisture will likely decrease. These climatic changes could lead to direct changes to the riparian biodiversity and exacerbate changes brought about by anthropogenic impacts to the water regime.

In this EPA-funded project we are integrating hydrologic, climatic, and vegetation modeling with wildlife habitat models in an attempt to predict the potential effects of future human population growth (and water extractions) and climate change on the riparian vegetation communities of the SPRNCA, and their ability to support important wildlife communities. In this paper, we describe how wildlife Habitat Suitability Index (HSI) models are being developed and how they are to be integrated with the overall modeling process, and the vegetation models in particular, to evaluate the risk of habitat change and loss.

Problem Identification and HSI Model Development

Previous applications of HSI models have included the prediction of the possible effects of particular land management alternatives on biota (Brand *et al.*, 1986; Schamberger and Farmer, 1978), the quantification of past injuries to ecosystem wildlife carrying capacities (Galbraith *et al.*, 1996; LeJeune *et al.*, 1996), and estimating the exposure to contaminants of wildlife species (Galbraith et al., 2001). If the structures, extents, and/or compositions of post-change vegetation communities can be predicted, it becomes possible to use HSI models to quantify and compare pre- and post-change habitat quality and potential carrying capacities. HSI models achieve this by:

1. Identifying the critical habitat variables that affect the carrying capacity of habitat.

2. Establishing quantitative relationships between the occurrence of these variables and the carrying capacity of habitat. Each variable is assigned a suitability index (SI). This is a score of between 0 and 1, where the former is completely unsuitable habitat (i.e., minimal carrying capacity) and the latter is optimal habitat (i.e., greatest carrying capacity).

3. Developing metrics that can be used in the field to quantify the occurrence of the critical habitat components (and, therefore, the carrying capacity of the habitat)

4. Developing algorithms that combine the variable scores (SIs) into an expression of the overall carrying capacity of the habitat. This final score is the HSI and can be between 0 (unsuitable for species or guild) and 1 (optimal habitat).

A fifth component, that is not often performed, should be the testing and validation of HSI models in the field.

To anticipate the effects of depth to groundwater changes on vegetation and wildlife habitat we are developing a hydrophytic model and a number of HSI models (Table 1). HSI models are being developed for species that are likely to lose habitat if the changes described above occur. However, they are also being developed for species that may not be adversely affected or, indeed, may benefit from the expected phytoecological changes.

Table 1. HSI models being developed and rationale for each.

Model species	Habitat gain or loss	Rationale for model
Southwestern willow flycatcher	Potential gain	Species, while riparian, is not dependent on native gallery forest but breeds successfully in the region in salt cedar and other riparian scrub
Yellow-billed cuckoo	Potential loss	In region this species is obligate of shady riparian gallery forest
Yellow and Wilson's warblers	Potential loss	Large migratory populations passing through the SPRNCA feed largely in willow and cottonwood canopy
Botteri's sparrow	Potential gain	Species characteristic of mesic or xerophytic grasslands bordering SPRNCA riparian forest. May benefit as former replaces latter.

The developmental and field-validation processes of two of these models are illustrated below by focusing on two individual models (southwestern willow flycatcher and yellow-billed cuckoo). Space does not permit the description of all five models.

Southwestern Willow Flycatcher.

The southwestern willow flycatcher, *Empidonax traillii extimus,* is a summer visitor to southern Arizona, wintering in Central and South America, south to Panama and northern Colombia (Finch et al., 2000; Sedgwick, 2000).

Throughout most of its breeding range the willow flycatcher is confined to brushy thickets associated with standing or slow-moving water (Sedgwick, 2000). The southwestern race is largely restricted as a breeding species to riparian shrubby thickets in an otherwise arid or semi-desert landscape (Sogge and Marshall, 2000; U.S. Fish and Wildlife Service, 1997; U.S. Fish and Wildlife Service, 2001; Arizona Game and Fish Department unpublished data; Paradzick and Woodward, in press; Allison et al., in press). In response to anthropogenic impacts to the subspecies' habitats (agriculture, mining, etc.) and to population declines, U.S. Fish and Wildlife Service listed the subspecies under the Endangered Species Act as "endangered" in 1995.

Southwestern willow flycatchers have bred or have been reported occupying territories in the SPRNCA in 1977 and 1989 (Krueper, 1997). At present, they may be considered intermittent and rare breeders in the area which may benefit if riparian forests are replaced by shrubbier habitats, including salt cedar stands.

No previous HSI model exists for the southwestern willow flycatcher in any part of its breeding range. Therefore, one was developed for this project. The components and structure of this model were initially based on a literature review of the habitat preferences and patterns of use of southwestern willow flycatchers. This resulted in a draft model, which provided a focus for discussions with southwestern willow flycatcher researchers in Arizona, and model testing and modification in the field (along the lower San Pedro and Gila Rivers at long-term Arizona Game and Fish Department study sites with known flycatcher densities). Based on the comments of species experts and the field test results, the draft model was modified, resulting in this version.

The final southwestern willow flycatcher HSI model incorporates eight variables (patch area and degree of isolation, width of habitat patch, shrub canopy cover, shrub foliage density at 3-5 meters, shrub canopy height, tree canopy cover, distance to standing or slow-moving water, soil moistness). The numerical relationships between each of the variables and habitat suitability are the core of the habitat model. These were developed from information in the scientific literature (e.g., U.S. Fish and Wildlife Service, 2001; Skaggs, 1996; Sedgwick, 2000; Marshall, 2000), and from conversations and field visits with Arizona Department of Game and Fish willow flycatcher researchers C. Paradzick and A. Woodward. They are described below and in Figures 1 through 8.

V1. Area and degree of isolation of riparian shrub patch. A riparian shrub or forest patch is defined as a shrub vegetation community (either with or without a tree canopy), generally less than 10 m in height, and dominated by willows and/or salt cedar. The patch size, patch isolation and habitat suitability scores that were developed are shown in Figure 1. It is assumed that patches smaller than 1 ha do not provide habitat for the species unless within a landscape of small patches, that small and medium size patches (1-2 and 2-5 ha, respectively) provide increasing levels of suitability and that patches larger than 5 ha are optimal for the species.

V2. Width of riparian patch. The patch width and habitat suitability scores that were developed are shown in Figure 2. It is assumed that the relationship between patch width and habitat suitability is approximately linear and that riparian strips narrower than 10 m do not provide habitat for the species.

V3. Percent shrub canopy cover in 20 m radius of sampling point. The shrub cover and habitat suitability scores that were developed are shown in Figure 3. It is assumed that the relationship between % cover and habitat suitability is approximately linear, except at very high percent shrub covers where continuous shrub cover eliminates the presence of canopy breaks, another important habitat feature for the species. It is also assumed that shrub cover that is less than 50% does not provide habitat for flycatchers.

V4. Shrub foliage density at 3-5 meters. The relationships between shrub foliage density at 3-5 m and habitat suitability are shown in Figure 4. Sparse equates with a site where it is possible to clearly see more than 20 m at 3-5 m elevation above ground level from the sampling point for the majority of 360°. Moderately dense equates with visibility between 10 and 20m.Dense equates with visibility between 5 and 10 m. Very dense equates with visibility less than 5m.

V5. Average shrub canopy height within 20 m radius of sampling point. The mean shrub canopy height and habitat suitability categorization are shown in Figure 5. It is assumed in this categorization that canopy heights of less than 4 meters do not provide habitat for flycatchers, and that optimal habitat is reached at 6 meters.

V6. Tree canopy cover within 20 m radius of sampling point. Based on the relationships among tree cover, shading and shrub growth, the tree canopy (woody vegetation > 10m in height) the habitat suitability scores shown in Figure 6 were developed. It is assumed that at high levels of tree canopy cover (>51%) shading is such that the surviving shrub layer will probably not be dense enough to provide high quality habitat.

V7. Distance to standing or slow-moving water. The distance to water (defined as standing or slow moving water greater than 5 meters in diameter or 2 meters in width) and habitat suitability scores shown in Figure 7 were developed. It is assumed that even sites distant from water may provide low quality flycatcher habitat. This relationship is presented graphically in Figure 7.

V8. Degree of soil waterlogging. On the lower San Pedro River and its confluence with the Gila River, southwestern willow flycatchers prefer nesting habitat that has at least moist soils, or, optimally, wet or waterlogged soils (Paradzick and Woodward pers comm.). Based on this the variable categorization shown in Figure 8 was developed:

Figure 1. Riparian shrub patch size (ha) and suitability index (SI) relationship

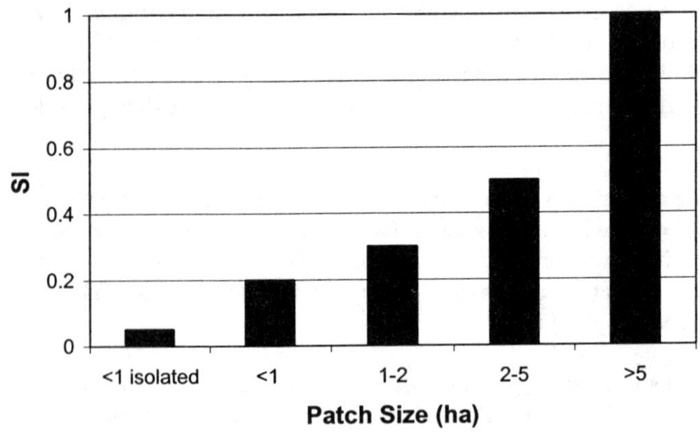

Figure 2. Riparian shrub patch width (m) and suitability index (SI) relationship

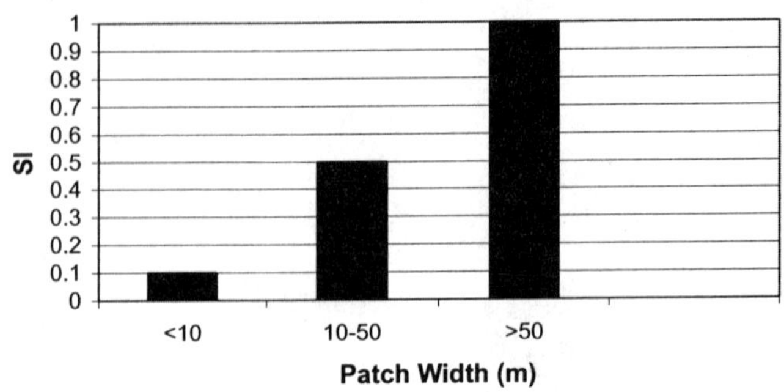

Figure 3. Percent shrub canopy cover and suitability index (SI) relationship

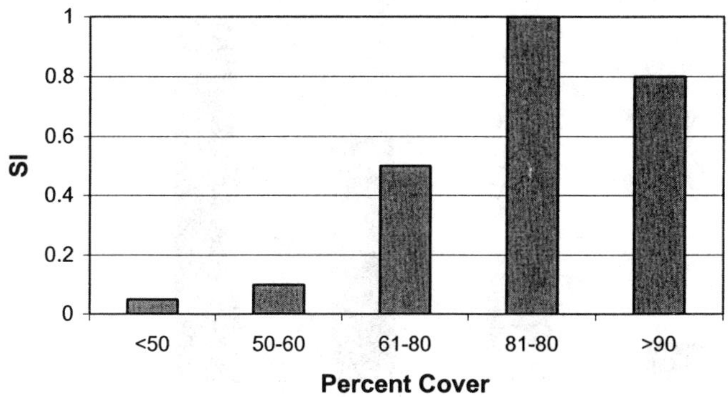

Figure 4. Shrub foliage density and suitability index (SI) relationship

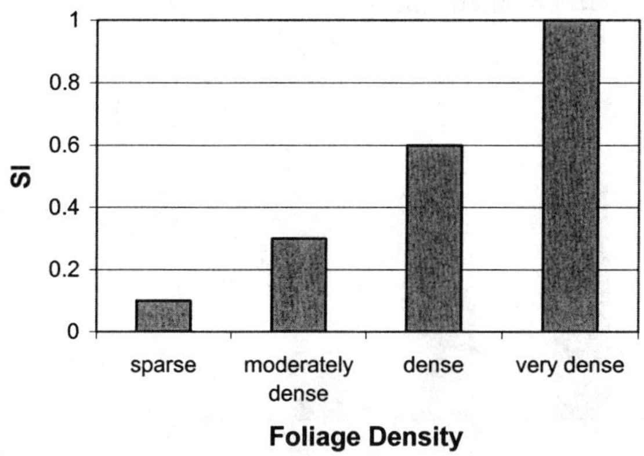

Figure 5. Mean shrub canopy height (m) and suitability index (SI) relationship

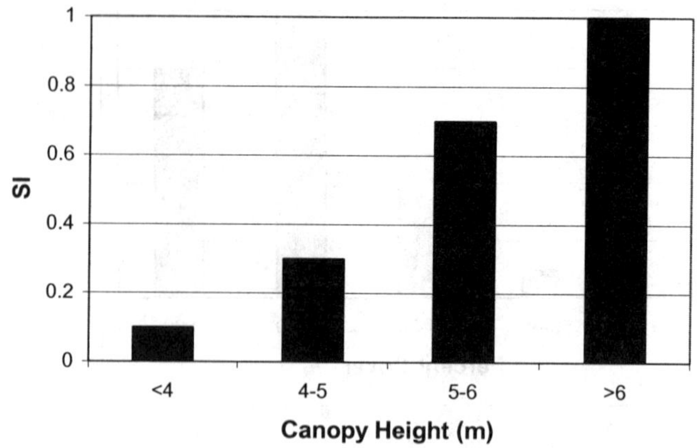

Figure 6. Tree canopy cover and suitability index (SI) relationship

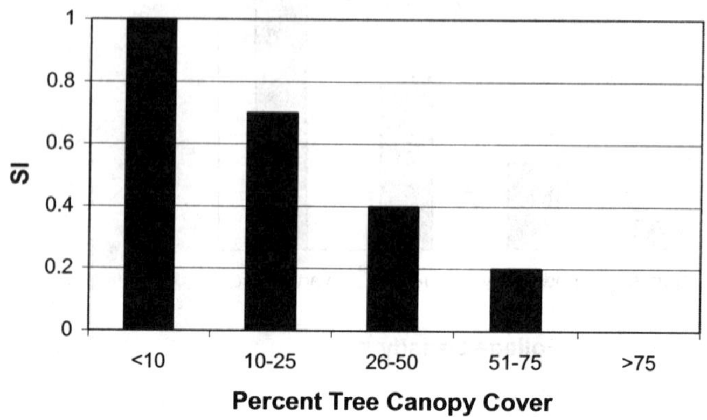

Figure 7. Distance (m) to water and suitability index (SI) relationship

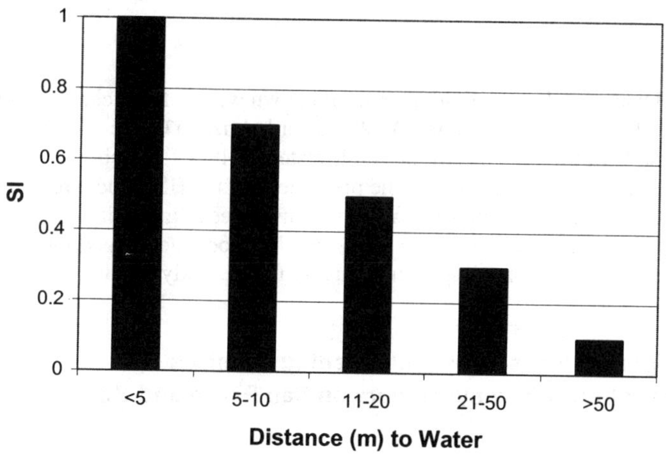

Figure 8. Soil moisture and suitability index (SI) relationship

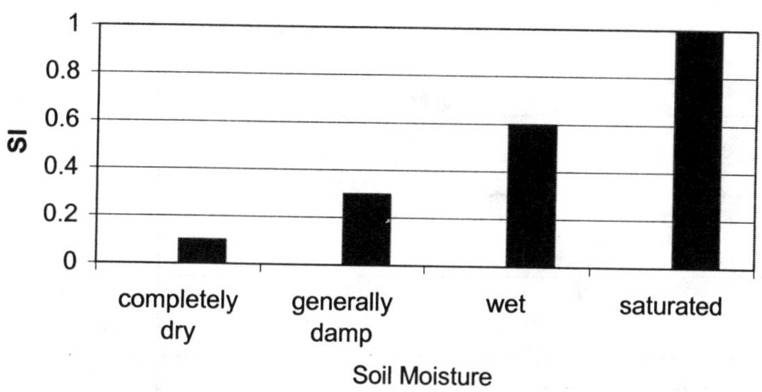

Model application

The eight variables described above were combined into an index of the overall assessment of the habitat suitability (HSI) of a particular patch of riparian habitat using the following 8th root algorithm (to limit the HSI scores to between 0 and 1):

$$HSI = (V1xV2xV3xV4xV5xV6xV7xV8)^{1/8}$$

The HSI values obtained using this equation range between 0 and 1 (lowest and highest estimates of suitability, respectively).

Field test results

The draft HSI model was tested in the field in areas of known willow flycatcher breeding density (unpublished data – C. Paradzick and A. Woodward, Arizona Game and Fish Department). The results of the field test of the southwestern willow flycatcher model are presented in Figure 9. These results show that the predictions of the HSI model regarding habitat suitability are generally accurate (if it is assumed that breeding density is a reflection of at least short term habitat quality). Thus the HSI model developed for this study is a reasonable predictor of breeding habitat quality for the study species.

Figure 9. HSI scores and breeding densities of southwestern willow flycatchers on San Pedro and Gila rivers

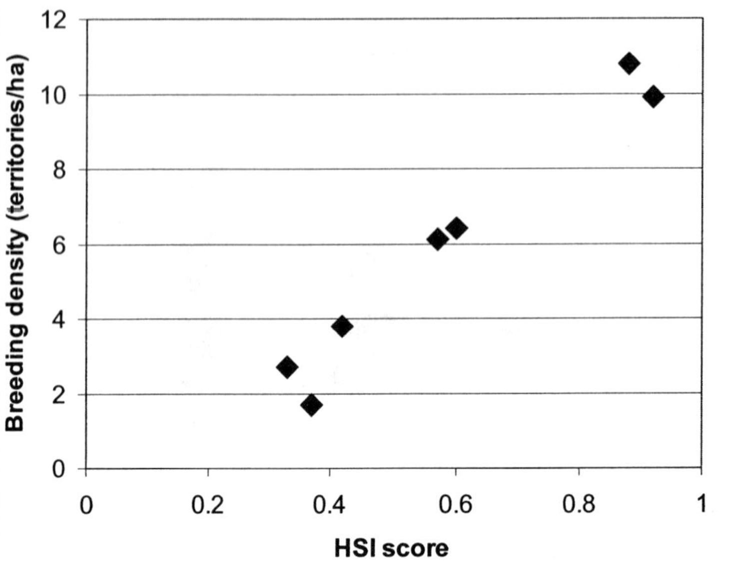

Yellow-billed Cuckoo

The yellow-billed cuckoo is a summer visitor to North America, wintering in Central and

South America (Kaufman, 1996; Hughes, 1999). The western race is largely confined to riparian broad-leaved woodlands, particularly those dominated by mature cottonwoods or willows (Gaines, 1974; Hamilton and Hamilton, 1965; Hughes, 1999). They apparently avoid riparian habitats dominated by invasive salt cedar, *Tamarix pentandra* (Laymon and Halterman, 1987).

It is likely that, at most, 600 pairs of yellow-billed cuckoos breed in Arizona, with 50-100 of these in the SPRNCA (Laymon and Halterman, 1987; Krueper, 1997). Rapid population decreases and range reductions of western yellow-billed cuckoos have recently prompted efforts (thus far unsuccessful) to have the race listed under the Endangered Species Act.

No previous HSI models existed for yellow-billed cuckoo in any part of its breeding range. Therefore, a model was developed to evaluate the potential effects of climate change to the biota of the riparian systems of the SPRNCA. The components and structure of this model were initially based on a literature review of the habitat preferences and patterns of use of yellow-billed cuckoos. This resulted in a draft model which provided a focus for discussions with researchers in Arizona, and model modification in the field. It was then tested in the SPRNCA at sites with known cuckoo breeding densities. Based on the comments of species experts and the field test results, the draft model was modified, resulting in this version.

The final yellow-billed cuckoo HSI model incorporates eight variables: continuity and width of habitat patch, shrub density (canopy cover), tree and shrub canopy heights, cottonwood/willow dominance in tree canopy, cottonwood/willow/ash/walnut/Baccharis dominance in shrub canopy, salt cedar and/or mesquite dominance in the shrub understory. The numerical relationships between each of the variables and habitat suitability are the core of the habitat model. These were developed from information in the scientific literature (e.g., Hughes, 1999; Gaines and Laymon, 1984; Laymon et al., 1997; Laymon and Halterman, 1989), and from conversations and field visits with yellow-billed cuckoo researchers working in the SPRNCA. They are described below and in Figures 10 through 17.

V1. Linear continuity of riparian forest or shrub habitat. In the SPRNCA, riparian shrub or tree linear patches that are relatively unbroken over at least 200 meters provide higher quality cuckoo habitat than more fragmented habitats. To accommodate this, the categorization shown in Figure 10 was developed.

V2. Width of riparian forest patch. Based on data presented in Laymon and Halterman (1989) and observations at sites of known breeding density in the SPRNCA, the patch width and habitat suitability categorizations shown in Figure 11 were developed. It is assumed that the relationship between patch width and habitat suitability is approximately linear and that riparian strips narrower than 50 m do not provide good habitat for the species.

V3. Percent shrub canopy cover in 50 m radius of sampling point. Based on the results reported in Gaines and Laymon (1984) and observations at sites with known breeding densities of cuckoos on the SPRNCA, the shrub cover and habitat suitability categorizations shown in Figure 12 were developed. It is assumed in this categorization that the relationship between % cover and habitat suitability is approximately linear except at very low percent shrub covers where a further reduction will result in a disproportionate effect on habitat suitability.

V4. Average shrub canopy height within 50 m radius of sampling point. Based on observations made at the SPRNCA in areas of known cuckoo breeding density, the shrub canopy height and habitat suitability categorizations shown in Figure 13 were developed.

V5. Average tree canopy height within 50 m radius of sampling point. Based on the results reported in Gaines (1977), Gaines and Laymon (1984), and observations made at the SPRNCA in areas of known cuckoo breeding density, the tree canopy height and habitat suitability categorization shown in Figure 14 were developed. It is assumed that the relationship between tree canopy height and habitat suitability is approximately linear. It is also assumed that even low canopy heights may have some habitat value.

V6. Cottonwood/willow dominance in tree canopy within 50 m radius of sampling point. Based on the results reported in Laymon and Halterman (1985), the cottonwood/willow dominance in the tree canopy and habitat suitability categorizations shown in Figure 15 were developed. It is assumed that the relationship between % dominance and habitat suitability is approximately linear. It is also assumed that very low representations of these species (<10%), provide at best marginal habitat for yellow-billed cuckoos.

V7. Cottonwood/willow/ash/walnut/seep willow percent dominance (relative cover) in shrub layer within 50 m radius of sampling point. Based on the results reported in Gaines and Laymon (1984) and observations made at the SPRNCA in areas of known cuckoo breeding density, the shrub cover and habitat suitability categorizations shown in Figure 16 were developed. It is assumed that the relationship between % dominance and habitat suitability is approximately linear except at very low percent shrub covers where a further reduction will result in a disproportionate effect on habitat suitability. It is also assumed that less than 10% dominance, provides only extremely marginal habitat for yellow-billed cuckoos.

V8. Percent dominance (relative cover) of salt cedar and/or mesquite in the shrub canopy within 50 m radius of sampling point. The relationships between representation of salt cedar and/or mesquite in the vegetation community and habitat suitability shown in Figure 17 are assumed. It is assumed in this categorization that the relationship between % dominance of salt cedar/mesquite and habitat suitability is not linear but that increasing dominance by salt cedar or mesquite has a disproportionate effect on habitat suitability. It is also assumed that the highest dominance by salt cedar and/or mesquite (>70%) will constitute extremely marginal habitat for yellow-billed cuckoos.

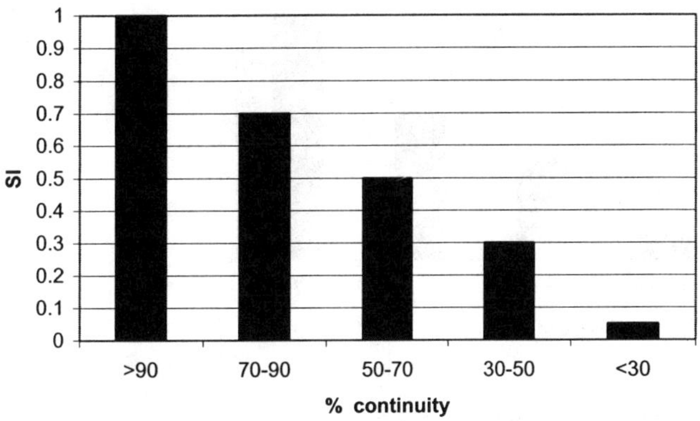

Figure 10. % Linear continuity of riparian forest or shrub over 200 m and SI relationships

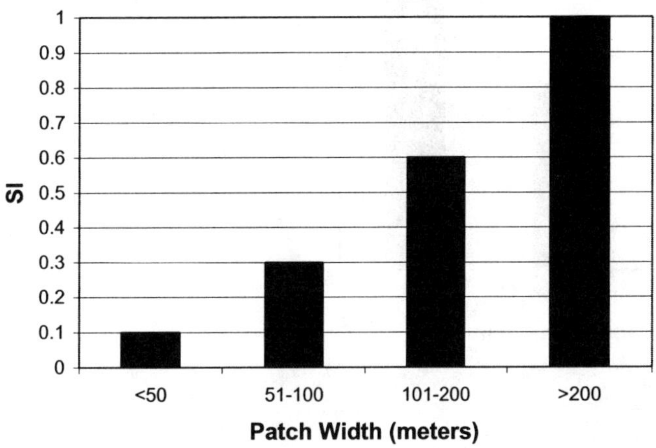

Figure 11. Riparian forest patch width and suitability index (SI) relationship

Figure 12. Percent shrub canopy cover and suitability index (SI) relationship

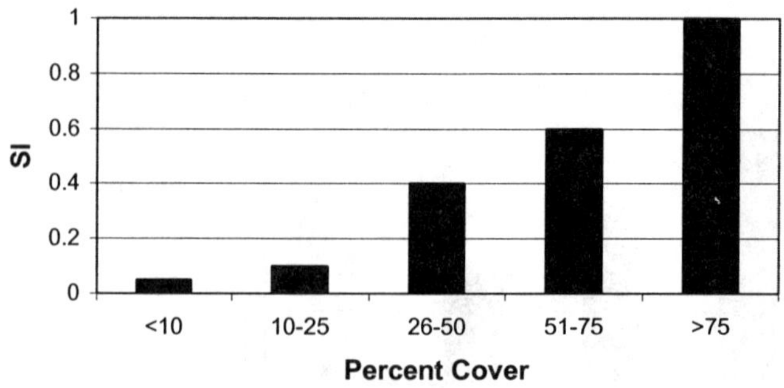

Figure 13. Average shrub canopy height and suitability index (SI) relationship

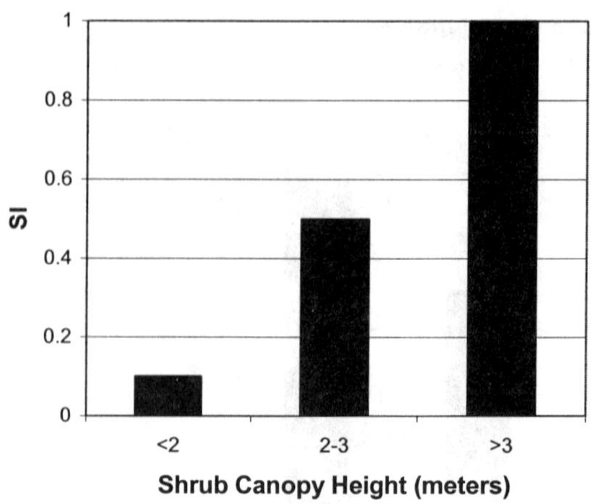

Figure 14. Average tree canopy height and suitability index (SI) relationship

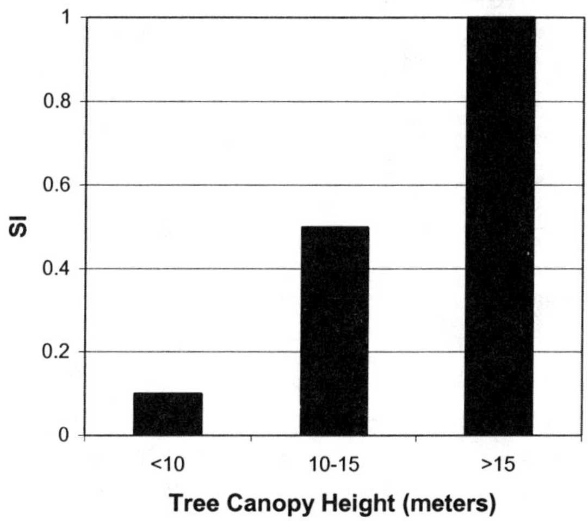

Figure 15. Cottonwood/willow dominance in tree canopy and suitability index (SI) relationship

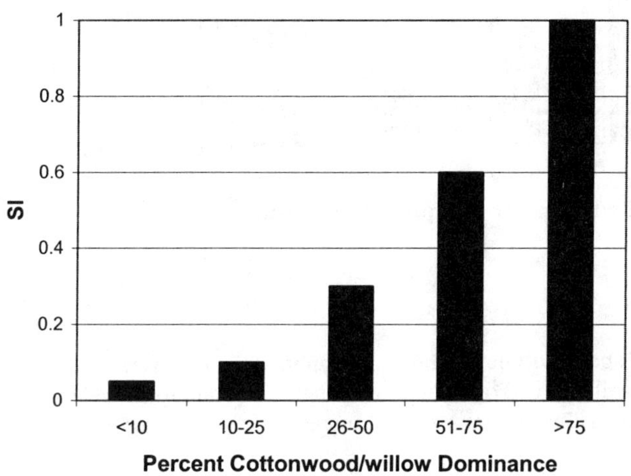

Figure 16. % dominance by willow and/or cottonwood in shrub canopy and suitability index (SI) relationship

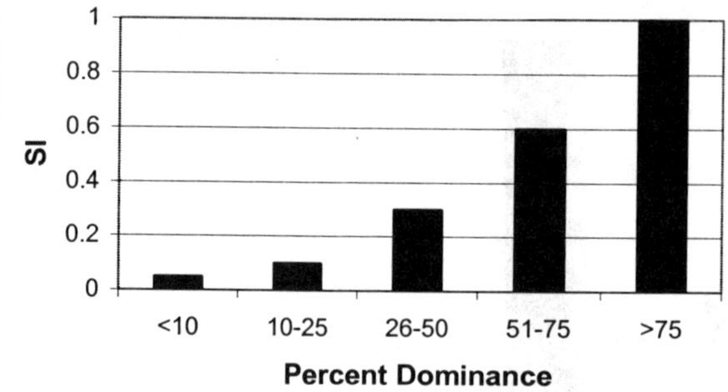

Figure 17. Percent dominance by salt cedar and/or mesquite in shrub canopy and suitability index (SI) relationship

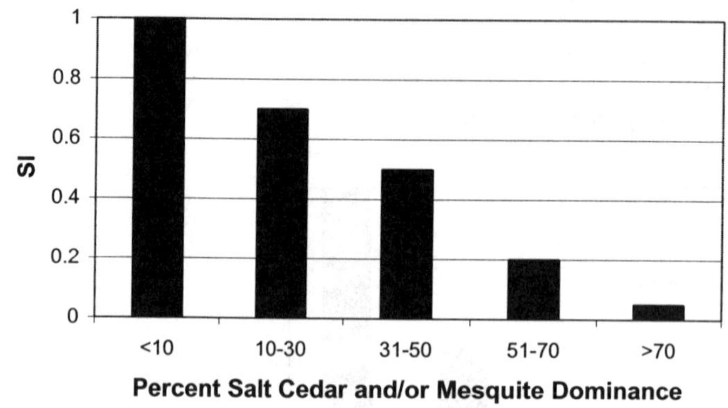

Model application

The eight variables described above will be combined into an index of the overall assessment of the habitat suitability (HSI) of a particular patch of riparian forest habitat using the following 8th root algorithm (to limit the HSI scores to between 0 and 1):

$$HSI = (V1xV2xV3xV4xV5xV6xV7xV8)^{1/8}$$

The HSI values obtained using this equation will range between 0 and 1 (lowest and highest estimates of suitability, respectively).

Field test results

The draft yellow-billed cuckoo HSI model was tested in the SPRNCA in areas of known yellow-billed cuckoo breeding density (unpublished data – M. Halterman, Bureau of Land Management). The results of the field test of the western yellow-billed cuckoo model are presented in Figure 18. These results show that the predictions of the HSI model regarding habitat suitability are generally accurate (if it is assumed that breeding density is a reflection of at least short term habitat quality). Thus the HSI model developed for this study is a reasonable predictor of breeding habitat quality for the study species.

Figure 18. Linear densities of pairs of yellow-billed cuckoos and HSI scores

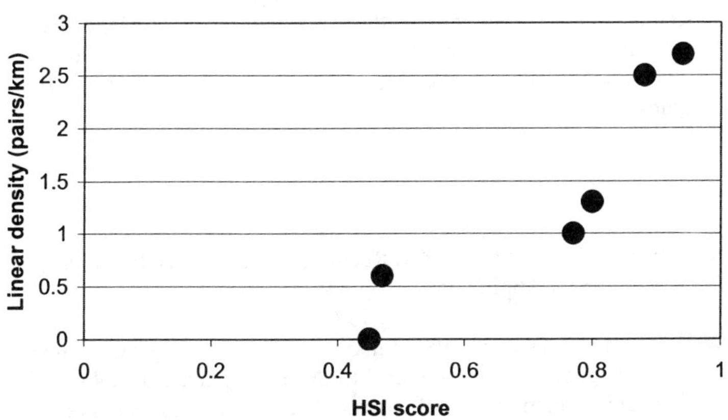

Conclusions

Hitherto, HSI models have largely been utilized to quantify the quality of existing habitat for wildlife species, without reference to how that habitat may have been altered in the past or how it might be altered in the future. In this study, we are using HSI models as part of an integrated modeling approach to estimate the habitat quality gain or loss for a variety of indicator species due to future climate change and aquifer management decisions. In effect, the HSI models are being utilized as part of an ecological risk assessment (ERA), except that the stressors are not the traditional ERA contaminants but are climatic changes and land-use policies, and the outcomes are not toxicity or physiological impairment but changes in habitat quality and carrying capacity. Other reports in this volume detail how HSI might be used as part of traditional ERA, however, the experience of this project has identified a number of areas in which these models could be profitable integrated into the process. These include:

- Determining the extent that wildlife may actually use, or avoid, the site or contaminated portions thereof, and therefore, their exposure risk.
- Determining the extent to which existing wildlife habitat may be altered by site remediation.
- Focusing remediation on areas of lower habitat value.
- Evaluating the extent to which proposed remediation activities might incur net benefits (due to contamination being removed) or costs (due to loss of habitat) through remediation.

One of the main limitations of the HSI models that are currently available is that relatively few have been developed. Also, most of those that have been developed have not been field-tested. Nevertheless, the development and testing of HSI models for ERA (focussed on species that typically occur in such assessments) could be accomplished relatively easily and quickly.

References

Auble, G.T., J.M. Friedman and M.L. Scott. 1994. Relating riparian vegetation to present and future streamflows. Ecological Applications, 4:544-554.

Brand, G.J., S.R. Shifley, and L.F. Ohman. 1986. Linking wildlife and vegetation models to forecast the effects of management. In: J. Verner, M.L.Morrison, and C.J. Ralph (eds), Wildlife 2000: modeling habitat relationships of terrestrial vertebrates. University of Wisconsin Press, Madison, Wisconsin.

Finch, D.M. 1989. Habitat use and habitat overlap of riparian birds in three elevational zones. Ecology, 70:866-880.

Gaines, D. 1974. Review of the status of the Yellow-billed Cuckoo in California: Sacramento Valley populations. Condor 76:204-209.

Galbraith, H., K. LeJeune, and J. Lipton. 1996. Metal and Arsenic Impacts to Soils, Vegetation Communities, and Wildlife Habitat in Southwest Montana Uplands Contaminated by Smelter Emissions: I Field Evaluation. *Environmental Toxicology and Chemistry*. 14:1895-1903.

Galbraith, H. Kapustka, L. and M. Luxon. 2001. Incorporating habitat quality in ecological risk assessments: selecting assessment species, SETAC Annual Conference, Baltimore, MD, November, 2001.

Halterman, M.D. 1991. Distribution and habitat use of the Yellow-billed Cuckoo on the Sacramento River, California, 1987-1990. MS Thesis, California State University, Chico.

Hamilton, W.J., and M.E. Hamilton. 1965. Breeding characteristics of Yellow-billed Cuckoos in Arizona. Proc. Calif. Acad. Sci. 32:405-432.

Hughes, J.M. 1999. Yellow-billed cuckoo. The Birds of North America. Life Histories for the 21st Century. No. 418 (A. Poole and F. Gill, eds). The Birds of North Ameica., Inc., Philadelphia, PA.

Kaufman, K. 1996. Lives of North American Birds. Houghton Mifflin Company. Boston.

Krueper, D. 1997. Annotated Checklist to the Birds of the Upper San Pedro River Valley. Unpublished report – Bureau of Land Management. San Pedro Riparian National Conservation Area.

Laymon, S.A. and M.D. Halterman. 1987. Can the western subspecies of Yellow-billed Cuckoo be saved from extinction? Western Birds 18:19-25.

Laymon, S.A., P.L. Williams, and M.D. Halterman. 1997. Breeding status of the yellow-billed cuckoo in the South Fork Kern River Valley, Kern County, California. Report to the USDA Forest Service Sequoia National Forest.

LeJeune, K., H. Galbraith, J. Lipton, and L.A. Kapustka. 1996. Effects of Metals and Arsenic on Riparian Soils, Vegetation Communities, and Wildlife Habitat in Southwest Montana. *Ecotoxicology*. 5:297-312.

Marshall, R.M. 2000. Population status on breeding grounds. In: *Status, Ecology, and Conservation of the Southwestern Willow Flycatcher*. USDA, Forest Service General Report RMRS-GTR-60.

Roja, H.A., J. Bredehoeft, R. Lacewell, J. Price, J. Stromberg, G.A. Thomas. 1998. *Sustaining and enhancing riparian migratory bird habitat on the Upper San Pedro River*. Report to the Commission for Environmental Cooperation.

Schamberger, M., and A. Farmer. 1978. The habitat evaluation procedures: their application in project planning and impact evaluation. North American Natural Resources Conference, 43:274-283.

Sedgwick, J.A. 2000. Willow Flycatcher. The Birds of North America. Life Histories for the 21st Century. No. 533 (A. Poole and F. Gill, eds). The Birds of North Ameica., Inc., Philadelphia, PA.

Skaggs, R.W. 1996. Population size, breeding biology, and habitat of willow flycatchers in the Cliff-Gila Valley, New Mexico. New Mexico department of Fish and game Report.

Sogge, M.K., and R.M. Marshall, 2000. A survey of current breeding habitats. In: *Status, Ecology, and Conservation of the Southwestern Willow Flycatcher*. USDA, Forest Service General Report RMRS-GTR-60.

Stromberg, J.C. 1998. Dynamics of Fremont cottonwood (*Populus fremontii*) and saltcedar (*Tamarix chinensis*) populations along the San Pedro River, Arizona. Journal of Arid Environments, 40:133-155.

Stromberg, J.C., R. Tiller, and B. Richter. 1996. Effects of groundwater decline on riparian vegetation of semiarid regions: the San Pedro, Arizona. Ecological Applications, 6:113-131.

U.S. Fish and Wildlife Service, 1997. Final determination of critical habitat for the southwestern willow flycatcher. Federal Register 62:39129-39146 (July 22, 1997).

U.S. Fish and Wildlife Service, 2001. Southwestern willow flycatcher – draft recovery plan. U.S. Fish and Wildlife Service, Albuquerque, New Mexico.

VEMAP, 1995. Vegetation/ecosystem modeling and analysis project: Comparing biogeography and biochemistry models in a continental-scale study of terrestrial ecosystem responses to climate change and CO_2 doubling. Global Biogeochemical Cycles, 9:407-437.

Vionnet, L.B., and T. Maddock. 1992. Modeling of ground-water flow and surface/ground-water interaction for the San Pedro Basin – Part 1- Mexican border to Fairbank, Arizona. Department of Hydrology and Water Resources, University of Arizona, Phoenix.

Lawrence A. Kapustka,[1] *Hector Galbraith,*[2] *Matt Luxon,*[3] *Joan M. Yocum,*[4] *William J. Adams*[5]

Application of Habitat Suitability Index Values to Modify Exposure Estimates in Characterizing Ecological Risk

REFERENCE: Kapustka, L. A., Galbraith, H., Luxon, M., Yocum, J. M., Adams, W. J., **"Application of Habitat Suitability Index Values to Modify Exposure Estimates in Characterizing Ecological Risk,"** *Landscape Ecology and Wildlife Habitat Evaluation: Critical Information for Ecological Risk Assessment, Land-Use Management Activities, and Biodiversity Enhancement Practices, ASTM STP 1458*, L. A. Kapustka, H. Galbraith, M. Luxon, G. R. Biddinger, Eds., ASTM International, West Conshohocken, PA 2004.

ABSTRACT: We developed an approach intended to improve the accuracy and relevancy of ecological risk assessments (EcoRA), facilitate communication of decisions, and control overall project costs. An Access® database containing data for 91 Habitat Suitability Index (HSI) Models species and a Correspondence Matrix, which cross-lists all wildlife species of the US in terms of similar habitat preferences and similar dietary patterns were prepared. An iterative approach was used to: (a) select candidate assessment species; (b) prescribe data collection needs for characterization of habitat quality; (c) generate spatially explicit descriptions of habitat quality for assessment species; and (d) allocate exposure estimates using habitat quality and distributions of chemical concentration. Information from a previous EcoRA conducted for Kennecott Utah Copper was used to evaluate the efficacy of the approach. Vegetation polygons across the 360-km^2 area of the site were generated from infrared aerial photographs. HSI values were calculated for each polygon for selected assessment species. HSI values then were used to modify exposure estimates for selected CoC. Qualitative comparisons of these modified exposure estimates to the findings of the original risk assessment illustrate that the approach would improve the accuracy and relevance of the risk estimates.

KEYWORDS: HSI, Habitat quality, ecological risk assessment, exposure assessment, vegetation, wildlife

---

[1] President and Senior Ecotoxicologist, ecological **p**lanning and **t**oxicology, inc, 5010 SW Hout street, Corvallis, OR 97330-9540

[2] Galbraith Environmental Sciences, 633 Furman Way, Boulder, CO 80303

3 Ecotoxicologist, Windward Environmental LLC, 200 West Mercer Street, Suite 401, Seattle, WA 98119

4 Environmental Scientist, ecological **p**lanning and **t**oxicology, inc, 5010 SW Hout street, Corvallis, OR 97330-9540

5 Principal Environmental Scientist, Rio Tinto Health Safety and Environment, 5295 South 300 West, Suite 300, Murray, UT 84107

Introduction

The function of ecological risk assessment (EcoRA) is to predict the potential effects of stressors (typically chemical) on ecological resources. One of the essential components of the EcoRA process is the identification of suitable assessment endpoint species (AES) that can be the focus of the exposure assessment and risk characterization phases. Risk assessment guidance specifies several criteria for assessment endpoint species (AES) selection (US EPA 1992, Suter 1993). These include: ecological relevance; susceptibility to known or potential stressors; relevance to management goals; accessibility to prediction and measurement; and reflection of societal values. Nevertheless, EcoRAs are often criticized for being limited to a few charismatic species and arbitrarily omitting too many others, for focusing on species unlikely to occur on the site because of habitat or range limitations, and for ignoring habitat quality in the exposure assessment. Ecological Risk Assessments (EcoRAs) can be made more relevant to stakeholders, including risk managers, if greater attention is given to characterization of landscape features that influence ecological resources. We have proposed modifications (Kapustka et al. 2001) of the US EPA EcoRA Framework (US EPA 1992) and Guidance (US EPA 1998) to:

- identify scenarios where habitat value is important in EcoRAs;
- guide selection of appropriate assessment species, keyed to wildlife distribution ranges;
  - keyed to a database of habitat suitability models;
  - cross-linked with the EPA exposure handbook species (US EPA 1993a,b)
  - referenced to wildlife distributions (e.g., breeding bird survey)
- define data collection needs for reconnaissance-, screening- and definitive-level characterization of habitat quality for potential assessment species;
- generate spatially explicit descriptions of habitat quality for various assessment species; and
- allocate exposure estimates using both habitat quality and spatial variations in chemical concentration.

In this paper, we examine the feasibility of implementing the modifications we proposed.

Materials and Methods

Information was developed as part of a series of studies conducted for a site-wide EcoRA. This background information is summarized to provide a context for the evaluation of feasibility of our proposed approach that uses landscape features to define habitat quality.

In 1994 and 1995 **ep** and **t** performed an EcoRA of the Kennecott Utah Coper Corporation (KUCC) site (**ep** and **t** 1994a,b, 1995a,b,c, 1996, 1997a). As part of that effort, we conducted and extensive vegetation sampling effort. More than 350 species of plants in the various vegetation zones were observed. Subsequent studies have tallied at least 410 plant species (Kapustka et al. 2004). Vegetation maps produced from aerial photographs were also prepared in Arcview® 6.0. As input to this feasibility assessment, we used the extensive collection of data on spatial distribution of Constituents of Potential Concern (CoPC) in soil, plant tissues, and small mammals as well as detailed descriptions of plant community composition across the 360 km^2 project area. Here, we have addressed selection of assessment species, development of a workplan, and application of habitat quality characteristics to modify exposure estimates.

Site Information and Risk Characterization Summarized from Prior EcoRA

The KUCC site west of Magma, Utah, and on the northern end of the Oquirrh Mountains, comprised nearly 460 km^2. The Project Area (40° 28' to 40° 45' N latitude; 112° 05' to 112° 15' W longitude) is in the Basin and Range Province, which is characterized by a series of rugged mountain ranges rising above the remnant lake beds. The current shoreline of the Great Salt Lake is at approximately 1300 m (4214 ft) above mean sea level. The Northern Oquirrh Mountains reach heights of 2800 m (>9000 ft) above mean sea level. The Oquirrh Mountains are at the eastern edge of the Great Basin Desert and include natural communities of desert shrub lands, submontane shrubs, and patches of juniper, aspen, or coniferous forests. Wetlands along the south shore of the Great Salt Lake are also included in the Core Project Area. About one dozen plant communities exist from the shoreline of the Great Salt Lake to the peaks of the Oquirrh Mountains. These include marshland, salt desert shrub, desert shrub, riparian forests, aspen/conifer, aspen, submontane shrub, juniper woodlands, and coniferous forest. Elk, raptors, and waterfowl are some of the more charismatic wildlife taxa in the area. Many species of songbirds, small mammals, and herpetofauna also reside in the Oquirrh Mountains. Numerous species of migratory birds and waterfowl use the Oquirrh Mountains and Salt Lake shoreline wetlands during spring and fall migrations or as a summer breeding area.

The initial EcoRA focused of eight CoPC: arsenic, cadmium, chromium, copper, fluorine, lead, selenium, and zinc. Concentrations of some of the CoPC in soils exceeded screening threshold phytotoxicity levels (**ep** and **t** 1995a). However, extensive site surveys found that the CoPC were not impeding successional development of complex plant communities as evidenced by continued increases in species richness and establishment of mid- and late successional plant species in areas that had been affected by smelter emissions in the 1960s (**ep** and **t** 1995b,c).

Similarly, CoPC concentrations in soil and plants exceeded effects thresholds for herbivorous and insectivorous animals and the carnivores that feed on them. Due to differences in distributions of CoPC and wildlife use patterns, upland areas were addressed separate from wetland areas. In this paper, we focus on copper and selenium.

Upland Areas

Concerns for herbivorous and insectivorous food chains were addressed directly by collecting soil, plants, invertebrates, and small mammals from a series of sampling sites along a gradient of environmental concentrations. At each sampling site, information on the abundance of small mammals, their rate of reproductive activity, and evidence of CoPC-related tissue lesions, were collected along with data on plant community structure at each sampling site in Coon, Little Valley, Kessler, and Black Rock Canyon and Pine Canyon (Figure 1).

The most commonly captured mammals were from the genus *Peromyscus*, the deer mouse (*P. maniculatus*) and piñon mouse (*P. truei*), ranging from 5 to 113 animals per sampling site. Small numbers of montane voles (*Microtus montanus*), Great Basin pocket mice (*Perognathus parvus*), western harvest mice (*Reithrodontomys megalotis*), and least chipmunks (*Tamias minimus*) also were captured. There was no evidence that the numbers of animals captured were reduced by the dietary concentrations of CoPC or the concentrations in the whole bodies of the mammals. Rates of reproductive activity varied considerably among the sampling sites, but there was no evidence that this was due to the CoPC. Seventy-one mammals (approximately 12 from each site) were necropsied for histological examination of liver and kidney tissues, but there were no CoPC-related tissue lesions.

Soil copper concentrations were elevated in the northern portion of the project area due to years of smelter emissions (Figure 2). Selenium soil concentrations were elevated in the northern canyons (i.e., Black Rock and Kessler), but were also high in Harkers Canyon (Figure 3). The distribution of selenium was interpreted by us and previous workers as naturally occurring background levels, and not due to industrial activities.

Site-specific information on CoPC concentrations was used to predict exposure profiles for different trophic levels of food webs in the Oquirrh Mountains. Estimation of effects was constrained by the limited toxicity profile information for the different wildlife. Trophic transfer factors were based on site-specific data collected on plant, insect, and small mammal tissues. The food web model for copper and selenium exposures are shown for Kessler Canyon copper (Figure 4, 5).

CoPC concentrations in the estimated diets of insectivores (e.g., shrews, chickadees, and bats) and carnivores were also near or occasionally exceeding projected effects thresholds at some sampling sites. The greatest potential for risk as measured in this study appears to be to some songbirds and some small mammals in Kessler and Black Rock Canyons from selenium in their (insectivorous) diets. The higher concentrations of selenium in the insectivore food chains were related to the dominance of selenium-occurring plants in these canyons. Foliar invertebrates feed on these plants and take up selenium into their bodies. Birds and mammals that feed upon these invertebrates might then be exposed to selenium concentrations higher than that found in the soil.

The estimated exposure levels slightly exceeded No Observed Adverse Effect Levels (NOAELs) for copper for both the most sensitive omnivores and herbivores, but did not exceed the more a broader group of animals (Table 1). Selenium levels slightly exceeded NOAEL values for Kessler Canyon, Black Rock Canyon, and Pine Canyon.

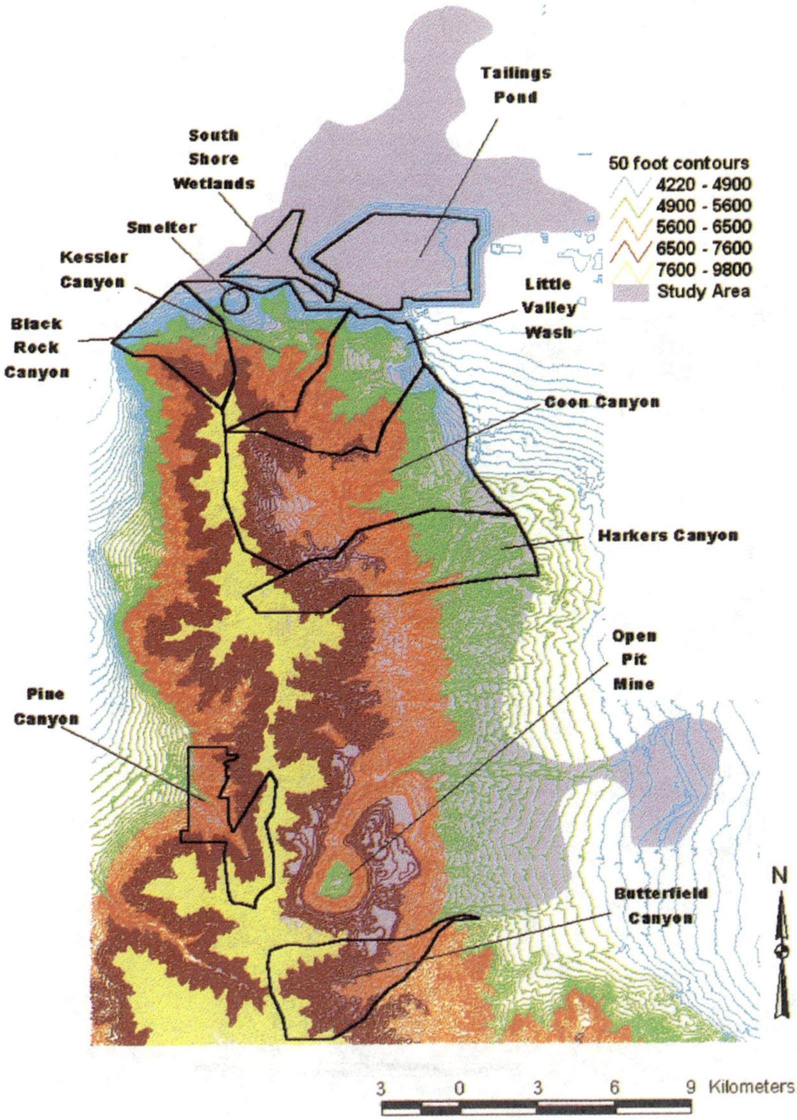

Figure 1. *Delineation of canyons and elevation zones of the project area.*

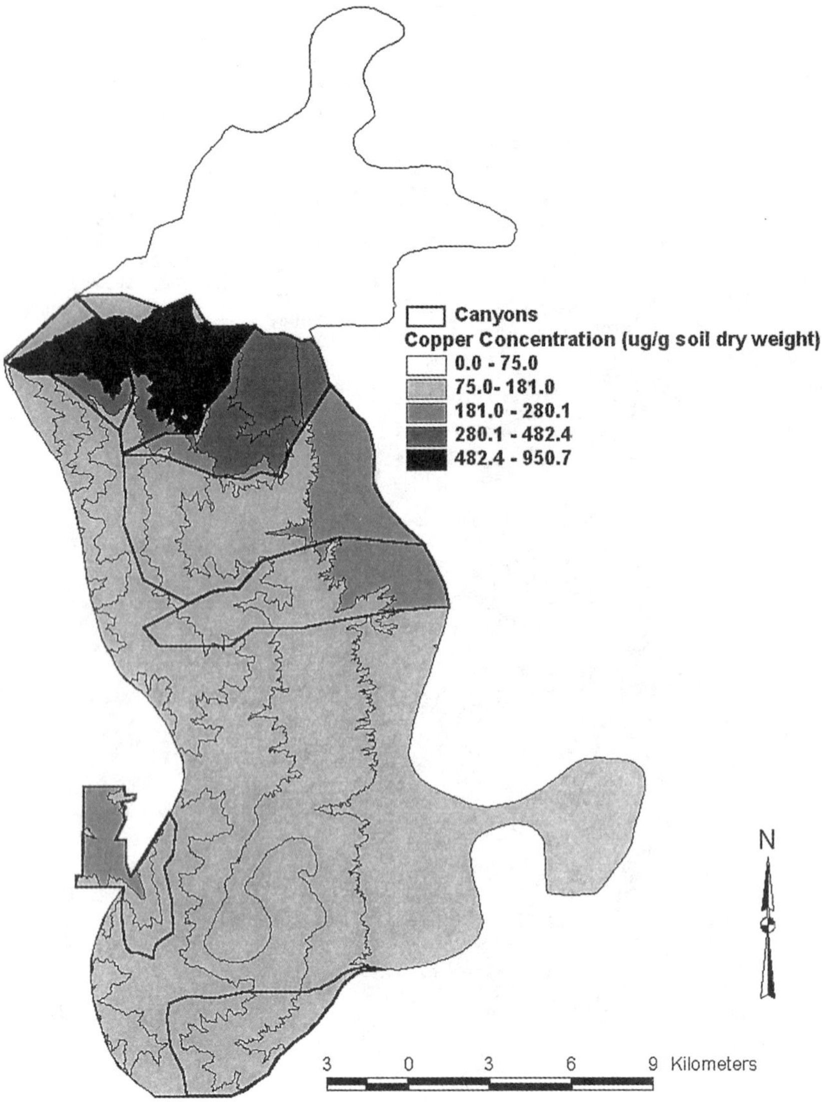

Figure 2. *Distribution of soil copper (ppm) on the KUCC Project Area.*

(Refer to Figure 1 for identification of canyon and elevation polygons.)

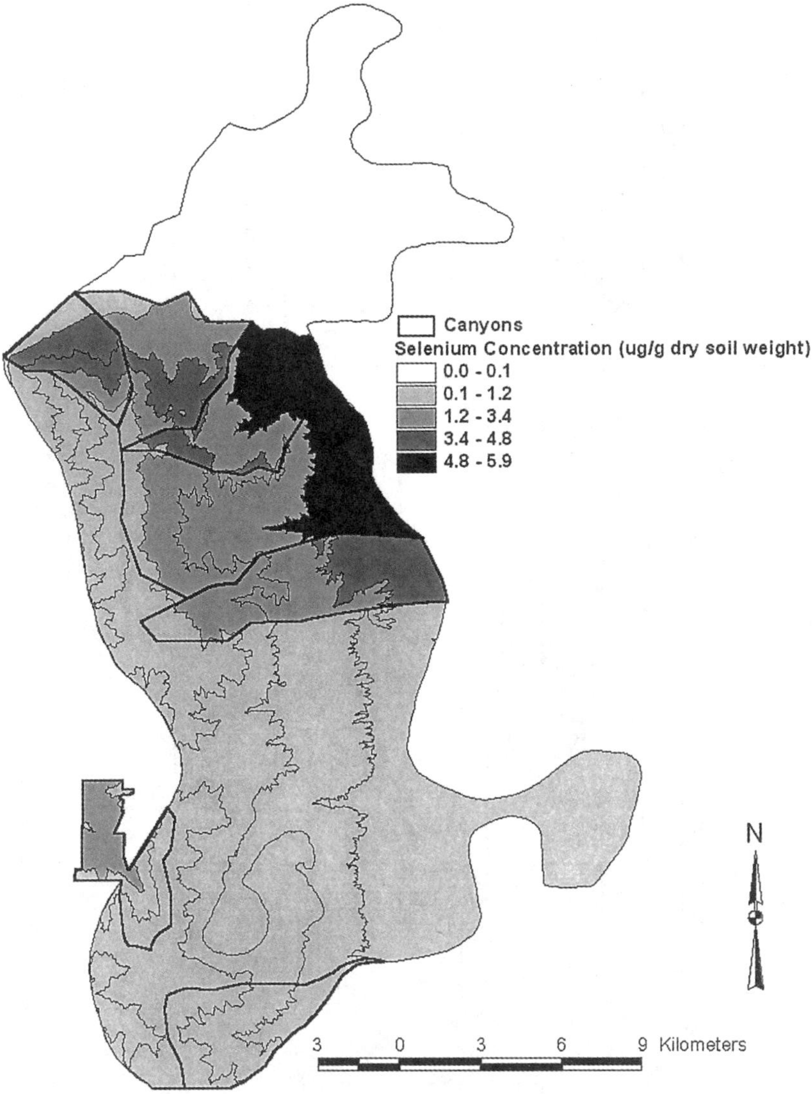

Figure 3. *Distribution of soil selenium (ppm) on the KUCC Project Area.*

(Refer to Figure 1 for identification of canyon and elevation polygons.)

SAMPLING SITE: KC2

CoPC: Copper

Values represent CoPC concentration (ppm, dry wt) in each environmental component
Shaded values are measured
Unshaded values are calculated

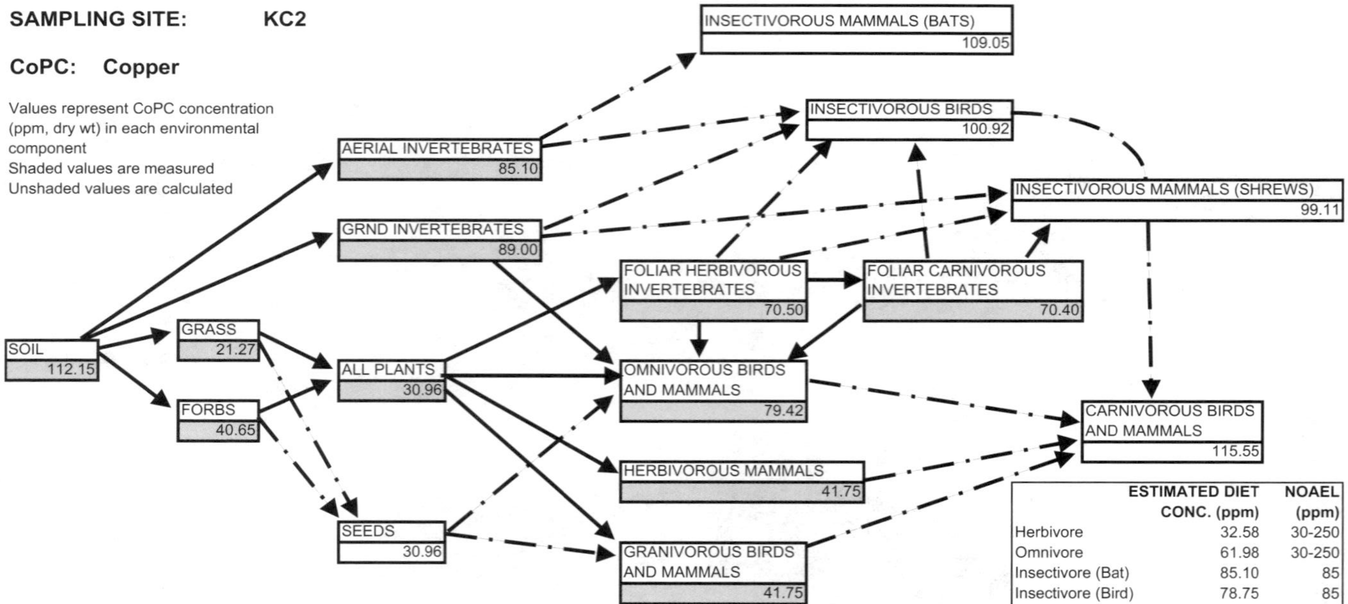

	ESTIMATED DIET CONC. (ppm)	NOAEL (ppm)
Herbivore	32.58	30-250
Omnivore	61.98	30-250
Insectivore (Bat)	85.10	85
Insectivore (Bird)	78.75	85
Insectivore (Shrew)	76.63	85
Carnivore	90.17	85

TERRESTRIAL FOOD WEB MODEL

TROPHIC TRANSFER FACTORS			
Soil to Grasses	0.19	Diet to Herbivorous/Granivorous Mammals	1.28
Soil to Forbs	0.36	Diet to Omnivorous Mammals	1.28
Soil to All Plants	0.28	Diet to Insectivorous Mammals	1.28
Soil to Ground Invertebrates	0.79	Diet to Carnivorous Mammals	1.28
Plants to Foliar Herbivorous Insects	2.28	Assumed Carnivore Diet: 50% Small Mammals	
Herbivorous to Carnivorous Invertebrates	1.00	50% Birds	

Figure 4. *Food web exposure model for copper in Kessler Canyon.*

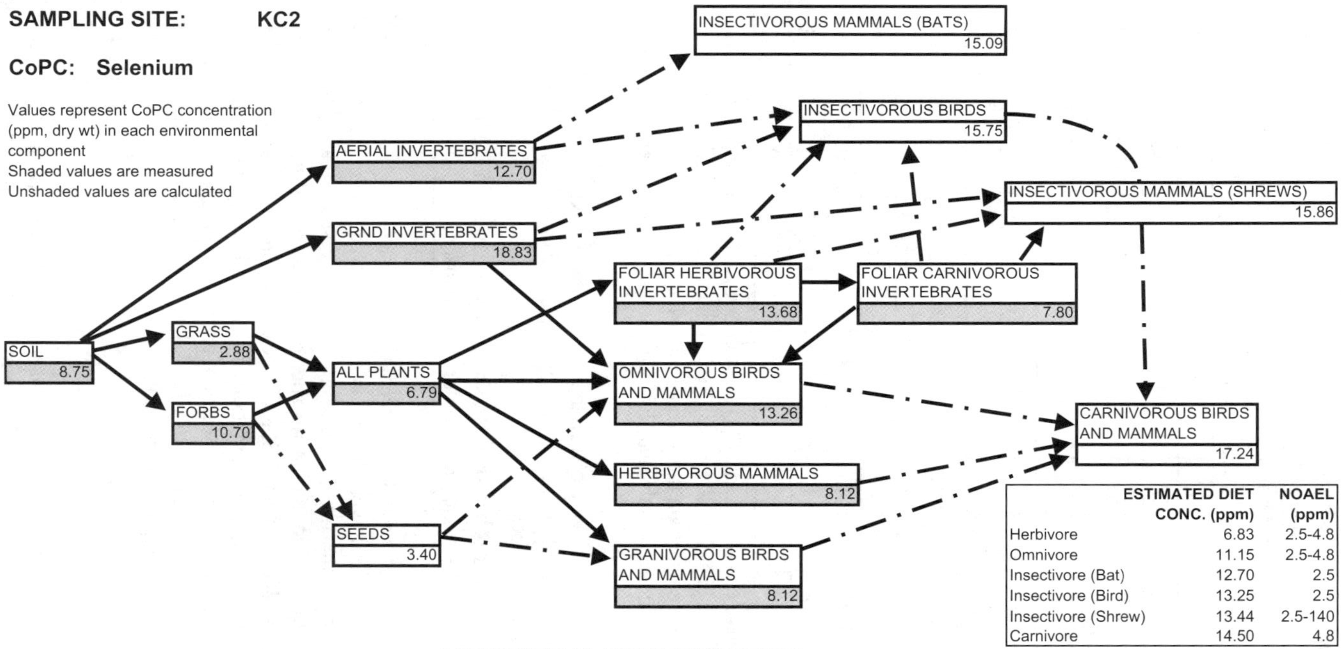

	ESTIMATED DIET CONC. (ppm)	NOAEL (ppm)
Herbivore	6.83	2.5-4.8
Omnivore	11.15	2.5-4.8
Insectivore (Bat)	12.70	2.5
Insectivore (Bird)	13.25	2.5
Insectivore (Shrew)	13.44	2.5-140
Carnivore	14.50	4.8

TERRESTRIAL FOOD WEB MODEL

TROPHIC TRANSFER FACTORS			
Soil to Grasses	0.33	Diet to Herbivorous/Granivorous Mammals	1.19
Soil to Forbs	1.22	Diet to Omnivorous Mammals	1.19
Soil to All Plants	0.78	Diet to Insectivorous Mammals	1.19
Soil to Ground Invertebrates	2.15	Diet to Carnivorous Mammals	1.19
Plants to Foliar Herbivorous Insects	2.01	Assumed Carnivore Diet:	50% Small Mammals
Herbivorous to Carnivorous Invertebrates	0.57		50% Birds

Figure 5. *Food web exposure model for selenium in Kessler Canyon.*

Table 1. Risk quotients for copper and selenium (based on diet concentration/NOAEL) for omnivores and herbivores by sampling sites.

	Omnivore[1]			Herbivore[2]		
Site	Copper[3]		Selenium	Copper[3]		Selenium
NOAEL (mg/kg)	30	250	4	30	250	4
LOAEL (mg/kg)	38	425	10	38	425	10
Coon Canyon 1	0.98	0.12	0.14	0.52	0.06	0.08
Coon Canyon 2	0.98	0.12	0.21	0.47	0.06	0.24
Little Valley 1	***1.80***	0.22	0.47	***1.14***	0.14	0.34
Little Valley 2	***1.91***	0.23	0.55	***1.27***	0.15	0.44
Kessler Canyon 1	***1.91***	0.23	***1.55***	***1.06***	0.13	0.80
Kessler Canyon 2	***2.01***	0.24	***2.72***	***1.09***	0.13	***1.71***
Black Rock Canyon 1	***3.12***	0.37	***1.10***	***2.26***	0.27	0.84
Black Rock Canyon 2	***3.15***	0.38	***1.03***	***1.61***	0.19	0.68
Pine Canyon 1	***1.07***	0.13	0.58	0.79	0.10	0.30
Pine Canyon 2	***1.38***	0.17	***1.35***	***1.36***	0.16	***2.44***
Pine Canyon 3	***1.39***	0.17	0.76	***1.76***	0.21	***1.13***

[1] Diets of omnivores were calculated as 61% invertebrates, 37% vegetation, and 2% soil. Vegetation concentrations were based on the average of all plant samples collected. Invertebrate concentrations were based on the average of surface-dwelling invertebrates and foliar herbivorous and carnivorous invertebrates. Risk quotients ≥1 are presented in bold italics.

[2] Diets of herbivorous mammals were calculated as 98% vegetation and 2% soil. Vegetation concentrations were based on the average of all plant samples collected. Risk quotients ≥1 are presented in bold italics.

[3] The left column under copper is based on the sensitivity observed in sheep. The right column under copper is based on monogastric herbivore mammals.

The greatest determinant of animal presence and abundance was the quality of the habitat. At sampling sites in the lower elevations of the five canyons, grazing by wild and domestic ungulates keeps the grasses and forbs relatively short with little, if any, standing dead vegetation to provide habitat for species such as voles and shrews. Less abundant woody cover in Kessler and Black Rock Canyons results in a reduced abundance of songbirds. Reduced grazing pressure in all of the canyons would increase the amount and quality of habitat for many species of animals. Management of selenium-concentrating plant species in Kessler and Black Rock Canyons could provide a means to reduce risks to herbivore and insectivore food chains.

Wetlands

Initial efforts in the 1994 and 1995 field seasons indicated potentially high CoPC (especially Se) exposure levels in wetlands. Sampling was conducted to quantify 1) the abundance and diversity of aquatic macroinvertebrates, 2) the abundance and diversity of wildlife, 3) the nesting success of birds, and 4) CoPC concentrations in water, sediment, aquatic invertebrates, and bird eggs (**ep** and **t** and Parametrix, 1997). They concluded that selenium may pose a limited risk to successful reproduction of some shorebirds that feed in specific portions of the wetland.

Subsequently, the Tailings Modernization Project, begun in 1996, led to many structural changes of the wetland areas. This included relocation of railroad tracks, repositioning culverts that lowered water levels of ponds associated with industrial activities, capping of wells and artesian flows (e.g., the Garfield Wells and Kessler Springs), and various other hydrologic changes related to industrial uses. These changes removed or reduced Selenium sources to the area and significantly altered the nature and extent of habitats attractive to birds. Consequently, additional field surveys and sample collections were performed in May and June 1999.

The objectives of the 1999 studies were to measure habitat quality and selenium concentration in bird eggs and aquatic invertebrates that are food resources of interest to the birds. Aquatic invertebrates are known to have the highest selenium accumulation relative to other avian dietary items. Risk from selenium to birds occurs primarily through ingestion of food having high selenium concentrations. Exposure potential is also affected by the attractiveness of the habitat to particular species or guilds. Habitat quality influences whether or not birds will use an area for nesting or as a stopover during migration.

Standardized Habitat Suitability Index Models for guilds (breeding waterfowl, migratory shorebirds) and individual species (avocets and red-winged blackbirds) were used to assess the habitat quality for these taxa. The area was divided into 12 subunits, identified by differences in cover and landuse activities, which were scored separately for suitability of habitat for these species. Each subunit was scored to parameterize the respective HSI models. Parameters included: 1) levels of human-associated disturbance; 2) percentage of an area unvegetated or sparsely vegetated; 3) proportion composed of islands, shallow water, open-water wetlands, or standing, non-woody vegetation of various heights; 4) permanence of water bodies; 5) the height at which vegetation becomes thick enough that one cannot see an object several meters away; and 6) the presence of other organisms such as carp, damselflies, and dragonflies. Scores for each of these attributes were used to calculate an index of habitat suitability, based on simple algebraic models that ranged between 0.0 (essentially unsuitable) and 1.0 (fully suitable).

The HSIs indicated relatively poor quality bird habitat for most of the wetland, (Table 2). The migratory shorebird model identified the I-80 Pond as the best habitat, but the score was only 0.25 (out of a possible 1.0). The habitat of the I-80 Pond, according to the avocet/stilt model, is somewhat better (HSI = 0.38). However, the Freeway Pond habitat for avocets and stilts was relatively poor (HSI = 0.26). The habitat on the sample unit is most suited for breeding ducks, particularly in the I-80 Pond, the Slag Pond, Pond A and the West Ponds, all of which scored a 1.0 in the HSI model. The Freeway Ponds and the Oolitic Sand Mining Area appeared to be suitable for ducks as well. Red-winged blackbird habitat was best at the Kessler Spring Area, around the Garfield Wells, and at the Oolitic Sand Mining Area. The remainder of the sample unit was not well suited for these songbirds.

Table 2. *Habitat suitability indices for the four taxa in 12 operational sub-areas of the wetland.*

Location/HSI model	Migratory Shorebird	Avocet/stilt	Duck breeding	Red-winged blackbird
Kessler Spring	0.00	0.00	0.08	0.50
Garfield Wells	0.03	0.06	0.20	0.30
Pond C	0.00	0.01	0.00	0.01
Pond B	0.00	0.00	0.00	0.04
Railroad Bridge Uplands	0.01	0.03	0.20	0.10
Oolitic Sand Area	0.07	0.12	0.80	0.30
East Ponds	0.05	0.08	0.20	0.10
Freeway Ponds	0.18	0.26	0.80	0.04
I-80 Pond	0.25	0.38	1.00	0.10
Pond A	0.09	0.14	1.00	0.10
Slag Pond	0.00	0.02	1.00	0.09
West Ponds	0.00	0.00	1.00	0.10

Migratory shorebirds, particularly those that move through or stage in the area in the fall, would likely find the Freeway Ponds and the I-80 Pond to be the most suitable habitat in the area. The remaining portions of the study area are completely unsuitable for this purpose (Table 2). These two areas are the only portions of the site with permanent water that are free enough from human disturbance to allow migratory shorebird use.

The heterogeneous bird habitat across the area was compared to invertebrate Se concentrations and bird egg selenium concentration. For example, in the Garfield Wells Area invertebrate selenium concentrations were elevated and presented some risk to nesting birds. The invertebrate selenium concentrations in the Kessler Springs Area suggested possible exposure problems to birds, but the bird eggs sampled at that location did not confirm the putative risk. Importantly, both the Garfield Wells and the Kessler Springs areas had low habitat quality scores for birds. And therefore, they were not attracting high numbers of birds. Another area known as the I-80 Pond was scored as having good habitat for the birds of interest. It had selenium concentrations in aquatic invertebrates that suggested a marginal risk to birds, but egg selenium concentrations were below toxicity levels. Two other areas, the Oolitic Sand Mining Area and Pond A, had good shorebird habitat with minimal risk from selenium exposure.

Subsequent modification of the wetlands was undertaken as part of an effort to expand the tailings pond. The HSI information was used in the design to identify areas for removal of soils and sediments having elevated CoPC. Habitat enhancement of areas with low CoPC concentrations remains under consideration as part of the corporate biodiversity program.

Habitat Suitability Index Database

Habitat evaluation procedures (Habitat Suitability Index Models; HSIs) have been developed for wildlife management activities (Schroeder and Haire 1993). We have located 62 Habitat Suitability Index (HSI) and Habitat Evaluation Procedure (HEP) models for bird species, 17 for mammals, and 6 for reptiles/amphibians (Appendix A). Each published HSI model includes a map of those areas of the species' range for which the model is applicable. Information from these publications has been encoded into an Access® database (Figure 6). Database fields include species distribution by EPA Region, State, and specific locality for which the model was produced; parameters required to compute the HSI; and prioritized methods that can be used to obtain data to parameterize the models. Equations to calculate relationships of parameters (e.g., percentage canopy cover) to variables and the algorithms that combine variables into HSI values have been encoded into Excel® spreadsheets.

Entries on all North American terrestrial and wetland wildlife species (avian, herpetofauna, or mammalian) are included. Notations are provided as to similarity of habitat requirements to HSI model species and to dietary preferences for US EPA exposure species (i.e., those listed in the US EPA *Wildlife Exposure Factors Handbook,* US EPA 1993). Built-in queries permit searches on any species or list of species to generate a compiled report of all potential species in a project area and the level of overlapping information for each taxon in terms of habitat and dietary preferences. For the bird HSI model species, we have identified an additional 107 "overlap species." These are species for which no HSI models exist but for which, because of close similarities in their habitat requirements, existing individual HSI models may be appropriate, perhaps after modification. In total, published HSI models exist for 169 primary and overlap bird species.

Assessment Species Selection Process

The steps to develop a candidate list of assessment species are:

1. Search database (Region, State, Eco-region applicability) to identify species having HSI models potentially relevant to the site.
2. Examine Breeding Bird Survey Database, Christmas Bird Count, range maps in atlases, other sources, and knowledge of the site to exclude unlikely species and add additional species known to occur there.
3. Use correspondence matrix to sort species into categories (I through IV) to obtain best match of surrogate species having both HSI and exposure parameter data.
4. Classify species in terms of habitat types, trophic levels, and exposure scenarios relevant for the site.
5. Submit this list of species for consideration as assessment species. Refinements based on the ease of obtaining critical habitat information for different species may be used to establish priorities among potential assessment species.
6. Document in the administrative record all supporting information used in making decisions regarding:
    - expansion or reduction of the candidate list
    - rationale for setting priorities among species.

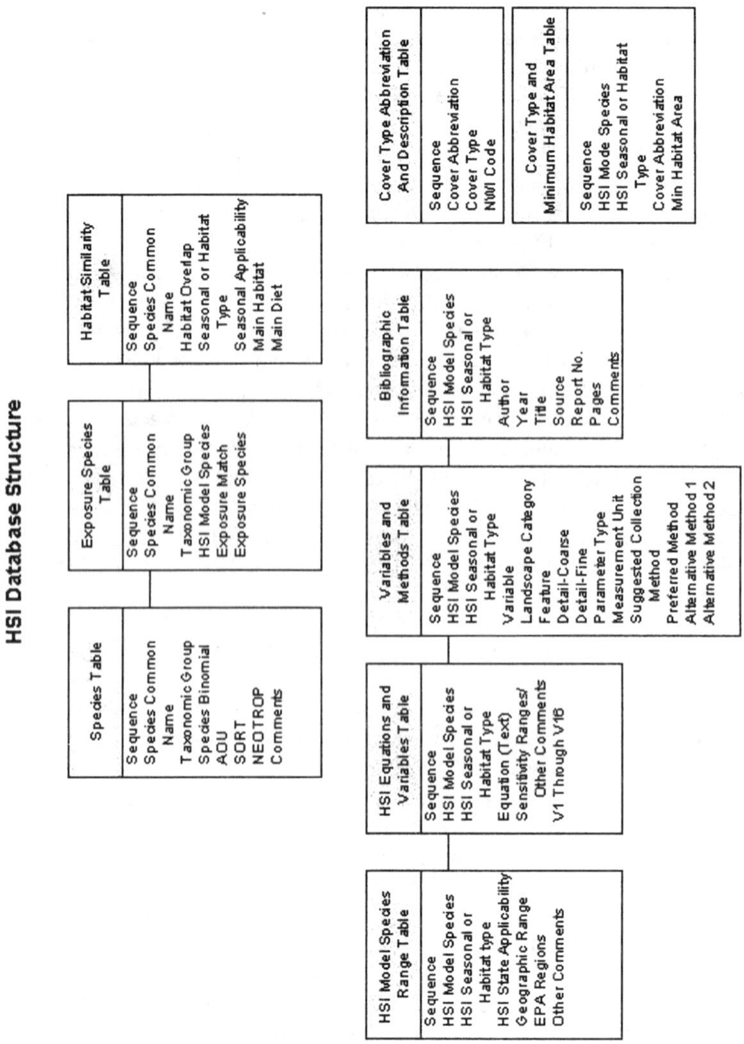

Figure 6. *Structure of Access Database.*

Workplan Development

After the candidate assessment species are agreed, the project workplan can be constructed to cover the full spectrum of potential site characterization required for the inclusion of habitat quality determination. The steps are:

1. Identify the complete list of candidate assessment endpoint species
2. Generate the corresponding list of HSI models
3. Compile the list of all variables needed to calculate HSI models for all the selected species
4. Qualitatively assess the level of effort needed to acquire the data
5. Examine the list of preferred and alternative methods capable of generating the required data
6. Review the sensitivity issues associated with the suite of variables
7. Evaluate which parameters
   - may be satisfied using aerial images
   - require routine on-site survey efforts
   - require specialized (detailed) on-site survey efforts
8. Structure a progressive sampling plan from reconnaissance level through definitive levels that maximizes the number of models satisfied with different levels of sampling effort

Application of Habitat Quality Characteristics to Modify Exposure Estimates

HSI values were developed for a subset of the candidate assessment species to provide an illustration of how habitat quality would alter the exposure estimates of animals of different trophic levels and of different home ranges (e.g., ones with home ranges equal to or smaller than the typical polygons and ones larger than the typical polygons). The methods presented in Kapustka et al. (2001) and data from the earlier EcoRA (**ep** and **t** 1995b) were used to calculate habitat-adjusted exposure estimates.

Results and Discussion

The results presented here pertain to the feasibility of performing an EcoRA using the approach presented earlier (Kapustka et al. 2001). The site contains a heterogeneous landscape both in terms of physical structure and vegetation cover. The distribution of CoPC also varies across landscape polygons. Therefore, the criteria for consideration of habitat characteristics of assessment species in an EcoRA are met.

Candidate Assessment Species

The Access® Database was queried to identify the HSI models that are applicable for Utah. This resulted in a list of 26 species (Table 3). We used Breeding Bird Survey data, personal observations at the site, and the correspondence matrix, to narrow the list of candidate assessment species. There are six bird species (American Coot, Yellow-headed Blackbird, Great Blue Heron, Yellow Warbler, Western Grebe, and Hermit Thrush) and two mammalian species (elk, red-backed vole) that have HSI models and for which exposure parameters are available in the US EPA *Wildlife Exposure Factors Handbook* (US EPA 1993).

In an EcoRA, exposure routes, including dietary considerations, bioavailability, and trophic transfer potential are important. On the KUCC site, with the CoPC being metals and metalloids, adverse effects are likely to be in the lower trophic levels of the food webs. The list of candidate species includes a mix of herbivores, insectivores, and carnivores.

Work Plan Analysis

A query of the HSI database identified 147 variables for the 26 candidate assessment species models (Table 3). These were classified under seven variable types. Cover types were expressed variously as area, basal area per area, connectivity, count, count/area, diameter at breast height (DBH), distance, height, kilometers, meters, canopy volume/area, percentage, predominance, presence, rank, and relative percentage. Faunal variables included fish kg/ha, potential nest sites/area, presence of carp, presence of odonates, and fish length.

Table 3. *HSI variables for candidate assessment species and tally of variables by categories.*

Number of Variables for the 26 Candidate Assessment Species			
American Coot	3	Lesser Scaup	5
Bald Eagle	4	Lewis' Woodpecker	7
Belted Kingfisher	7	Mink	6
Black-capped Chickadee	4	Osprey	5
Blue Grouse	7	Red-winged Blackbird	8
Brewer's Sparrow	6	Ruffed Grouse	5
Downy Woodpecker	5	Southern Red-backed Vole	4
Elk	15	Veery	6
Ferruginous Hawk	6	Western Grebe	8
Gray Partridge	8	Western Meadowlark	2
Great Blue Heron	6	Williamson's Sapsucker	4
Hairy Woodpecker	5	Yellow Warbler	3
Lark Bunting	4	Yellow-headed Blackbird	4
Distribution of Variable Types			
Cover	103	Soil moisture	1
Fauna	5	Water	18
Human use or development	6	Wetlands	7
Physiognomy	7	Total	147

We ranked each variable as to the level of effort that would be required to obtain the input data to calculate the HSI values. Four categories used were 1) the data can be obtained entirely from aerial imagery; 2) data collection requires ground reconnaissance of the site; 3) data collection requires on-site sampling or survey; and 4) detailed or otherwise highly specialized on-site data collection is required (e.g., tally the number of nesting cavities in trees). Some models are relatively easy to parameterize and others require considerable effort. For example, the American coot model can be parameterized entirely from aerial imagery (scored 1 in Table 4). The bald eagle and elk models require reconnaissance-level observations (scored 2 in Table 4) for at least one parameter. Twelve other models require routine site survey data (scored 3 in Table 4) for at least one parameter and 11 require detailed data for at least one parameter (scored 4 in Table 4).

HSI Calculations

In this paper, we have focused on three assessment species to illustrate the application of the approach. In a real-time EcoRA, a similar strategy would be employed, but would likely begin with species identified above (Table 4) as those easiest to parameterize (i.e., scoring 1 or 2). It would then proceed to those requiring greater effort or expertise. The three described here are the southern red-backed vole, the western meadowlark, and the ferruginous hawk.

Southern Red-backed Vole

Southern Red-backed vole HSI values were calculated for each habitat type in the KUCC GIS vegetation coverage (see Kapustka et al. 2004 in review). HSI values for each KUCC region were generated by averaging HSI scores across all vegetation polygons within each KUCC region. We assumed that an average HSI score of less than 0.1 represents unsuitable habitat. The mean home range area for the southern red-backed vole in Colorado varied from 0.01 to 0.5 ha (0.02 to 1.25 acres; Merritt and Merritt 1978). Maximum and minimum home ranges for the species in Michigan were 1.4 ha (3.56 acres) and 0.20 ha (0.49 acre), respectively (Blair 1941). The home ranges of both males and females overlapped the ranges of other individuals of both sexes.

HSI values for cover types (polygons) were used to calculate home range areas proportional to the spread of published home ranges. For example, for cover types having an HSI score of 1.0, the home range was designated as the minimum value (here 0.20 ha). The area of the home range was increased to the maximum published area (here 1.4 ha) for the lowest quality habitat cover type.

Western Meadowlark

Western Meadowlark HSI scores were calculated for each habitat type within the KUCC GIS vegetation coverage. HSI values for each KUCC region were generated by averaging HSI scores across all habitat polygons within each KUCC region. We assumed that an average HSI score of less than 0.1 represents unsuitable habitat. In Wisconsin, breeding territories ranged from 1.2 to 6.1 ha (3 to 15 ac), with mean of 3 ha (~7.5 ac; Lanyon 1956).

Table 4. *Effort/expertise required to parameterize HSI variables.*

HSI Species	V1	V2	V3	V4	V5	V6	V7	V8	V9	V10	V11	V12	V13	V14	V15
American Coot	1	1	1												
Bald Eagle (Breeding Season)	1	2	1	1											
Belted Kingfisher	3	3	1	3	1	4	1								
Black-Capped Chickadee	1	3	4	4											
Blue Grouse	1	3	3	3	3	3	1								
Brewer's Sparrow	1	1	1	3	1	3									
Downy Woodpecker	4	4													
Eastern Meadowlark[1]	3	3	3	3	3										
Elk	1	1	2	1	1	1	1	1	1	1	1	2	2	1	1
Ferruginous Hawk	3	1	1	1	3	3									
Gray Partridge	1	1	1	1	1	3	3	1							
Great Blue Heron	1	3	1	1	1	1									
Hairy Woodpecker	4	4	4	1	1										
Lark Bunting	3	3	3	3											
Lesser Scaup (Breeding)	3	3	1	1	1										
Lewis' Woodpecker	1	3	1	4	1	1	4								
Mink	1	1	3	1	1	1									
Osprey	4	1	3	1	4										
Red-Winged Blackbird	3	1	3	3	3	3	3	3							
Ruffed Grouse	4	4	4	4	1										
Southern Red-Backed Vole (Western U.S.)	4	4	3	3											
Veery	4	3	3	3	3	3									
Western Grebe	1	4	1	3	1	1	3	3							
Williamson's Sapsucker	1	1	3	4											
Yellow Warbler	3	3	3												
Yellow-Headed Blackbird	3	1	3	3											

V1 through V15 are variables defined for the specific HSI model. Each model has unique variables (e.g., American Coot model has 3 variables and the Elk model has 15 variables).

1 = Possible to quantify or reasonably estimate value from aerial imagery.

2 = Requires reconnaissance-level observations to obtain site-specific data.

3 = Requires routine on-the-ground site survey to obtain the data.

4. = Entails detailed or sophisticated survey data.

[1] We used the HSI model for the Eastern Meadowlark for the Western Meadowlark found at the site.

Ferruginous Hawk

The ferruginous hawk HSI was calculated over the entire site using three different methods. We calculated a site-wide HSI by 1) using data only from suitable habitats and 2) using data from all habitats within the site. For the third method, we calculated HSI values independently for each region of KUCC. Site-wide HSI values of 1.0 and 0.6 were generated using suitable habitat data and all data, respectively. When HSI values were calculated separately for each KUCC region, HSIs were 1.0 for all regions, with the exception of the wetlands region (0.65) and Kessler Canyon (0.99).

Ferruginous hawk home ranges are reported to range from 5 to 6346 km^2 for a nesting pair (BLM 2003. Snake River Birds of Prey conservation area website http://www.id.blm.gov/bopnca/ferrug.htm). Thus, (although it would be conservative to assume a smaller home range), for the sake of model exploration, KUCC could represent as little as 6% of the home range for a pair of hawks. Based on the apparently high habitat suitability of the KUCC site, a ferruginous hawk pair whose home range extends beyond the KUCC site would be expected to forage extensively at KUCC.

The HSI values were calculated for each vegetation type (Table 5) and applied to the respective polygons. The meadowlark distribution (Figure 7) was predominantly in the lower elevations and in the northern canyons dominated by grasslands. The distribution of the southern red-backed vole (Figure 8) was confined to the higher elevation forests that would provide the mycorrhizal fungi that constitute the majority of the vole's diet. Based on the areas involved and the different home ranges for the suitable habitats, the estimated populations of meadowlark would be highest in Little Valley (Table 6). The highest populations of southern red-backed vole would be in the spine and in the upper elevations of Harkers Canyon (Table 6). The ferruginous hawk ranged over the lower elevation areas and would forage considerable distances off the Project Area boundaries (Figure 9).

Habitat-adjusted Exposure Estimates

In the previous EcoRA, estimates of exposure were made using assumptions of entire populations encountering the levels of CoPC measured for the various canyons and elevation zones. Calculations were performed using mean, median, and 95% Upper Confidence estimates of soil and dietary concentrations. By using habitat quality approaches described in this paper, the spatially explicit exposure estimates can be made to reflect the size of the local population for the various polygons. For the meadowlark, the largest populations would be found in Little Valley where copper and selenium concentrations were intermediate. The previous EcoRA demonstrated only moderate exceedence of NOAELs of the most sensitive species (Table 1). The value of the habitat information would have been to provide a stronger ecological basis for the conclusions of negligible risk. For the southern red-backed vole, the relationship between CoPC distribution and suitable habitat indicates that the animals would be exposed primarily to background levels of CoPC, with the possible exception of portions of high elevation areas in Pine Canyon.

Table 5. *HSI values for each cover type for two species.*

Cover Type	Western Meadowlark	Red-backed Vole
Agricultural	0.00	0.00
Aspen	0.00	0.70
Aspen/Conifer	0.00	1.00
Barren	0.00	0.00
Conifer Forest	0.00	1.00
Conifer/Submontane Shrub	0.00	0.50
Desert Shrub	0.00	0.00
Developed	0.00	0.00
Floodplain Grassland	0.72	0.00
Grassland	1.00	0.00
Grassland/Desert Shrub	0.13	0.00
Juniper	0.00	0.00
Marshland	0.00	0.00
No Data	0.00	0.66
Recently Re-vegetated	0.00	0.00
Riparian	0.00	1.00
Russian Olive Savannah/Woodland	0.00	0.04
Salt Desert Shrub	0.00	0.00
Submontane Shrub	0.00	0.04
Submontane Shrub/Aspen	0.00	0.00
Submontane Shrub/Desert Shrub	0.00	0.00
Submontane Shrub/Grassland	0.13	0.00
Townsite Woodland	0.50	0.00

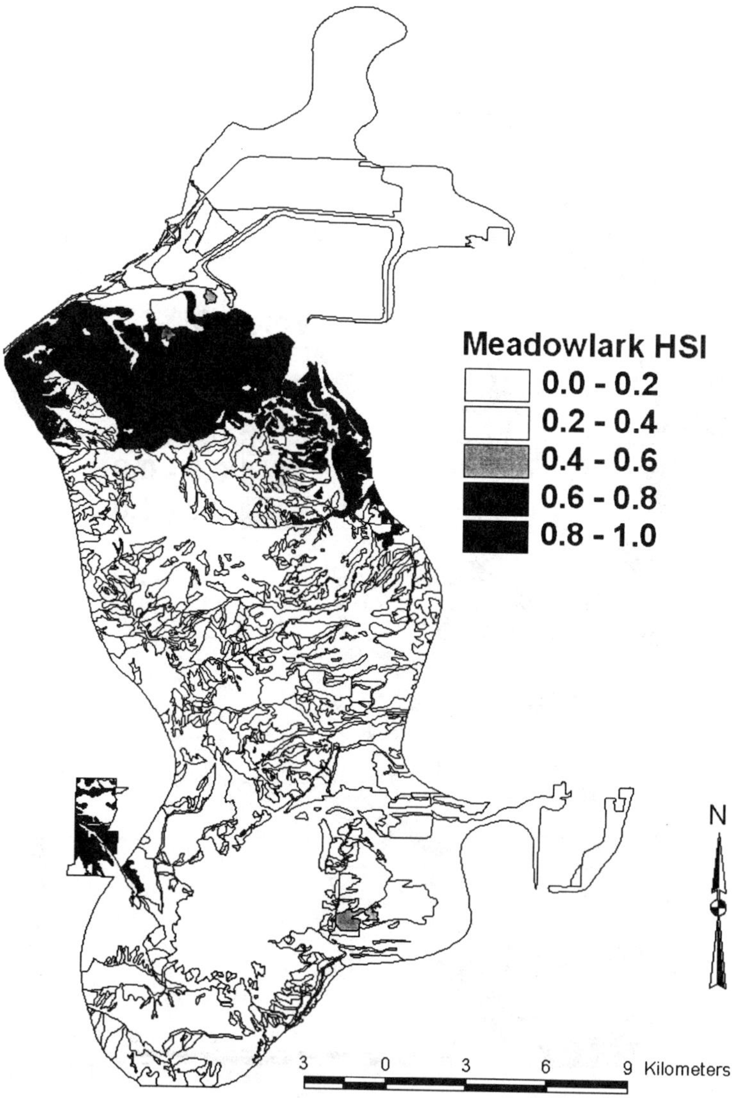

Figure 7. *Representation of habitat quality for the western meadowlark across the project area. (Refer to Figure 1 for general orientation of canyon and elevation polygons.)*

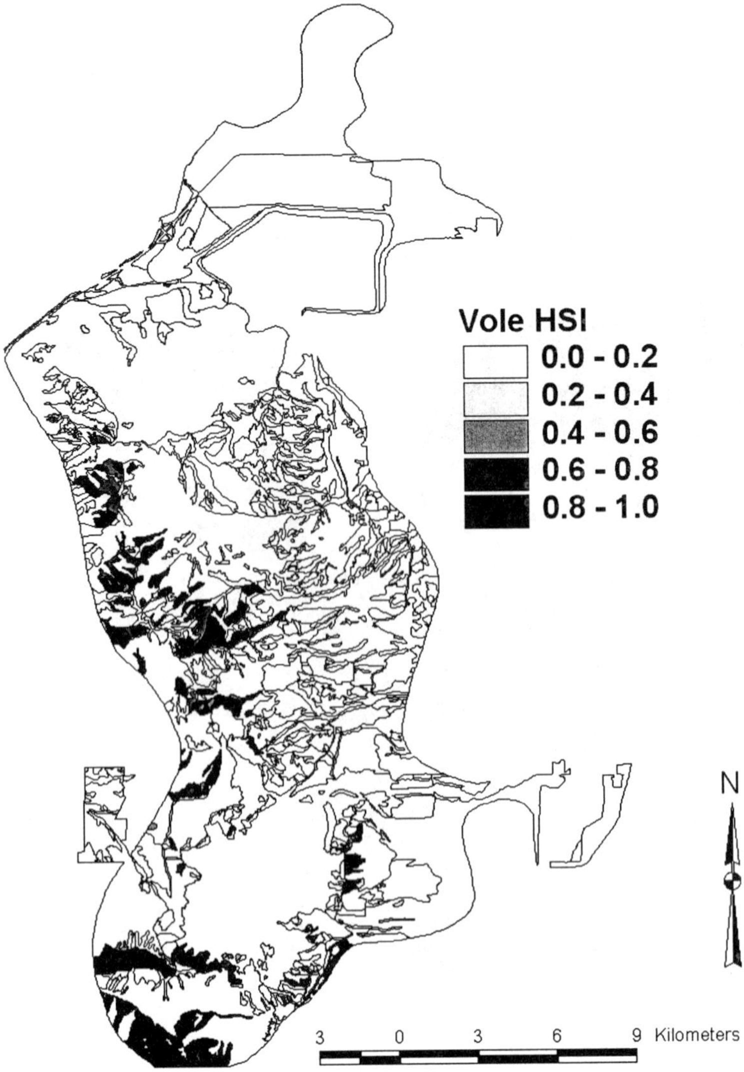

Figure 8. *Representation of habitat quality for the southern red-backed vole across the project area. (Refer to Figure 1 for general orientation of canyon and elevation polygons.)*

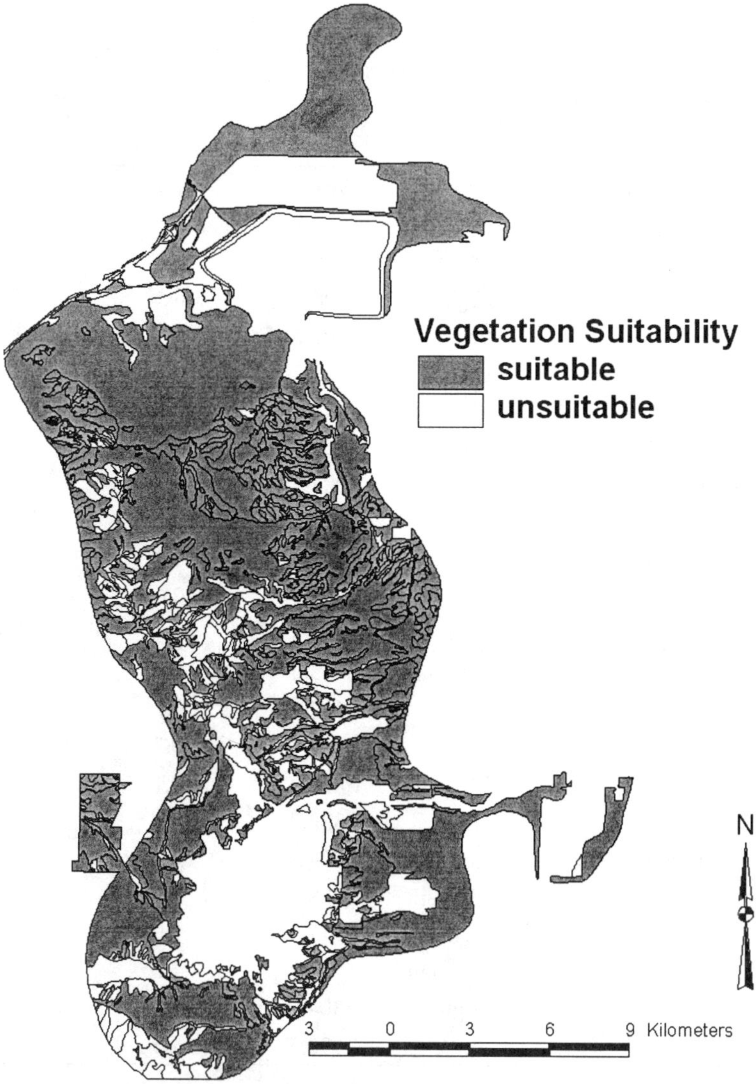

Figure 9. *Representation of suitable habitat areas for the ferruginous hawk across the project area. . (Refer to Figure 1 for general orientation of canyon and elevation polygons.)*

Table 6. *Projected populations (percentage of estimated site-wide population) for each elevation zone by canyon for the Western Meadowlark and theSouthern Red-back Vole.*

Sample Unit	Elevation	Area (ha)	Western Meadowlark	Southern Red-backed Vole
Black Rock Canyon	Base	961.6	6.3	0
	Low	1,011.1	6.5	0
	Medium	946.3	6.3	0
	High	908.1	6.3	0
Kessler Canyon	Base	907.7	5.0	0
	Low	1,126.7	5.1	0
	Medium	772.2	5.0	0
	High	757.8	5.0	0
Little Valley	Base	2,007.9	10.5	0
	Low	2,493.5	12.2	0.2
	Medium	2,246.2	11.3	0.6
	High	1,899.9	10.6	0.5
Coon Canyon	Low	2,705.8	0.4	2.4
	Medium	3,543.0	4.5	3.5
	High	1,412.1	0.2	3.3
Harkers Canyon	Low	1,487.6	0.7	0.4
	Medium	1,858.0	0.1	3.3
	High	1,322.8	0.0	6.6
Butterfield Canyon	Low	824.9	0.3	0.1
	Medium	1,328.2	0.2	4.0
	High	1,217.3	0.0	3.3
Pine Canyon	Medium	731.2	1.0	0.2
	High	304.0	0.1	0.1
Spine	Spine	6,990.5	2.5	71.6

Conclusions

The previous EcoRA studies began nearly a decade ago at a time when there were few examples of comprehensive assessments. We were fortunate to be given the latitude to build the assessment upon a solid foundation of descriptive and quantitative ecology. Having characterized the vegetation across the project area, quantified concentrations of

CoPC in soils and dietary items (e.g., different types of plants, invertebrates, and small mammals), we were able to produce site-specific exposure estimates. Moreover, the ecological dynamics of plant succession, biodiversity, and wildlife demographics enabled weight-of-evidence assessments of the relationships between exposure levels such as exceedences of NOAELs and ecological effects. This retrospective exploration of the previous EcoRA was done to gauge the feasibility of adopting our proposed approach that focuses on landscape characteristics to define habitat suitability and to let the ecological relationships become important drivers of the EcoRA process. A decade ago, this would have been a very difficult approach to undertake. However, with the remarkable advances in computer technology and software, all of the tasks are feasible. Though in the end, we would have reached very similar conclusions regarding the extent of risk posed to wildlife, it appears that the EcoRA would have had greater ecological relevance, would have been easier to explain to stakeholders, would have generated a more complete administrative record in terms of assessment species, and could have been done with the same level of funding or less.

Acknowledgements

This work is funded through the American Chemistry Council, Long-Range Research Initiative, Project Reference Number 0236; Project Title: Evaluating habitat use to improve exposure assessment in ecological risk assessments. Previous EcoRA work was funded by Kennecott Utah Copper Corporation. Many colleagues contributed to the studies and reports prepared for the EcoRA. Elaine-Dorward-King was Project Manager for KUCC; Anne Fairbrother, Rick Bennett, Joe Garrison, Missy Fix, Bryan Wemple, and Alesha Dern were key participants in the EcoRA studies; Ron Kass, Nancy VanDyke, Rick Black, and Geoff Dawson prepared the initial vegetation maps and were key members of plant survey teams.

References

Blair, W. F. 1941. Some Data on Home Ranges and General Life History of the Short-tailed Shrew, Red-backed Vole, and Woodland Jumping Mouse in Northern Michigan. American Midland Naturalist. 25: 681-685.

ep and **t**. 1994a. *Work plan for the ecological risk assessment of Kennecott Utah Copper, Salt Lake City.* Prepared for Kennecott Utah Copper Corporation, Salt Lake City, UT.

ep and **t**. 1994b. *Sampling and analysis plan for the 1994 screening level field studies for the ecological risk assessment of Kennecott Utah Copper, Salt Lake City.* Prepared for Kennecott Utah Copper Corporation, Salt Lake City, UT.

ep and **t**. 1995a. *Problem formulation report: Screening level for the ecological risk assessment of Kennecott Utah Copper, Salt Lake City.* Prepared for Kennecott Utah Copper Corporation, Salt Lake City, UT.

ep and **t**. 1995b. *Analysis report: Screening level for the ecological risk assessment of Kennecott Utah Copper, Salt Lake City.* Prepared for Kennecott Utah Copper Corporation, Salt Lake City, UT.

ep and **t**. 1995c. *Risk characterization report: Screening level for the ecological risk assessment of Kennecott Utah Copper, Salt Lake City, UT.* Prepared for Kennecott Utah Copper Corporation, Salt Lake City, UT.

ep and **t**. 1996. *Final Report: Ecological Risk Assessment, Northern Oquirrh Mountains*. Prepared for Kennecott Utah Copper Corporation, Salt Lake City, UT.

ep and **t**. 1997a. *Final Report: Ecological risk assessment, South Shore Wetlands.* Prepared for Kennecott Utah Copper Corporation, Magna, UT.

Fairbrother A., K. V. Brix, J. E. Toll, S. McKay, and W. J. Adams. 1999. Egg Selenium Concentrations as Predictors of Avian Toxicity. *Human Ecological Risk Assessment.* 5: 1229-1253.

Kapustka, L. A., Galbraith, H., Luxon, M., and Yocum, J. 2001. Using Landscape Ecology to Focus Ecological risk Assessment and Guide Risk Management Decision-Making. *Toxicology and Industrial Health* 17: 236-246.

Kapustka, L. A., Galbraith, H., Luxon, M., Yocum, J., Adams, B. (in review) Predicting Biodiversity Potential Using a Modified LAYERS of Habitat Model, *Landscape Ecology and Wildlife Habitat Evaluation: Critical Information for Ecological Risk Assessment, Land-Use Management Activities, and Biodiversity Enhancement Practices, ASTM STP 1458*, L. A. Kapustka, H. Galbraith, M. Luxon, G. R. Biddinger, Eds., ASTM International, West Conshohocken, PA.

Lanyon, W. E. 1956. Territory of the Meadowlarks, Genus *Sturnella* in Wisconsin. *Ibis* 98: 485-489.

Schroeder RL and Haire SL. 1993. *Guidelines for the Development of Community-level Habitat Evaluation Models*. U. S. Department of Interior, Fish and Wildlife Service, Biological Report 8. Washington DC 20240.

Suter, G. W. II. 1993. *Ecological Risk Assessment*, Lewis Publishers, Boca Raton, FL.

US EPA. 1992. *Framework for Ecological Risk Assessment.* EPA/630/R-92-001. US Environmental Protection Agency, Washington, DC 20240.

US EPA. 1993a. *Wildlife exposure factors handbook, Vol.* I. EPA/600/R-93/187a. US Environmental Protection Agency, Washington, DC 20240.

US EPA. 1993b. *Wildlife exposure factors handbook, Vol.* I. EPA/600/R-93/187a. US Environmental Protection Agency, Washington, DC 20240.

US EPA. 1998. *Guidelines for Ecological Risk Assessment.* EPA/630/R-95/002F. US Environmental Protection Agency Risk Assessment Forum, Washington, DC 20240.

Appendix A. List of HSI models abstracted into the Access® database.

American Alligator.
Newsom, J. D., T. Joanen, and R. J. Howard. 1987. Habitat suitability index models: American alligator. U.S. Fish and Wildlife Service. Biological Report 82(10.136).

American Black Duck (Wintering).
Lewis, J. C. and R. L Garrison. 1994. Habitat suitability index models: American black duck (wintering). U.S. Fish and Wildlife Service. FWS/OBS-82/10.68

American Coot.
Allen, A. W. 1985. Habitat suitability index models: American Coot. U.S. Fish and Wildlife Service. Biological Report 82(10.115).

American Eider (Breeding).
Blumton, A. K., R. B. Owen, Jr., and W. B. Krohn. 1988. Habitat suitability index models: American eider (breeding). U.S. Fish and Wildlife Service. Biological Report 82(10.149).

American Woodcock (Wintering).
Cade, B. S. 1985. Habitat suitability index models: American woodcock (wintering). U.S. Fish and Wildlife Service. Biological Report 82(10.105).

Arizona Guild and Layers of Habitat Models.
Short, H. L. 1984. Habitat suitability index models: The Arizona guild and layers of habitat models. U.S. Fish and Wildlife Service. FWS/OBS-82/10.70

Baird's Sparrow.
Sousa, P. J. and W. N. McDonal. 1983. Habitat suitability index models: baird's sparrow. U.S. Fish and Wildlife Service. FWS/OBS-82/10.44

Bald Eagle (Breeding Season).
Peterson, A. 1986. Habitat suitability index models: Bald eagle (breeding season). U.S. Fish and Wildlife Service. Biological Report 82(10.126).

Barred Owl.
Allen, A. W. 1987. Habitat suitability index models. barred owl. U.S. Fish and Wildlife Service. Biological Report 82(10.143).

Beaver.
Allen, A. W. 1983. Habitat suitability index models: beaver. U.S. Fish and Wildlife Service. FWS/OBS-82/10.32

Belted Kingfisher.
Prose, B. L. 1985. Habitat suitability index models: belted kingfisher. U.S. Fish and Wildlife Service. Biological Report 82(10.87).

Black Bear (Upper Great Lakes Region).
Rogers, L. L. and A. W. Allen. 1987. Habitat suitability index models: Black Bear, Upper Great Lakes Region. U.S. Fish and Wildlife Service. Biological Report 82(10.144).

Black Brant.
Schroeder, R. L. 1984. Habitat suitability index models: Black brant. U.S. Fish and Wildlife Service. FWS/OBS-82/10.63

Black-bellied Whistling-Duck (Breeding).

McKenzie, P. M. and P. J. Zwank,. 1988. Habitat suitability index models: Black-bellied whistling-duck (breeding). U.S. Fish and Wildlife Service. Biological Report 82(10.150).

Black-Capped Chickadee.

Schroeder, R. L. 1983. Habitat suitability index models: black-capped chickadee. U.S. Fish and Wildlife Service. FWS/OBS-82/10.37

Black-Shouldered Kite.

Faanes, C. A. and R. J. Howard. 1987. Habitat suitability index models: black-shouldered kite. U.S. Fish and Wildlife Service. Biological Report 82(10.130).

Black-tailed Prairie Dog.

Clippinger, N. W. 1989. Habitat suitability index models: black-tailed prairie dog. U.S. Fish and Wildlife Service. Biological Report 82(10.156).

Blue Grouse.

Schroeder, R. L. 1984. Habitat suitability index models: Blue grouse. U.S. Fish and Wildlife Service. FWS/OBS-82/10.81

Blue-Winged Teal (Breeding).

Sousa, P. J. 1985. Habitat suitability index models: Blue-winged teal (breeding). U.S. Fish and Wildlife Service. Biological Report 82(10.114).

Bobcat.

Boyle, K. A. and T. T. Fendley. 1987. Habitat suitability index models: bobcat. U.S. Fish and Wildlife Service. Biological Report 82(10.147).

Brewer's Sparrow.

Short, H. L. 1984. Habitat suitability index models: Brewer's sparrow. U.S. Fish and Wildlife Service. FWS/OBS-82/10.83

Brown Thrasher.

Cade, B. S. 1986. Habitat suitability index models: Brown thrasher. U.S. Fish and Wildlife Service. Biological Report 82(10.118).

Bullfrog.

Graves, B. M. and S. W. Anderson. 1987. Habitat suitability index models: bullfrog. U.S. Fish and Wildlife Service. Biological Report 82(10.138).

Cactus Wren.

Short, H. L. 1985. Habitat suitability index models: Cactus wren. U.S. Fish and Wildlife Service. Biological Report 82(10.96).

Canvasback (Breeding Habitat).

Schroeder, R. L. 1984. Habitat suitability index models: Canvasback (breeding habitat). U.S. Fish and Wildlife Service. FWS/OBS-82/10.82

Clapper Rail.

Lewis, J. C., and R. L. Garrison. 1983. Habitat suitability index models: clapper rail. U.S. Fish and Wildlife Service. FWS/OBS-82/10.51

Diamondback Terrapin (Nesting-North Atlantic Coast).

Palmer, W. M. and C. L. Cordes. 1988. Habitat suitability index models: diamondback terrapin (nesting)—Atlantic coast. U.S. Fish and Wildlife Service. Biological Report 82(10.151).

Downy Woodpecker.
Schroeder, R. L. 1982. Habitat suitability index models: Downy woodpecker. U.S. Fish and Wildlife Service. FWS/OBS-82/10.38

Eastern Brown Pelican.
Hingtgen, T. M., R. Mulholland, and A. V. Zale. 1985. Habitat suitability index models: eastern brown pelican. U.S. Fish and Wildlife Service. Biological Report 82(10.90).

Eastern Cottontail.
Allen, A. W. 1984. Habitat suitability index models: Eastern cottontail. U.S. Fish and Wildlife Service. FWS/OBS-82/10.66

Eastern Meadowlark.
Schroeder, R. J. and P. J. Sousa. 1982. Habitat suitability index models: Eastern meadowlark. U.S. Fish and Wildlife Service. FWS/OBS-82/10.29

Eastern Wild Turkey.
Schroeder, R. L. 1985. Habitat suitability index models: Eastern wild turkey. U.S. Fish and Wildlife Service. Biological Report 82(10.106).

Elk (Winter Range).
Buckmaster, G. et al. 1995. Elk Winter Range Draft Habitat Suitability Index (HSI) Model. University of Alberta.

Elk.
Galbraith, H. et al. 1992. Elk Model. RCG/Hagler, Bailey, Inc.

Ferruginous Hawk.
Jasikoff, T. M. 1982. Habitat suitability index models: Ferruginous hawk. U.S. Fish and Wildlife Service. FWS/OBS-82/10.10

Field Sparrow.
Sousa, P. J. 1983. Habitat suitability index models: Field sparrow. U.S. Fish and Wildlife Service. FWS/OBS-82/10.62

Fisher.
Allen, A. W. 1983. Habitat suitability index models Fisher. U.S. Fish and Wildlife Service. FWS/OBS-82/10.45

Forster's Tern (Breeding-Gulf and Atlantic Coast).
Martin, R. P., and P. J. Zwank. 1987. Habitat suitability index models: Forster's tern (breeding) Gulf Atlantic Coasts. U.S. Fish and Wildlife Service. Biological Report 82(10.131).

Fox Squirrel.
Allen, A. W. 1982. Habitat suitability index models: fox squirrel. U.S. Fish and Wildlife Service. FWS/OBS-82/10.18

Gadwall (Breeding).
Sousa, P. J. 1985. Habitat suitability index models. Gadwall (breeding). U.S. Fish and Wildlife Service. Biological Report 82(10.100).

Gray Partridge.
Allen, A. W. 1984. Habitat suitability index models: Gray partridge. U.S. Fish and Wildlife Service. Biological Report 82(10.73).

Gray Squirrel.
Allen, A. W. 1982. Habitat suitability index models: gray squirrel. U.S. Fish and Wildlife Service. FWS/OBS-82/10.19

Great Blue Heron.
Short, H. L. and R. J. Cooper. 1985. Habitat suitability index models. Great blue heron. U.S. Fish and Wildlife Service. Biological Report 82(10.99).

Great Egret.
Chapman, B. R., and R. J. Howard. 1984. Habitat suitability index models: great egret. U.S. Fish and Wildlife Service. FWS/OBS-82/10.78

Greater Prairie-Chicken (Multiple Levels of Resolution).
Prose, B. L. 1985. Habitat suitability index models: Greater Prairie-chicken. (Multiple levels of resolution). U.S. Fish and Wildlife Service. Biological Report 82(10.102).

Greater Sandhill Crane.
Armbruster, M. J. 1987. Habitat suitability index models: greater sandhill crane. U.S. Fish and Wildlife Service. Biological Report 82(10.140).

Greater White-Fronted Goose (Wintering).
Kaminski, R. M. 1986. Habitat suitability index models: greater white-fronted goose (wintering). U.S. Fish and Wildlife Service. Biological Report 82(10.116).

Hairy Woodpecker.
Sousa, P. J. 1987. Habitat suitability index models: hairy woodpecker. U.S. Fish and Wildlife Service. Biological Report 82(10.146).

Indiana Bat.
Rommé, R. C., K. Tyrell, and V. Brack Jr. 1995. Literature Summary and Habitat Suitability Index Model Components of Summer Habitat for the Indiana Bat, Myotis Sodalis. Indiana Dept. of Natural Resources.

Lark Bunting.
Finch, D. M., S. H. Anderson, and W. A. Hubert. 1987. Habitat suitability index models: lark bunting. U.S. Fish and Wildlife Service. Biological Report 82(10.137).

Laughing Gull.
Zale, A. V. and R. Mulholland. 1985. Habitat suitability index models: laughing gull. U.S. Fish and Wildlife Service. Biological Report 82(10.94).

Least Tern.
Carreker, R. G. 1985. Habitat suitability index models; Least tern. U.S. Fish and Wildlife Service. Biological Report 82(10.103).

Lesser Scaup (Breeding).
Allen, A. W. 1986. Habitat suitability index models: Lesser scaup (breeding). U.S. Fish and Wildlife Service. Biological Report 82(10.117).

Lesser Scaup (Wintering).
Mulholland, R. 1985. Habitat suitability index models: lesser scaup (wintering). U.S. Fish and Wildlife Service. Biological Report 82(10.91).

Lesser Snow Goose (Wintering).

Leslie, J. C., and P. J. Zwank. 1985. Habitat suitability index models: lesser snow goose (wintering). U.S. Fish and Wildlife Service. Biological Report 82(10.97).

Lewis' Woodpecker.

Sousa, P. J. 1982. Habitat suitability index models: Lewis' woodpecker. U.S. Fish and Wildlife Service. FWS/OBS-82/10.32

Mallard (Winter Habitat, Lower Mississippi Valley).

Allen, A. W. 1986. Habitat suitability index models: mallard (winter habitat, lower Mississippi Valley. U.S. Fish and Wildlife Service. Biological Report 82(10.132).

Marsh Wren.

Gutzwiller, K. J. and S. H. Anderson. 1987. Habitat suitability index models: marsh wren. U.S. Fish and Wildlife Service. Biological Report 82(10.139).

Marten.

Allen, A. W. 1982. Habitat suitability index models: Marten. U.S. Fish and Wildlife Service. FWS/OBS-82/10.11

Mink.

Allen, A. W. 1986. Habitat suitability index models: mink, revised. U.S. Fish and Wildlife Service. Biological Report 82(10.127).

Moose (Lake Superior Region).

Allen, A. W., P. A. Jordan, and J. W. Terrell. 1987. Habitat suitability index models: moose, Lake Superior region. U.S. Fish and Wildlife Service. Biological Report 82(10.155).

Mottled Duck.

Rorabaugh, J. C. and P. J. Zwank. 1983. Habitat suitability index models: mottled duck. U.S. Fish and Wildlife Service. FWS/OBS-82/10.52

Muskrat.

Allen, A. W. and R. D. Hoffman. 1984. Habitat suitability index models: Muskrat. U.S. Fish and Wildlife Service. FWS/OBS-82/10.46

Northern Bobwhite.

Schroeder, R. L. 1985. Habitat suitability index models: Northern bobwhite. U.S. Fish and Wildlife Service. Biological Report 82(10.104).

Northern Pintail.

Suchy, W. J. and S. H. Anderson. 1987. Habitat suitability index models: Northern pintail. U.S. Fish and Wildlife Service. Biological Report 82(10.145).

Northern Pintail (Gulf Coast Wintering).

Howard, R. J., and H. A. Kantrud. 1986. Habitat suitability index models: Northern pintail (gulf coast wintering). U.S. Fish and Wildlife Service. Biological Report 82(10.121).

Osprey.

Vana-Miller, S. L. 1987. Habitat suitability index models: osprey. U.S. Fish and Wildlife Service. Biological Report 82(10.154).

Pileated Woodpecker.

Schroder, R. L. 1982. Habitat suitability index models: Pileated woodpecker. U.S. Fish and Wildlife Service. FWS/OBS-82/10.39

Pine Warbler.

Schroeder, R. L. 1982. Habitat suitability index models: pine warbler. U.S. Fish and Wildlife Service. FWS/OBS-82/10.28

Plains Sharp-Tailed Grouse.

Prose, B. L. 1987. Habitat suitability index models: plains sharp-tailed grouse. U.S. Fish and Wildlife Service. Biological Report 82(10.142).

Pronghorn.

Allen, A. W., J. G. Cook, and M. J. Armbruster. 1984. Habitat suitability index models: Pronghorn. U.S. Fish and Wildlife Service. FWS/OBS-82/10.65

Redhead (Wintering).

Howard, R. J., and H. A. Kantrud. 1983. Habitat suitability index models: redhead (wintering). U.S. Fish and Wildlife Service. FWS/OBS-82/10.53

Red-Spotted Newt.

Sousa, P. J. 1985. Habitat suitability index models: Red-spotted newt. U.S. Fish and Wildlife Service. Biological Report 82(10.111).

Red-Winged Blackbird.

Short, H. L. 1985. Habitat suitability index models: red-winged blackbird. U.S. Fish and Wildlife Service. Biological Report 82(10.95).

Roseate Spoonbill.

Lewis, J. C. 1983. Habitat suitability index models: roseate spoonbill. U.S. Fish and Wildlife Service. FWS/OBS-82/10.50

Ruffed Grouse.

Cade, B. S. and P. J. Sousa. 1985. Habitat suitability index models: Ruffed grouse. U.S. Fish and Wildlife Service. Biological Report 82(10.86).

Slider Turtle.

Morreale, S. J. and J. W. Gibbons. 1986. Habitat suitability index models: Slider turtle. U.S. Fish and Wildlife Service. Biological Report 82(10.125).

Snapping Turtle.

Graves, B. M. and S. H. Anderson. 1987. Habitat suitability index models: snapping turtle. U.S. Fish and Wildlife Service. Biological Report 82(10.141).

Snowshoe Hare.

Carreker, R. G. 1985. Habitat suitability index models: Snowshoe hare. U.S. Fish and Wildlife Service. Biological Report 82(10.101).

Southern Red-backed Vole (Western United States).

Allen, A. W. 1983. Habitat suitability index models: Southern red-backed vole (Western United States). U.S. Fish and Wildlife Service. FWS/OBS-82/10.42

Spotted Owl.

Laymon, S. A., H. Salwasser, and R. H. Barrett. 1985. Habitat suitability index models: Spotted owl. U.S. Fish and Wildlife Service. Biological Report 82(10.113).

Swamp Rabbit.
Allen, A. W. 1985. Habitat suitability index models: Swamp rabbit. U.S. Fish and Wildlife Service. Biological Report 82(10.107).

Veery.
Sousa, P. J. 1982. Habitat suitability index models: Veery. U.S. Fish and Wildlife Service. FWS/OBS-82/10.22

Western Grebe.
Short, H. L. 1984. Habitat suitability index models: Western grebe. U.S. Fish and Wildlife Service. FWS/OBS-82/10.69

White Ibis.
Hingtgen, T. M., R. Mulholland, and R. W. Repenning. 1985. Habitat suitability index models: white ibis. U.S. Fish and Wildlife Service. Biological Report 82(10.93).

White-Tailed Deer (Gulf of Mexico and South Atlantic Coastal Plains).
Short, H. L. 1986. Habitat suitability index models: White-tailed deer in Gulf of Mexico and South Atlantic coastal plains. U.S. Fish and Wildlife Service. Biological Report 82(10.123).

Wildlife Species Richness in Shelterbelts.
Schroeder, R. L. 1986. Habitat suitability index models: wildlife species richness in shelterbelts. U.S. Fish and Wildlife Service. Biological Report 82(10.128).

Williamson's Sapsucker.
Sousa, P. J. 1983. Habitat suitability index models: Williamson's sapsucker. U.S. Fish and Wildlife Service. FWS/OBS-82/10.47

Wood Duck (Breeding).
Sousa, P. J. and A. H. Farmer. 1983. Habitat suitability index models: Wood duck. U.S. Fish and Wildlife Service. FWS/OBS-82/10.43

Wood Duck (Wintering).
Sousa, P. J. and A. H. Farmer. 1983. Habitat suitability index models: Wood duck. U.S. Fish and Wildlife Service. FWS/OBS-82/10.43

Yellow Warbler.
Schroeder, R. L. 1982. Habitat suitability index models: yellow warbler. U.S. Fish and Wildlife Service. FWS/OBS-82/10.27

Yellow-Headed Blackbird.
Schroeder, R. L. 1982. Habitat suitability index models: yellow-headed blackbird. U.S. Fish and Wildlife Service. FWS/OBS-82/10.26

Greg Linder,[1] *Elizabeth Harrahy,*[2] *Lynne Johnson,*[1] *Larry Gamble,*[3] *Kevin Johnson,*[4] *Joy Gober,*[5] and *Stephanie Jones*[6]

Sunflower Depredation and Avicide Use: A Case Study Focused on DRC-1339 and Risks to Non-target Birds in North Dakota and South Dakota

---

Reference: Linder, G., Harrahy, E., Johnson, L., Gamble, L., Johnson, K., Gober, J., and Jones, S., **"Sunflower Depredation and Avicide Use: A Case Study Focused on DRC-1339 and Risks to Non-target Birds in North Dakota and South Dakota,"** *Landscape Ecology and Wildlife Habitat Evaluation: Critical Information for Ecological Risk Assessment, Land-Use Management Activities, and Biodiversity Enhancement Practices, ASTM STP 1458*, L. A. Kapustka, H. Galbraith, M. Luxon, and G. R. Biddinger, Eds., ASTM International, West Conshohocken PA, 2004.

ABSTRACT: Agricultural land-use, cropping practice, agrichemical use, and wildlife interactions have long provided conflicts between wildlife needs and human uses of habitat. For example, sunflower seeds ripening in late summer and early autumn throughout agricultural areas of North Dakota and South Dakota are highly sought food items for red-winged blackbirds, common grackles, and yellow-headed blackbirds. Unfortunately, loss of sunflower seeds prior to fall harvest has been attributed to these birds, with crop losses estimated at greater than $5 million per year. An avicide, DRC-1339 (3-chloro-p-toluidine hydrochloride), has been proposed for use in spring baiting programs in North Dakota and South Dakota to control fall depredation of the sunflower crop. An estimated 60 species of non-target birds with varying sensitivities to DRC-1339 occur near spring baiting sites, with nearly half of these species being granivores that might feed on the DRC-1339-treated bait. At least nine species are birds of management concern. Our work evaluated risks to non-target birds that are potentially associated with DRC-1339 spring baitings. From the current analysis, spring baiting presents risks to non-target birds, especially small-bodied species characterized by marked responsiveness to DRC-1339, e.g., ingestion of a single baited grain will likely yield mortality in a small-bodied bird. A simple comparison of hazard quotients for small-bodied nontarget birds and target birds suggests these species have similar risks

---

[1] U.S. Geological Survey, Biological Resources Division (USGS/BRD), Columbia Environmental Research Center, HeronWorks Field Office, Brooks OR 97305.

[2] U.S. Fish and Wildlife Service, Department of Fishery and Wildlife Biology, Colorado State University, Ft. Collins CO 80523.

[3] U.S. Fish and Wildlife Service, Region 6, Denver CO 80228.

[4] U.S. Fish and Wildlife Service, Region 6, North Dakota Field Office, Bismarck ND 58501.

[5] U.S. Fish and Wildlife Service, Region 6, South Dakota Field Office, Pierre SD 57501.

[6] U.S. Fish and Wildlife Service, Region 6, Nongame Migratory Bird Coordinator, Denver CO 80225.

for dietary exposures to DRC-1339. At present mitigation practices are unproven with respect to minimizing non-target bird loss. Potential losses from non-target populations presently thought to be declining suggests that risks vary across the relatively simple "non-target" category, and risk managers must be wary of oversimplifying management plans based on a "non-target"and "target" categorization of species at-risk. In view of the uncertainties apparent in the present analysis, as well as other risk assessments focused on the issue, decisions regarding DRC-1339's use hinge on differing management perspectives of "acceptable risk" and resource valuation. Only when these issues are resolved can resource management plans benefit the long-term sustainability of resources at risk.

KEYWORDS: red-winged blackbird, yellow-headed blackbird, common grackle, sunflower, wetlands, northern Great Plains, DRC-1339

Introduction

Sunflower seeds ripening in late summer and early autumn throughout agricultural areas of North Dakota and South Dakota are highly sought food items for red-winged blackbirds (*Ageldius phoenicaus*), common grackles (*Quiscalus quiscula*), yellow-headed blackbirds (*Xanthocephalus xanthocephalus*), and to a more limited extent for other birds such as brown-headed cowbirds (*Molothrus ater*). Unfortunately, loss of sunflower seed attributed to these birds (Family Icteridae) has been estimated at greater than $5 million per year (USDA 2000). Generally, blackbird damage is predominated by red-winged blackbirds and common grackles and occurs in late summer to early autumn as the sunflower crop ripens. As a chemical control measure, DRC-1339 (3-chloro-4-methyl benzamine hydrochloride, or 3-chloro-p-toluidine hydrochloride) has been proposed for use throughout sunflower-planted fields of North Dakota and South Dakota under a FIFRA (Federal Insecticide, Fungicide, and Rodenticide Act) Section 24 Special Use Permit.

DRC-1339 is the only lethal chemical agent currently registered in the U.S. for managing blackbird damage in sunflower fields during both the spring and fall migrations, and it has been used by the U.S. Department of Agriculture-Animal and Plant Health Inspection Service (USDA-APHIS) in an experimental blackbird control program in eastern South Dakota. Since 1994, USDA-APHIS has been evaluating the effectiveness of spring baiting near blackbird roosts in South Dakota to reduce blackbird breeding populations. Their goal of the spring baiting program is the anticipated decrease in the fall depredation of sunflowers by blackbirds throughout the Dakotas. However, at least 60 species of non-target birds with varying sensitivities to DRC-1339 occur near spring baiting sites. Approximately half of these species are granivorous birds which may feed on the DRC-1339-treated rice bait, and at least nine are species of management concern.

Here, we summarize an analysis of risks associated with DRC-1339 spring baitings to non-target birds, including an analysis of effects based on existing toxicity data, and an analysis of exposure based on existing field studies and non-target information.

Problem Formulation and Development of Conceptual Models: DRC-1339 Risks to Non-target Birds

In the present implementation, the process of problem formulation, analyzing exposure and effects, and the subsequent characterization of risks to non-target birds exposed to DRC-1339 is consistent with the current practice of ecological risk assessment as summarized in Suter (1993), EPA (1992, 1998), and draft guidance developed to support FIFRA (EPA 1999).

Assessment Endpoints and Species of Concern

Assessment endpoints focused on individual responses of non-target birds to exposures to DRC-1339 during spring baitings, with a primary interest in individual-level responses that potentially affected the abundance of species normally occurring in the northern Great Plains (i.e., North Dakota, South Dakota, and Minnesota) as residents or spring migrants. Bird species serving as receptors of concern in the current analysis have been observed at or near baited plots located near sunflower fields or roosts during experimental spring baitings with DRC-1339 in North Dakota and South Dakota (Table 1). These species are primarily granivores or will take sunflower seeds when available. Granivores will potentially be exposed to DRC-1339 through primary routes of exposure, which given the baiting scenarios guiding this assessment translates to dietary exposures through consumption of brown-rice baits. Species of concern served as representatives of all non-target birds potentially exposed to DRC-1339 as a consequence of spring baitings, and are regarded as species most likely exposed to the avicide during the baiting period.

TABLE 1--*Summary tabulation of receptors of concern*

Nontarget species		Target species
Primary exposure Large-bodied granivore	Primary exposure Small-bodied granivore	Primary exposure
Mallard	Chipping sparrow	Red-winged blackbird
	Clay-colored sparrow	Yellow-headed blackbird
Bobwhite quail	Field sparrow	Common grackle
	American tree sparrow	
	Bobolink	
	Horned lark	
	Eastern meadowlark	
	Western meadowlark	

Beyond diet as the primary route of exposure, body size was an additional attribute that influenced selection of representative species. Small-bodied birds (< 100 grams) and large-bodied birds (≥ 100 grams) were identified, given the importance of body size in characterizing dose of DRC-1339 associated with adverse effects in equally sensitive birds. The selected non-target granivores are representative of birds at risk, given their life history traits (e.g., granivores), resident period, seasonal patterns of nesting, or migratory passage through the areas of concern.

Measures of Adverse Effects and Pathway Analysis

Measures of effects for this initial iterate of risk analysis are focused on toxicity endpoints such as median effects estimates (e.g., as median level doses, LD50s), and similar measures of toxicity, including estimates of benchmark dose (BMD) as available. These measures of adverse effects depend on laboratory toxicity tests, while similar measures derived from field studies have been incorporated as part of exposure analysis, especially as those relate to observations of mortality potentially linked to chemical exposure. As a potential source of uncertainty captured in risk characterization, these measures of adverse effects were also considered within the context of confounding factors, wherein adverse effects observed (for example) in the field are misdiagnosed as causally linked to DRC-1339 exposure. Given the multiple stressors characteristic of exposures in the field, misdiagnosis and the potential for "false positive" (misassignment of cause to DRC-1339 exposure) and "false negative" (failure to assign cause when linkage to DRC-1339 exists) findings will be critical to risk management decisions.

Identification of Exposure Pathways

The analysis of exposure pathways combines spatial and temporal sources and receptors, which in this instance simplifies to spring baitings with DRC-1339 that occur over a 30- to 40-day period in March and April. The chemical's routine application is critical in the evaluation of pathways within both temporal and spatial context, and is briefly summarized below.

As suggested by the available literature (e.g, Eisemann et al. 2000), DRC-1339 applications during spring baiting will ideally follow a reproducible process, given the chemical's physicochemical behavior in the environment and provided permit specification are consistently implemented. DRC-1339-treated brown rice will be formulated at a concentration of 2% and diluted with untreated rice the day of application at a ratio of 1:25. This mixture will then be broadcast in bait plots with an all-terrain-mounted seed-spreader at a rate of 12 to 23 kg/ha. Treated plots are anticipated to be about 0.8 ha and would be located near roads under roost-to-field flight paths. Plots will be pre-baited with untreated rice for a period sufficient to habituate target species (e.g., foraging blackbirds) to bait sites and to monitor non-target activity. Up to four subsequent applications of baits tainted with DRC-1339 can be made after 75% of the previous application has been consumed or 10 mm of precipitation has fallen (EPA Registration Numbers 56228-30 and SD-980005). Decoy birds may be used to attract greater numbers of blackbirds to bait sites.

Conceptual Model for Assessing Risks Associated with Exposure of Non-target Birds to Spring Baitings of DRC-1339

Pathways linking sources with receptors, and subsequently describing potential risks associated with dietary exposures to avicide exposure to non-target birds are illustrated in conceptual models adapted from EPA (1999; Figure 1). These are primary exposures which are the focus of this analysis and reflect potential dietary intake of brown-rice baits.

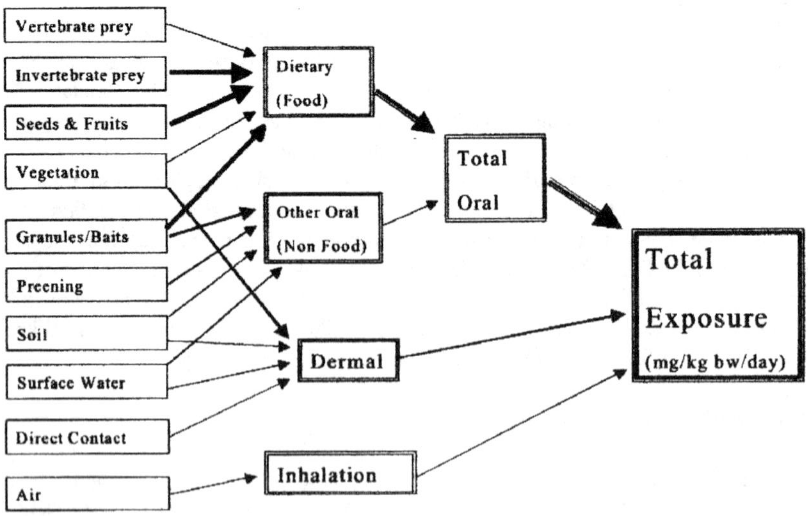

Fig. 1--*Conceptual model illustrating dietary pathways likely dominating exposure to DRC-1339 in non-target and target birds (from EPA 1999).*

Analysis: Acute Toxicity Profile for DRC-1339

DRC-1339, or 3-chloro-p-toluidine hydrochloride (chemical synonym, 3-chloro-4-methyl benzamine hydrochloride) was developed as an avicide in the early 1960s. The material is palatable to target birds, but highly toxic and slow-acting which minimizes the likelihood that other target birds would avoid exposure by developing avoidance behaviors. European starlings (*Sturnus vulgaris*) were the original target species for the control agent (see Feare 1984 for background on European starlings).

The toxicity of DRC-1339 to birds has received much review since the chemical agent's development over 40 years ago (EPA 1995). Relative to existing data on other environmental chemicals, that available for pesticides and other regulated chemicals (e.g., veterinary pharmaceuticals) is often better developed than those encountered in other risk assessment arenas (e.g., see Suter 1993). Overall, the existing data suggests that DRC-1339 is highly toxic to a few bird species (e.g., Icteridae), and ranges from toxic to relatively non-toxic to other birds. The variability of acute data, however, must be considered in light of the number of species tested and the test methods used in evaluating the chemical, since the range of test species is highly restricted relative to the range of species potentially exposed to the toxicant in the field.

The acute toxicity of DRC-1339 has been widely characterized, with some toxicity data (mostly acute) being available for over 40 of the more than 800 North American bird species and some Old World bird species (Figure 2). Interpretation of early laboratory studies suggested that DRC-1339 exhibited differential toxicity among taxonomic families, e.g., sparrows, finches, and raptors appeared to be relatively insensitive to DRC-1339 with LD50s (median lethal doses) greater than 100 mg/kg (Eisemann et al. 2001), but recent results indicate that species sensitivity to DRC-1339 does not consistently follow taxonomic lines (e.g., Borchert 2001a, 2001b; Mach 2001; Sayre 2001a, 2001b). Casual observation of Figure 2, for example, indicates a range in toxicity among the emberizids that covers nearly two orders of magnitude. Not surprisingly, the chemical's mode of action appears to differ across species and is dose-dependent (e.g., high dose and low dose exposures may achieve adverse effects along different modes of action). Target species such as blackbirds, starlings, and species of corvids are highly sensitive to DRC-1339, having LD50s less than 10 mg/kg. Columbiforms, galliforms, and many passerines are also acutely sensitive to DRC-1339 (LD50s < 20 mg/kg). Given the relatively small percentage of North American species tested with respect to acute toxicity (40 of 800, or ca 5%), generalizations at higher taxonomic categories should be guardedly developed for DRC-1339.

From studies focused on the toxicant's mechanism of action, DRC-1339 is considered nephrotoxic to sensitive species. Proximal convoluted tubules are target tissues within the kidney, and their destruction results in increased levels of uric acid in the blood (e.g., see Apostolou 1969; Mull 1971; Westberg 1973; Schafer 1981; EPA 1995). Metabolism studies have shown that as much as 90% of a dose administered to birds is excreted in the form of parent compound or metabolite within 30 minutes (EPA 1995). Non-responsive birds do not present renal pathologies or show increased levels of methemoglobin in the blood, especially at expected environmental concentrations. In contrast to responsive animals, non-responsive birds and mammals excrete acetylated metabolites of the toxicant in the urine, although at high exposure concentrations, nonsensitive vertebrates present depression of the central nervous system and respiratory failure which generally causes death (see Apostolou, 1969; Mull, 1971; Westberg 1973, EPA 1995).

Studies on acute toxicity dominate the literature for DRC-1339, since the original criteria leading to the development of the control agent focused on selective toxicity to target species achieved through short-term exposures yielding delayed mortalities some distance from foraging areas to reduce development of aversive behaviors. Acute studies (oral exposures) driven by testing required under FIFRA present toxicity tests yielding data for bobwhite quail (*Colinus virginianus*; Fletcher and Pedersen 1991a), mallards (*Anas platyrhynchos*; Fletcher and Pedersen 1991b), and comparable tests with non-target species have been completed with American tree sparrows (*Spizella arborea*; Mach 2001, Borchert 2001a), horned larks (*Eremophila alpestris*; Sayre 2001a), dark-eyed junco (*Junco hyemalis*, Sayre 2001b) and western meadowlarks (Borchert 2001b). These studies yielded no-observed effect levels (NOELs) and LD50s for each species. DRC-1339 LD50s for bobwhite quail and American tree sparrows are similar to those observed for red-winged blackbirds and starlings. When considered in light of the supporting literature focused on the chemical agent, the data for evaluating acute effects are well developed for evaluating risks.

Analysis of Effects: Derivation of Benchmark Doses (BMDs)

Existing toxicity values (as LD50s) reflect a wide range of studies characterized by various levels of data quality. Historic data may be limited with respect to statistical rigor (e.g., design considerations of replication and sample size; see Harrahy 2001, 2002), but published historic data generally conforms to draft guidelines for data and information quality (DOI 2002). Given the increasing use of benchmark dose (BMD) in risk assessment, those studies with data sufficient to the derivation were used to develop BMDs for DRC-1339.

For data sufficient to the calculation, U.S. EPA Benchmark Dose Software (BMDS) was used, following EPA (2000). BMDS (version 1.3.1) offers 16 different models that are appropriate for the analysis of dichotomous (quantal) data, continuous data, and nested developmental toxicology data. While numerous curve fitting mathematical models were available to evaluate dose-response data from FIFRA studies, Log-Probit was used in the present analysis owing to its common application in reports of historic studies. Differences between historic and present outputs could then be considered in the absence of model differences confounding the picture. Results from model fitting included BMD and the estimate of the lower-bound confidence limit on the BMD (BMDL).

Using available toxicity data, BMDs and BMDLs were derived for Bobwhite, Mallard, American Tree Sparrow, Dark-eyed Junco, Western Meadowlark, and Horned Lark and have been summarized in Table 2 with other toxicity values previously reported (e.g., LD50s). BMDLs ranged over an order of magnitude from 0.7 mg/kg for the American Tree Sparrow to 76.1 mg/kg for the Horned Lark. For studies presenting sufficient data, the current analysis yielded BMDs and BMDLs consistent with median effect estimates (Table 2).

TABLE 2--*Benchmark dose values for selected species.*

Species	*BMD (mg/kg body weight)*	*BMDL (mg/kg body weight)*	*LD50* (mg/kg body weight)*	*LLD* (mg/kg body weight)*	*NOEL* (mg/kg body weight)*	*Data source*
Bobwhite quail	*1.7*	*1.2*	*2.6*	*NR**	*1.5*	*Fletcher and Pedersen, 1991a*
Mallard	*51.0*	*31.6*	*100*	*NR*	*NC**	*Fletcher and Pedersen, 1991b*
Western meadowlark	*1.9*	*1.4*	*4.0*	*2.2*	*1.1*	*Borchert, 2001a*
American tree sparrow	*1.1*	*0.7*	*3.5*	*4.0*	*NC*	*Borchert, 2001b*
Horned lark	*103.3*	*76.1*	*232*	*101.3*	*NC*	*Sayre, 2001a*
Dark-eyed junco	*90.5*	*53.7*	*162*	*100.0*	*2-20 (from range-finding test)*	*Sayre, 2001b*

** LD50 = median lethal dose, LLD = lowest level dose, NOEL = no observable effects level, NR = not reported, NC = not calculable*

Analysis of Exposure: Food-chain Analysis

Exposed dose was calculated by applying a dietary exposure model identified in Pastorok et al. (1996) and EPA (1999) as:

$$\text{Exposed dose} = \frac{[[(C_{food(s)}\ (ug/gm)\ x\ FR\ (gm/day)]}{BW\ (kg)}$$

where,

$C_{food(s)}$	=	concentration of chemical in food items (plant and animal food items in total);
FR	=	foraging rate (feed ingestion rate); and
BW	=	body weight of receptor species.

For an analysis focused on dietary exposure to DRC-1339-coated brown rice baits, the $C_{food(s)}$ term in the equation would be further decomposed to the amount of poison on each tainted rice grain and the number of rice grains (tainted and untainted) consumed in a total daily food ration. Other factors influencing exposed dose could also be incorporated into the dietary exposure model, including behavioral modifiers (e.g., factors for food preference or food avoidance or discriminate feeding) and physiological modifiers (e.g., gut absorption factors that influence realized dose), if available.

The calculation of exposed dose relied on body weights for receptors of concern (Dunning 1993), food ingestion rates derived from those body weights (Nagy 1987), and weights of individual rice grains and the amount of DRC-1339 on treated rice grains (Eisenmann 2001). Exposed doses varied from 40-220 mg DRC-1339/kg body weight, with small-bodied species consistently presenting greater doses than large-bodied birds (see Table 3). All calculations of exposed doses reflect a single dietary source–brown rice bait presented at field dilution without food discrimination (e.g., tainted brown rice was neither avoided or preferred).

Analysis of Effects: Field Studies and Bird Surveys

A number of field studies have been completed to evaluate the efficacy and risks of DRC-1339's use in controlling pest bird species in agricultural areas. These studies have addressed various aspects of the issues identified during problem formulation, and include recent field efforts to identify non-target bird use of sunflower fields and other agricultural lands at various times of the year (Smith 1999, Schaaf et al. 2001, Custer 2002, Knutsen 1998, Kostecke 1998, Kostecke 2001, Kenyon 1996 and Barras 1996).

Similarly, long-term bird surveys contribute to the analysis of effects potentially expressed at a population level. Evaluation of risks to species of concern requires a wider frame of reference such as the characterization of population trends of these species of concern across North America. Local events focused on use of DRC-1339 must have an ecological context, and evaluating the potential risks of blackbird control practices should consider a larger population setting, if possible. USGS/BRD and Canadian Wildlife

Service coordinate the North American Breeding Bird Survey (BBS), which is a primary source of population trend and distribution information for most species of North American birds (Robbins et al. 1986, Droege and Sauer 1990).

TABLE 3--*Exposed doses of DRC-1339 for target and non-target birds.*

	Exposed dose (mg/kg body weight)		
Species	Males (based on Mean body weight)		Females (based on Mean body weight)
Bobwhite quail*		66.27	
American tree sparrow*		203.00	
Western meadowlark	156.89		162.28
Mallard*		42.28	
Dark-eyed junco	202.55		205.05
Horned lark*		189.41	
Bobolink	178.71		185.17
Chipping sparrow*		218.52	
Clay-colored sparrow*		219.33	
Field sparrow*		217.99	
Red-winged blackbird	170.79		182.08
Yellow-headed blackbird	165.10		159.58
Common grackle	153.96		159.58
European starling	163.60		NA

*Exposed doses based on body weight data for male and female birds combined; exposed doses for other species reported for both male and female, since gender-specific body weights were available.

Population Trend Estimation and Trend Maps

For our present use of BBS data, population trends for selected species likely to occur in the northern Great Plains in spring were characterized for the available survey interval (1966 - 2000; see Sauer et al. 2001). Trends were estimated using the route-regression method in which population trends and annual indices are described (Geissler and Sauer, 1990). Trend, or a consistent change in counts of birds on a route, is the quantity estimated, and annual indices of abundance are used to assess higher levels of pattern in these data in the context of trend. Regional trends are estimated as a weighted average of trends on individual routes. Route trends are estimated using the estimating equations estimator described by Link and Sauer (1994), in which a multiplicative trend is estimated. Peterjohn et al. (1997) and Link and Sauer (1997) should be consulted for more details regarding the BBS analysis supporting the trend maps used in this risk evaluation.

Trend Maps, 1966-1996

Trend maps generated by BBS for those species are included in their survey, and these maps provide a “best guess” of population change for the species over its range in the time period considered. As illustrated for the trend maps for Horned Lark and Western Meadowlark (Figure 3 and Figure 4, respectively), areas of population increase and areas of population decline are illustrated through gray-scale images ranging from gray through black for declining and increasing trends, respectively. These maps are intended to provide a general view of population change for the long-term, but do not provide insight into short-term changes within the 1966-1996 period.

For this analysis focused on species potentially at-risk, the trend at any point was estimated as a weighted average of trend information from nearby survey routes containing information from the species. Trend on these routes was estimated as a yearly change (Link and Sauer 1994; see also Geissler and Sauer 1990). Given our ecological context of the northern Great Plains, population trends for bird species selected as representative of birds at-risk present a mixed signature of current estimates of population status (i.e., increasing, decreasing, or unchanged during the survey period). For example, population trends for Mallard are uniformly increasing throughout the area of concern (North Dakota, South Dakota), while various species of sparrow display trends that are relatively mixed with respect to their spatial distribution of population estimates. Horned Lark population trends consistently present decreasing estimates, as did Western Meadowlark in areas of North Dakota and South Dakota covered by trend estimates.

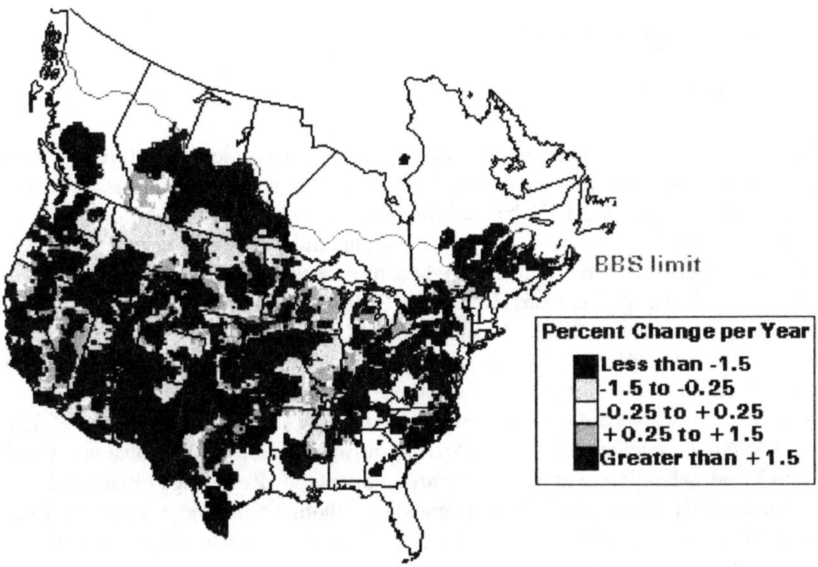

Fig. 3--*Horned lark trend map.*

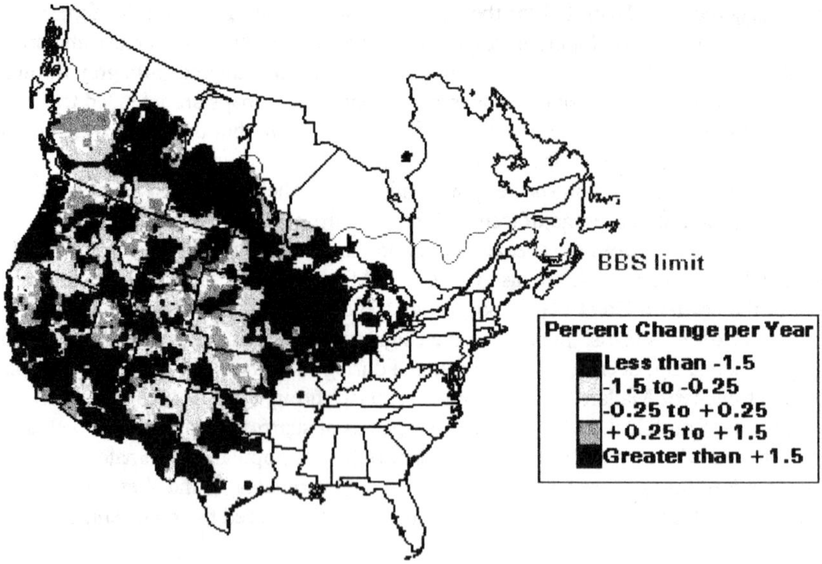

Fig. 4--*Western meadowlark trend map.*

Characterization of Risk

Hazard quotients are summarized in Table 4, and reflect simple ratio estimators based on exposed doses derived from a conservative dietary exposure model that assumed a single feed source, 2% DRC-1339 baited brown rice, with no behavioral or other modifiers (e.g., physiological or metabolic) that would influence realized dose. These exposed doses were compared to increasingly conservative benchmark toxicity values that ranged from median effective doses (LD50s) to BMDs to BMDLs. Hazard quotients based on BMDLs would be the most conservative. Depending on the available data, hazard quotients were derived for each sex. If data by sex were not available, simple ratio estimators were calculated for the species without discriminating for potential sex differences in responsiveness to dietary exposures to DRC-1339. In the absence of data sufficient to target-specific BMDs and BMDLs, surrogate values from similar sized birds (Western Meadowlark) were applied to hazard quotients for Red-winged Blackbird, Yellow-headed Blackbird, and Common Grackle. Although not directly a concern to the current analysis, hazard quotients were also derived for European Starling, given its historic role in the development of DRC-1339 as a bird control agent.

Two attributes of birds at risk strongly influence hazard quotients derived in the current analysis. A bird's sensitivity to DRC-1339 and its body size are the most critical individual factors influencing risk, and life history attributes linked to these attributes (e.g., food preference and foraging behaviors) are critical to evaluating potential effects in the

TABLE 4--*Hazard quotients for DRC-1339 in target and non-target birds.*

		Sexes combined (unless noted)	
Species	Exposed dose/BMD	Exposed dose/BMDL	Exposed dose/LD50
Bobwhite quail	39.0	55.2	25.5
American tree sparrow	184.5	290.0	58.0
Western meadowlark			
Males	82.6	112.1	39.1
Females	85.4	115.9	40.5
Mallard	0.8	1.3	0.4
Dark-eyed junco			
Males	2.2	3.8	1.3
Females	2.3	3.8	1.3
Horned lark	1.8	2.5	0.8
Bobolink			
Males	94.1	127.7	44.6
Females	97.5	132.3	46.2
Chipping sparrow	198.7	312.2	62.4
Clay-colored sparrow	199.4	313.3	62.7
Field sparrow	198.2	311.4	62.3
Red-winged blackbird			
Males	89.9	122.0	42.6
Females	95.8	130.1	45.4
Yellow-headed blackbird			
Males	86.9	117.9	41.2
Females	84.0	114.0	39.8
Common grackle			
Males	81.0	110.0	38.4
Females	84.0	114.0	39.8
European starling			
Males	86.1	116.9	40.8

field. All other factors being equal, small-bodied birds (< 100 g body weight) will be at greatest risk to exposure in the field, which was a desired attribute ("1-particle lethal bait" for target species) during the development DRC-1339. Hazard quotients reflecting BMDL comparisons ranged from near unity (1.34 for Mallard) to greater than 300 (for American Tree Sparrow, Chipping Sparrow, Clay-colored Sparrow, and Field Sparrow). The role that body size plays in influencing hazard quotients is illustrated by simple ratio estimators for a large-bodied bird (e.g., Northern Bobwhite) and relatively small-bodied birds (e.g., Western Meadowlark and sparrows) having BMDLs near 1.0 mg/kg (1.2 mg/kg for Northern Bobwhite, 1.4 mg/kg for Western Meadowlark, and 0.7 mg/kg for sparrows). Here, derived hazard quotients are clearly influenced by the bird's body weight, which suggests that field exposures will allow little margin for error when a bird is foraging on baited rice; ingestion of a single baited grain will likely yield mortality in a small-bodied bird. A simple comparison of hazard quotients for small-bodied non-target birds and target birds suggests these species have similar risks for dietary exposures to DRC-1339. The use of surrogate benchmark values for target species likely has little influence in biasing such comparisons, given the comparative estimates of risks derived independently by other workers (e.g., Eisemann et al. 2001).

Risks of DRC-1339-treated Brown Rice to Non-target Birds

Previously completed field studies reinforced the selection of species of concern identified in problem formulation. For example, Kenyon (1996), Knutsen (1998), and Linz et al. (1995) reported non-target species that corresponded with those birds identified as species of concern. Granivores such as American tree sparrow, song sparrow, clay-colored sparrow, dark-eyed junco, horned lark, and mourning dove were commonly observed in field surveys intended to identify non-target species likely to be present at the time of spring baiting.

Species such as meadowlark, American tree sparrow, and quail are sensitive to DRC-1339 and make use of fields likely baited with DRC-1339-treated brown rice; hence, their exposures are likely to occur and adverse effects potentially expressed. These species serve as surrogates for other non-targets sharing common taxonomic grounds (e.g., same genus, sub-family, or family). Relatively larger-bodied granivores (e.g., Northern Bobwhite, Ring-necked Pheasant) are likely at-risk but exposed dose is effectively reduced as a function of body mass unless foraging behaviors alter exposure (e.g., preferential feeding on bait grains). Many small-bodied granivores (e.g., some sparrows) frequently observed at or around bait sites may be at relatively low risk, if their insensitivity to DRC-1339 mirrors that of the dark-eyed junco or horned lark (e.g., Emberizidae and Alaudidae, respectively). The relatively wide range in species sensitivity within the Emberizidae, however, clearly suggests that caution be exercised when categorical characterizations of risk are developed, especially in view of the apparent sensitivity of some species (e.g., American Tree Sparrows) within a given taxonomic group such as the Emberizidae. As Eisemann, et al. (2001) observed, life history factors related to diet and food preference will influence the actual risk realized by any of these "at-risk" species.

Focus on Potential Ecological Adversity

Risk characterization should discuss whether ecological receptors exposed to chemical stressors are likely to reflect impacts to the ecosystem or to populations of the particular valued species within that ecosystem (assessment endpoint). Risk characterization should also consider whether ecological receptors may be adversely effected in the future (EPA 1992, EPA 1998, Suter 1993). Hence, the potential impacts associated with changes in local population are considered within the context of regional populations characterized by BBS trends.

Estimating actual population declines can be technically difficult and potentially associated with various sources of uncertainties. While efforts have been made to evaluate population-level impacts potentially linked to DRC-1339 exposure, little empirical data are available for the projection of reliable risk estimators at a population-level. The BBS trends brought forward as part of this analysis at best capture a snapshot of populations "at-risk," and the potential role of DRC-1339 in changing those trends is speculative. Estimates of numbers of target (e.g., Barras 1996) and non-target birds exposed and dying as a consequence of DRC-1339 exposure share common problems, e.g., "false negative" readings of pathology, local movements of flocks having varying numbers of individuals, seasonal differences in migration behaviors that potentially influence efficacy (e.g., males arriving earlier than females), that complicate population estimation and ultimately limit the interpretation of risks within the "level of comfort of risk managers." Those levels of comfort will vary, given the perspectives of the risk managers and their "willingness to pay" for miscalculated risks or risk interpretations gone awry (Belzer 2001, Costanza, et al. 1997, Field 1996, Field 2000, Hartwick and Olewiler 1998).

The ecological context should be considered when efficacy and the long-term consequences of spring baiting are considered. While estimates of numbers of target birds killed in a single baiting operation provide some short-term measure of efficacy (e.g., Barras 1996), decreased numbers of target birds arriving at nesting areas after exposures in sunflower fields in areas of spring baiting periods may not yield the desired population reductions in those same species when depredation of sunflowers occurs. For example, target species such as red-wing blackbirds are polygynous, and during spring migration males arrive before females en route to nesting areas north of baiting areas (Orians 1980; 1985); hence, most of the birds likely to succumb to spring baiting with DRC-1339 tainted rice will be males. Other migratory routes that include males of other populations not exposed to DRC-1339 will still enter the breeding populations at the nesting habitats further north, and the effective reduction in birds returning in the fall has not been characterized. Long-term consequences of spring baiting with DRC-1339 have not been considered within the context of impacts to the resource or economic costs associated with the program.

Risk Analysis and Risk Management Implications

Interactions among agricultural land-use and cropping practice, agrichemical use, and wildlife have long provided examples of conflicts between wildlife needs and human uses of habitat. As in all regional environmental issues, these interactions occur within the context of cumulative risks associated with multiple stressors, including larger-scale processes such as climate change and its consequence impacts on habitat; hence, the northern Great Plains provides a current example of how competing uses of spatially-limited land resources sets the stage for policy- and decision-makers to develop management plans that assure the long-term sustainablity of habitat and the resources dependent on those habitats. From the current analysis, the spring baiting program targeted on blackbirds presents potential risks to non-target birds, especially small-bodied species characterized by marked responsiveness to DRC-1339 (acutely toxic). To reduce risks to non-target birds, mitigation measures have been incorporated into spring baiting operations in the field, but these mitigation practices are presently unproven with respect to minimizing non-target bird loss. The impact any loss from non-target populations presently thought to be declining suggests that risks vary across the relatively simple "non-target" category, and risk managers must be wary of oversimplifying management plans based on a "non-target" versus "target" categorization of species at-risk. Management decisions regarding DRC-1339's use should reflect practices that benefit the long-term sustainability of the resources at risk. In view of the uncertainties apparent in the present analysis decisions regarding DRC-1339's use hinge on differing management perspectives of "acceptable risk" and resource valuation. Only when these issues are resolved can resource management plans benefit the long-term sustainability of resources at risk.

References

Apostolou, A., 1969, *Comparative Toxicology of the Avicides 3-chloro-p-toluidine and 2-chloro-4-acetotoluidide in Birds and Mammals*, Ph.D. Dissertation, University of California, Davis. 167 pp.

Avery, M. L., Kenyon, M. J., Linz, G. M., Bergman, D. L., Decker, D. G., and Humphrey, J .S., 1998, "Potential Risk to Ring-necked Pheasants from Application of Toxic Bait for Blackbird Control in South Dakota," *Journal of Wildlife Management*, Vol. 62, pp. 388-394.

Barras, A. E., 1996, *Evaluation of Spring Baiting with an Avicide, DRC-1339, and Habitat Preferences of Migratory Blackbirds*, M.S. Thesis, North Dakota State University of Agriculture and Applied Science, Fargo, ND. 108 pp.

Belzer, R. B., 2001, "Using Economic Principles for Ecological Risk Management," In Stahl, Jr., R. G., Bachman, R. A., Barton, A. L., Clark, J. R., deFur, P. L., deFur, Ells, S. J., Pittinger, C. A., Slimak, M.W., and Wentsel, R.S. (Editors), *Risk Management: Ecological Risk-based Decision Making*, SETAC Press, Pensacola FL., pp. 75-90.

Borchert, J., 2001a, *DRC-1339: An acute oral toxicity study with the Western Meadowlark (Sturnella neglecta)*, FIFRA Guideline, Subdivision E: 71-1, Genesis Laboratories, Incorporated, Wellington, CO.

Borchert, J., 2001b, *DRC-1339: An acute oral toxicity study with the American Tree Sparrow (Spizella arborea)*, FIFRA Guideline, Subdivision E: 71-1, Genesis Laboratories, Incorporated, Wellington, CO.

Costanza, R., Cumberland, J., Daly, H., Goodland, R., and Norgard, R., 1997, *An Introduction to Ecological Economics*, International Society for Ecological Economics/St. Lucie Press, Boca Raton FL.

Custer T. W., Custer, C. M., Linz, G. M., Sileo, L., and Johnston, J. J. 2002. *Exposure of Nontarget Birds to DRC-1339 Avicide in Fall Baited Sunflower Fields*, U.S. Geological Survey, Upper Midwest Environmental Sciences Center, LaCrosse WI.

Droege, S. and Sauer, J. R. (Editors), 1990, *North American Breeding Bird Survey Annual Summary 1989*. U. S. Fish and Wildlife Service, Biological Report 90(8):1-22.

Dunning, Jr., J. B. (Editor), 1993, *CRC Handbook of Avian Body Masses*, CRC Press, Boca Raton FL.

Eisemann, J.D., Linz, G. M., and Johnston, J. J., 2001, "Non-target Hazard Assessment of Using Drc-1339 Avicide to Manage Blackbirds in Sunflower," In Johnston, J.J. (Editor), *Pesticides and Wildlife*, American Chemical Society Symposium Series 771, American Chemical Society, Washington, D.C. pp. 197-211.

Environmental Protection Agency (EPA), 2000, *Benchmark Dose Technical Guidance Document*, External Review Draft, EPA/630/R-00/001, Risk Assessment Forum, U. S. Environmental Protection Agency, Washington, DC 20460.

EPA, 1999, *ECOFRAM Terrestrial Draft Report. Ecological Committee on FIFRA Risk Assessment Methods (ECOFRAM)*, Office of Pesticide Programs, U.S. Environmental Protection Agency, Washington, D.C.

EPA, 1998, *Guidelines for Ecological Risk Assessment*, EPA/630/R-95/002F, Risk Assessment Forum, U.S. Environmental Protection Agency, Washington, D.C.

EPA, 1995, *Reregistration Eligibility Decision (RED): Starlicide (3-chloro-p-toluidine hydrochloride)*, EPA-738-R-96-003, Office of Prevention, Pesticides and Toxic Substances, Washington, D.C.

EPA, 1992, *Framework for Ecological Risk Assessment*, EPA/630/R-92/001, Risk Assessment Forum, U.S. Environmental Protection Agency, Washington, D.C.

Feare, C., 1984, *The Starling*, Oxford University Press, Oxford, England. 315pp.

Field, B. C., 2000, *Natural Resource Economics*, McGraw-Hill, New York.

Field, B. C., 1996, *Environmental Economics*, Third Edition, McGraw-Hill, New York.

Fletcher, D. W. and Pedersen, C. A., 1991a, *Volume 4, Compound 1339: 21-day Acute Oral LD50 Study in Bobwhite Quail*, FIFRA Guideline, Subdivision E: 71-1, BLAL/89/QD 135, Bio-Life Associates, Limited, Neillsville, WI. 176pp.

Fletcher, D. W. and Pedersen, C. A., 1991b, *Volume 3, Compound 1339: 21-day Acute Oral LD50 Study in Mallard Ducks*, FIFRA Guideline, Subdivision E: 71-1, BLAL/89/DD 71, Bio-Life Associates, Limited, Neillsville, WI. 148pp.

Geissler, P. H. and Sauer, J. R., 1990, "Topics in Route-regression Analysis," In Sauer, J. R. and Droege, S. (Editors), *Survey Designs and Statistical Methods for the Estimation of Avian Population Trends*, U. S. Fish and Wildlife Service, Biological Report 90(1). pp 54-57.

Harrahy, E. A., 2002, Response to comments on "*A Critical Review of the Literature on the Effects of DRC-1339 on Nontarget Birds with Special Emphasis on Experimental Design, Analysis and Inference*," U.S. Fish and Wildlife Service. Colorado Cooperative Fish and Wildlife Research Unit, Colorado State University, Fort Collins, CO.

Harrahy, E. A., 2001, *A Critical Review of the Literature on the Effects of DRC-1339 on Nontarget Birds with Special Emphasis on Experimental Design, Analysis and Inference*, Unpublished report prepared for the U.S. Fish and Wildlife Service. Colorado Cooperative Fish and Wildlife Research Unit, Colorado State University, Fort Collins, CO. 63 pp.

Hartwick, J. M. and Olewiler, N. D., 1998, *the Economics of Natural Resource Use*, Addison Wesley & Benjamin Cummings, New York.

Kenyon, M. J.,1996, *Nontarget Avian Risks Associated with the Avicide DRC-1339*, M.S. Thesis, North Dakota State University, Fargo, ND.156 pp.

Knutsen, G. A., 1998, *Avian Use of Rice-baited and Unbaited Stubble Fields During Spring Migration in South Dakota*, M.S. Thesis, North Dakota State University of Agriculture and Applied Science, Fargo, ND.160 pp.

Kostecke, R. M., 1998, *Evaluation of Risks to Secondary Poisoning to Birds and Mammals Following the Use of a Pesticide*, M.S. Thesis, North Dakota State University, Fargo, ND.

Kostecke, R. M., Linz, G.M., and Bleier, W.J., 2001, "Survival of Avian Carcasses and Photographic Evidence of Predators and Scavengers," *Journal of Field Ornithology*, Vol. 72, pp. 439-447.

Link, W. A. and Sauer, J. R., 1994, "Estimating Equations Estimates of Trend," *Bird Populations*, Vol. 2, pp. 23-32.

Link, W. A. and Sauer, J. R., 1997, "New Approaches to the Analysis of Population Trends in Land Birds: Comment,"*Ecology*, Vol. 78, pp. 2632-2634.

Linz, G. M., Bergman, D. L., and Bleier, W. J., 1995, "Birds Associated with Blackbird Spring Feeding Sites in South Dakota," *Prairie Naturalist*, Vol. 27, pp. 11-15.

Mach, J. J., 2001, *DRC-1339: An acute oral toxicity study with the American Tree Sparrow (Spizella arborea)*, FIFRA Guideline, Subdivision E: 71-1, Genesis Laboratories, Incorporated, Wellington, CO. 91pp.

Mull, R. L., 1971, *Comparative Studies of Acute and Sub-acute Toxicity of 3-chloro-p-toluidine, Suties of the Effects of an Acutely Toxic Dose in Chickens on Certain Clinical Biochemical Parameters, and Comparative Studies of a Renal Aromatic-n-deacylase*, Ph.D. Dissertation, University of California, Davis CA. 105pp.

Nagy, K., 1987, "Field Metabolic Rate and Food Requirement Scaling in Mammals and Birds," *Ecological Monographs*, Vol. 57, pp. 11-128.

Pastorok R. A., Butcher, M. K., and Nielsen, R. D., 1996, "Modeling Wildlife Exposure to Toxic Chemicals: Trends and Recent Advances," *Human and Ecological Risk Assessment*, Vol. 2, pp. 444-480.

Peterjohn, B. G., Sauer, J. R., and Link, W. A., 1997, "The 1994 and 1995 Summary of the North American Breeding Bird Survey," *Bird Populations*, Vol. 3, pp. 48-66.

Peterjohn, B. G. and Sauer, J. R., 1993, "North American Breeding Bird Survey Annual Summary 1990-1991," *Bird Populations*, Vol. 1, pp. 1-15.

Orians, G. H., 1980, *Some Adaptations of Marsh-nesting Blackbirds*, Princeton University Press, Princeton NJ. 295pp.

Orians, G. H., 1985, *Blackbirds of North America*, University of Washington Press, Seattle WA. 163pp.

Robbins, C. S., Bystak, D., and Geissler, P. H., 1986, *The Breeding Bird Survey: its First Fifteen Years, 1965-1979*, U.S. Fish and Wildlife Service, Resource Publication 157. 196pp.

Sauer, J. R., Hines, J. E., and Fallon, J., 2001, *The North American Breeding Bird Survey, Results and Analysis 1966 - 2000.* Version 2001.2, USGS Patuxent Wildlife Research Center, Laurel, MD

Sayre, R.W., 2001a, *DRC-1339: An Acute Oral Toxicity Study with Grassland Birds: the Horned Lark (Eremophila alpestris)*, FIFRA Guideline, Subdivision E: 71-1, Genesis Laboratories, Incorporated, Wellington, CO. 67 pp.

Sayre, R.W., 2001b, *DRC-1339: An Acute Oral Toxicity Study with the Dark-eyed Junco (Junco hyemalis)*, FIFRA Guideline, Subdivision E: 71-1, Genesis Laboratories, Incorporated, Wellington, CO.

Schaaf, D., Linz, G. M., Bleier, W. J., and Homan, H. J., 2001. "Avian Use of Ripening Sunflower Fields," *Proceedings of 23rd Sunflower Research Workshop*, January 17-18, 2001, Fargo, ND. pp. 177-179.

Smith, J. A., 1999, *Nontarget Avian Use of DRC-1339 Treated Plots During an Experimental Blackbird Control Program in Eastern South Dakota*, M.S. Thesis, Wildlife and Fisheries, South Dakota State University, Brookings, SD. 54pp.

Stahl, Jr., R. G., Bachman, R. A., Barton, A. L., Clark, J. R., deFur, P. L., Ells, S. J., Pittinger, C. A., Slimak, M. W., and Wentsel, R. S., 2001, *Risk Management: Ecological Risk-based Decision Making*, SETAC Press, Pensacola FL.

Stahl, Jr., R. G., Pittinger, C. A., Barton, A. L., Clark, J. R., deFur, P. L., Ells, S. J., Slimak, M. W., Wentsel, R. S., Belzer, R. B., and Bachman, R. A., 2001, "Introduction and Background to the Development of a Framework for Ecological Risk Managment," In Stahl, Jr., R.G., Bachman, R.A., Barton, A.L., Clark, J.R., deFur, P. L., Ells, S. J., Pittinger, C. A., Slimak, M. W., and Wentsel, R. S., 2001, *Risk Management: Ecological Risk-based Decision Making*, SETAC Press, Pensacola FL. pp.1-20.

Suter, G.W., 1993, *Ecological Risk Assessment*, Lewis Publishers, Boca Raton FL.

U.S. Department of Agriculture (USDA), Animal and Plant Health Inspection Service (APHIS), Wildlife Services, 2000, *Spring blackbird damage management project for the protection of ripening sunflowers*, Pre-Decisional Environmental Assessment. USDA/APHIS, Washington, D.C.

Westberg, G. L., 1973, *Comparative Studies of the Metabolism of 3-Chloro-P-Toluidine and 2-Chloro-4-Acetotoluidine in Rats and Chickens and Methodology for the Determination of 3-Chloro-P-Toluidine and Metabolites in Animal Tissues*. Ph.D. dissertation, Comparative Pharmacology & Toxicology, University of California, Davis CA.

Benjamin L. Preston[1]

GIS-Based Localization of Impaired Benthic Communities in Chesapeake Bay: Associations with Indicators of Anthropogenic Stress

REFERENCE: Preston, B. L., **"GIS-Based Localization of Impaired Benthic Communities in Chesapeake Bay: Associations with Indicators of Anthropogenic Stress,"** *Landscape Ecology and Wildlife Habitat Evaluation: Critical Information for Ecological Risk Assessment, Land-Use Management Activities, and Biodiversity Enhancement Practices, ASTM STP 1458*, L. A. Kapustka, H. Galbraith, M. Luxon, and G. R. Biddinger, Eds., ASTM International, West Conshohocken, PA, 2004.

ABSTRACT: Current ecological risk assessment methodologies have limited capacity to account for the spatial heterogeneity in stressors and physical/chemical conditions characteristic of aquatic and estuarine ecosystems. The Shannon-Weaver index was utilized to construct spatial models of benthic macroinvertebrate biodiversity in Chesapeake Bay over several time periods between 1987-2001. These models were subsequently compared to monitoring data for sediment contaminant concentrations and physical/chemical water quality conditions. A series of GIS exercises demonstrated that low values for species diversity were associated with higher concentrations of a diverse array of contaminants as well as physical/chemical water quality conditions. Multivariate regression analysis among a range of contaminant and water quality variables accounted for up to 61% of the observed variation in benthic biodiversity. Collectively, these results demonstrate the numerous challenges for conducting ecologically relevant risk assessments at the ecosystem level, which can be partially ameliorated with quality monitoring data and geographic approaches to environmental assessment.

KEYWORDS: benthic, biodiversity, water quality, contaminant, ecological risk assessment, GIS

Introduction

Many factors interact to affect ecological systems. Some of these factors are natural, such as gradients in physical/chemical environmental conditions and species interactions. However, a broad range of anthropogenic perturbations have significantly altered ecosystems in the past and are projected to continue doing so well into the future (Vitousek et al. 1997; Sala et al. 2000). Land-use change associated with both agriculture and urbanization along with the externalities of population growth and industrial activity collectively expose ecosystems to diverse perturbations such as novel chemical stressors (Dyer and Wang 2002; Kolpin et al. 2002), changes in nutrient loads and cycling (Paerl 1997; Tilman et al. 2001), and habitat loss (Allan et al. 1997; Harding et al. 1998). Furthermore, complex interactions may occur among contaminants and/or natural factors such as water quality conditions (Lozano and Pratt 1994; Folt et al. 1999). Thus, the amelioration of anthropogenic impacts is complicated by the challenge of attributing

[1] Senior Research Fellow, Pew Center on Global Climate Change, 2101 Wilson Boulevard, Suite 550, Arlington, VA 22201.

observed environmental changes to specific stressors that can be addressed through environmental management (Barnthouse et al. 2000).

The predominate method for identifying potential ecological hazards is the ecological risk assessment (ERA) framework (U.S. EPA 1998). ERA has proven to be a useful tool for estimating the likelihood of adverse ecological effects resulting from exposure to stressors. However, ERA is designed to estimate the risk of direct effects from single exposures, making it difficult to implement under more realistic scenarios where multiple stressors co-occur and interact with other drivers of ecology (Barnthouse et al. 2000; Dyer et al. 2000; Moore and Bartell 2000; Preston 2002b, 2002c). This is particularly true at the whole ecosystem or regional level, where conventional risk assessment approaches may, in fact, be impractical (Barnthouse et al. 2000). In such situations, it may be necessary to consider a broad range of both natural and anthropogenic factors, which must be assessed collectively for their relationship with ecological effects of interest. This information can subsequently be used to construct process models that link a suite of ecological drivers to specific or general ecological effects (Barnthouse et al. 2000). These drivers can then be targeted through environmental management to enhance or reduce their influence on ecological receptors.

A critical necessity for the development of such process models is extensive data regarding both stressors and ecological effects, which necessitates moderate- to long-term environmental monitoring. Unfortunately, few environmental monitoring programs are sufficiently robust for rigorous risk assessment (Preston and Shackelford 2002b), and there may be substantial differences and inconsistencies both within and among monitoring programs (Hughes et al. 2000). Chesapeake Bay is one of the largest and most extensively studied estuarine ecosystems in the United States (Boesch 1996), and environmental monitoring of the watershed over the past two decades has led to the development of an extensive database containing data on biota, contaminants, and physical/chemical water quality conditions. As such, the Chesapeake Bay serves as a useful field laboratory for developing a comprehensive understanding of drivers of ecological change. Although the environmental monitoring data from the bay are not fully integrated in that significant spatial and temporal heterogeneity exists with respect to sampling for individual variables (Preston 2002c), geographic information systems (GIS) and multivariate statistics have been applied to these data to assist in overcoming some of these limitations and to permit more integrated data analysis (Preston 2002a 2002c; Preston and Shackelford 2002a).

The current study represents an update of previous studies utilizing geographic information systems (GIS) in the analysis of environmental monitoring data from Chesapeake Bay (Preston 2002c; Preston and Shackelford 2002a). The overall objectives of the current study were to identify the principle drivers of impairment of benthic macroinvertebrate communities in Chesapeake Bay, assess the spatial autocorrelation among various drivers, and determine whether temporal changes have occurred in benthic biodiversity that are coincident with changes in contaminant concentrations or water quality conditions.

Materials and Methods

Data Sources

Digital geographic data for surface features were obtained from several sources. Land and water features were obtained as ArcView® shapefiles from the U.S. Environmental Protection Agency's (EPA) Better Assessment Science Integrating Point and Non-point Sources (BASINS) geographic information system (available over the internet at http://www.epa.gov/OST/BASINS). These features were based upon digitized images of U.S. Geologic Survey (GS) base maps (1:250,000 scale). These features were used to construct a polygon theme representing the primary basin of the Chesapeake Bay and the lower regions of major tributaries comprising an area of 11,170 km^2 (Fig. 1). All subsequent analyses were restricted to this study area. State boundaries were also obtained as ArcView® shapefiles from U.S. EPA's BASINS system, based upon U.S. GS digital line graphs (1:2,000,000 scale).

The Shannon-Weaver biodiversity index for the benthic macroinvertebrate community was used as an effect indicator. Shannon's index describes the equitability with which individuals in a community are allocated among species and was calculated using equation 1 (Newman 1995):

$$SBI=\Sigma p_i \ln p_i \tag{1}$$

where, SBI=Shannon Biodiversity Index and p_i=percentage of all individuals in the ith species. Shannon's index is a useful metric in regional biological assessments where spatial heterogeneity in the distribution of specific species may be significant (Weisberg et al. 1997). Biodiversity data for summer months (May-September) between 1987 and 2001, inclusive, were obtained from the Chesapeake Bay Program's (CBP) data clearinghouse (available over the internet at http://www.chesapeakebay.net). Data were originally derived from benthic community samples collected from 2147 fixed and random monitoring stations located throughout the basin resulting in a total of 2687 individual observations (Fig. 1). The locations of monitoring stations in decimal degrees were obtained by Loran-C (accurate to ± 500 m) using the North American Datum 1927 between 1987-1996, after which locations were based upon global positioning system receivers using the North American Datum 1983. Methods for sample collection have been previously reported (CBP 2000; Preston 2002c).

Eleven water quality parameters were analyzed including physical/chemical conditions such as temperature, dissolved oxygen, pH, and total suspended solids as well as nutrient concentrations. Water quality data for summer months between 1987-2001 were also obtained from the CBP. Data were originally collected by monthly sampling from the basin floor at 152 fixed and random monitoring stations located throughout the basin (Fig. 2). Geographic locations for monitoring stations were obtained in a similar fashion as described above. Specific methods used in the analysis of individual water quality parameters have been previously reported (CBP 1993). The total number of observations for each parameter ranged from 205 to 2295.

Sediment concentrations of 60 contaminants were analyzed, including total concentrations for 10 metals, 12 pesticides, 20 polycyclic aromatic hydrocarbons (PAHs),

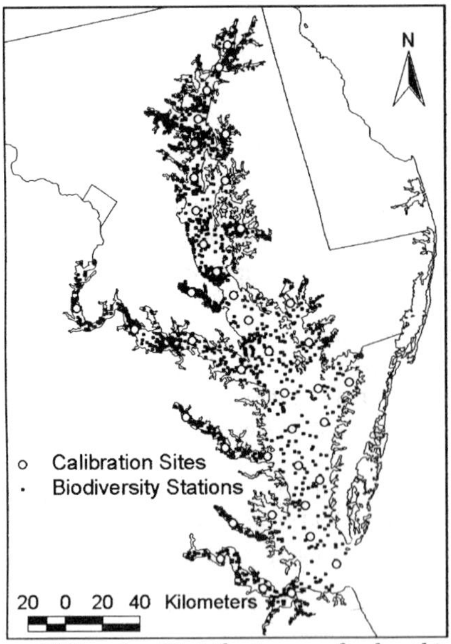

FIG. 1—*Distribution of benthic biodiversity monitoring stations (n=2147) and calibration points for effect modeling (n=40) within Chesapeake Bay (1987-2001).*

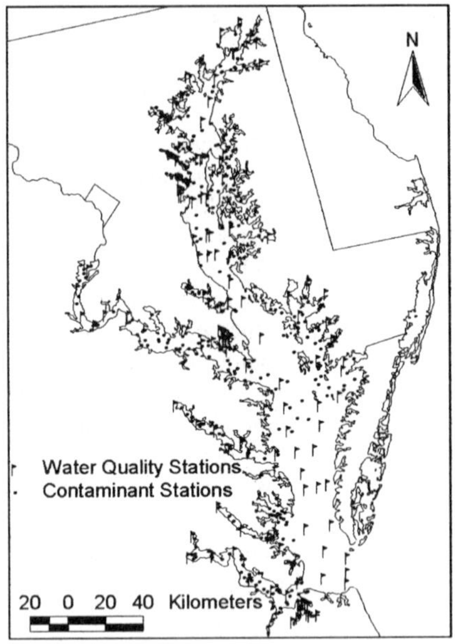

FIG. 2—*Distribution of water quality monitoring stations (n=152) and sediment contaminant monitoring stations (n=485) within Chesapeake Bay (1987-2001).*

and 18 polychlorinated biphenyl (PCB) congeners (Appendix 1). Sediment contaminant data for summer months between 1984-1999 were obtained from the CBP. Data were originally collected by yearly sediment sampling at 485 fixed monitoring stations located around the perimeter of the basin and in most of its major tributaries (Fig. 2). Geographic locations for monitoring stations were obtained in a similar fashion as described above. The total number of observations made throughout the basin for each contaminant varied, ranging from 192 to 756. Data on sediment concentrations of contaminants available from CBP were synthesized from approximately 50 independent assessment reports (CBP 1999).

Effect Modeling

Four basin-wide models of benthic biodiversity were constructed, with each model corresponding with a different time period. The first model was based upon temporally aggregated biodiversity data over the time period 1987-2001. The remaining three models subdivided this 15-year time period into three intervals: a) 1987-1991, b) 1992-1996, c) 1997-2001. Models were constructed by interpolating values for benthic biodiversity at locations where samples were not collected based upon observed values where samples were collected. Prior to interpolation, values for Shannon-Weaver biodiversity were each assigned a probability, based upon the cumulative probability distribution for the entire 15-year data set. Data were then projected using an Albers

Equal Area-Conic projection to account for curvature of the earth. Surface interpolation was based upon a grid system comprised of cells of equal area. Biodiversity values were assigned to each grid cell using an inverse distance-weighted (IDW) average of observed biodiversity values within or neighboring the cell. In the current study, an IDW model with the weights determined by the inverse square law (i.e., value of the weights decrease with the square of distance) was utilized. Biodiversity values were assigned to each grid cell using equation 2 (Fischer et al. 1996):

$$X_0=\Sigma w_i X_i \qquad (2)$$

where X_0=interpolated value for a grid cell, X_i= transformed value for Shannon's index and $w_i= D(X_0,X_i)^{-2/W}$ where $D(X_0,X_i)$=distance from X_0 to X_i and W=a normalization factor that enables $\Sigma w_i=1$. Validation of the use of the IDW model in the spatial analysis of biodiversity data from Chesapeake Bay has been previously reported (Preston 2002c).

Both the size (i.e., area) of the grid cells as well as the neighborhood size (i.e., number of observations used to interpolate the value of a grid cell) were important considerations (Preston 2002c). Therefore, a sensitivity analysis was conducted to optimize grid cell and neighborhood size. Forty calibration points (independent of biotic sampling points) distributed throughout the basin were selected (Fig. 1). The interpolated value for Shannon's biodiversity index at each of these points was calculated assuming a range of cell sizes (0.01-625 km^2) and neighborhood sizes (1-40). Optimal cell and neighborhood sizes were determined by quantifying the variation associated with each interpolated cell value with stepwise decreases in cell size or increases in neighborhood size and selecting neighborhood and cell sizes that minimized this variation. Using data from the time period 1987-1999, at cell sizes of 0.0625 km^2, further reductions in cell size caused no further significant decreases in within-cell variation. No significant variation was defined as a less than 1% change in mean values among the 40 calibration points and a less than 5% change in value at 95% of individual calibration points. Due to variations in sample size, interpolation parameters for the three time-series models were optimized for the time period 1987-1991 (the time period with the fewest samples). Subsequent time periods were based upon the same parameters. For the time period 1987-1991, at cell sizes of 1 km^2, further reductions in cell size caused no further significant decreases in within-cell variation. For both the time periods 1987-2001 and 1987-1991, at neighborhood sizes of 10, further increases in neighborhood size caused no further significant decrease in within-cell variation. Thus, effect modeling of benthic biodiversity was performed using grid cell and neighborhood sizes of 0.0625 km^2 and 10, respectively for the time period 1987-1999 and 1 km^2 and 10, respectively, for all other time periods.

Upon the completion of the IDW interpolations, areas corresponding with the upper and lower 20th percentiles for the cumulative probability distribution for benthic biodiversity (based upon all observations between 1987 and 2001) were identified and designated low- and high-impact zones, respectively. The upper and lower 20th percentiles were selected because the modeled biodiversity values for these areas were greater than one standard deviation from the basin-wide observed mean for benthic biodiversity. Water quality and sediment contaminant monitoring stations intersecting with low- and high-impact zones were selected and their associated data were organized into separate data sets for statistical comparison. The total area of the low- and high-impact zones for each model was estimated by image analysis.

An uncertainty analysis was subsequently conducted to assess confidence in each of the biodiversity models. The uncertainty associated with the interpolated value for Shannon's index was quantified for the subset of 40 grid cells associated with the 40 calibration points used in the sensitivity analysis (see above). For each of these 40 grid cells, the coefficient of variation (CV) was calculated for the ten observed values for Shannon's index used to interpolate the value for that grid cell.

Stressor Identification

As noted previously, a broad range of water quality parameters and contaminants were assessed for their strength of association with modeled patterns of benthic biodiversity. Two general approaches were utilized for assessing the strength of association. First, aggregate analyses were conducted to determine whether the spatial and temporal distributions of modeled high and low-impact zones were consistent with concentrations of potentials stressors averaged over all water quality and contaminant samples taken within those zones. Second, site-specific analyses were conducted to ensure spatial aggregation of sampling sites did not bias results and to determine the combination of stressors that best accounted for estimates of benthic biodiversity among a number of specific locations.

Aggregate Analyses—Aggregate analyses consisted of statistical comparisons of water quality parameters and sediment contaminant concentrations between the low- and high-impact zones (i.e., low-impact zones were effectively used as a reference site). First, comparisons were made among all data between 1987-1999 for contaminants and 1987-2001 for water quality parameters, following the method of Preston (2002c). All comparisons were made among mean values using a two-tailed t-test (α=0.05) to account for both variance and sample size, as sample sizes differed among different variables and impact zones. Subsequently, similar comparisons were made among three different time intervals within the 1987-2001 sequence (1987-1991; 1992-1996; 1997-2001), using the 1987-2001 means as a common basis for comparison (for analysis of contaminants, the third time period and long-term mean ended in 1999). However, sample sizes for contaminant and water quality parameters varied substantially among these three time intervals, which necessitated the use of water quality and contaminant data from throughout the basin, rather than only in the low- and high-impact zones. In addition, for the analysis of time intervals (as well as all site-specific analyses described below) mean concentrations of individual pesticides, PAHs, and PCBs were summed and each of these classes of contaminants were analyzed as a group.

Site-Specific Analyses—Although spatial aggregation of data may allow some broad associations to be established among benthic biodiversity, water quality, and contaminant concentrations, there may be substantial variability in the stressors present at any particular location within a region, which may be masked through spatial aggregation. Thus, a number of additional analyses were conducted on a site-specific basis to determine a) how well individual water quality or contaminant characteristics correlated with benthic biodiversity among a suite of 485 study sites, and b) the combination of stressors that could best account for the values of benthic biodiversity among those sites.

The value of Shannon's index, each water quality variable (n=11), and each contaminant (n=13) was estimated at each of the 485 study sites using different methods. Benthic biodiversity at each site was estimated from the benthic biodiversity model, aggregated over 1987-2001 (see above). Contaminant concentrations were measured directly at each of the 485 sites (i.e., the selected study sites corresponded with those stations where contaminant monitoring had been conducted). Water quality parameters were estimated for each site based upon values obtained at the nearest water quality monitoring station (mean distance and standard error between study sites and the closest water quality monitoring station was 4.8±0.2 km). As the biodiversity model was basin-wide, an estimate of biodiversity was obtained for all 485 study sites. However, different contaminant and water quality parameters were reported among the monitoring stations, and thus not all of the 485 study sites had complete information for all contaminants and water quality variables. Each of the 13 independent variables described above were parameterized for each study site in three different ways: minimum, maximum, and mean (or total for organic contaminants). This was done to account for the fact that the distribution and abundance of species may be affected by both mean concentrations of stressors as well as by anomalous extremes in concentrations. Furthermore, as the relationship between Shannon-Weaver biodiversity and various water quality and contaminant variables was not assumed to be linear, statistical calculations were performed on untransformed as well as log-transformed dependent and independent variables.

Statistical analyses of the relationship between water quality and contaminant variables and benthic biodiversity were performed in two phases, following the method of Preston and Shackelford (2002a). First, the correlation between each independent water quality or contaminant variable and the interpolated value for Shannon-Weaver biodiversity at each of the 485 study sites was determined by calculation of the Pearson's product-moment correlation coefficient and Fisher's test (α=0.05) (Sokal and Rohlf 1995). Variables were subsequently ranked according to the strength of their association with Shannon's index. Only those parameterizations (i.e., minimum, mean, maximum, total, transformed and untransformed) that yielded the highest correlation between independent variables and biodiversity were reported. Second, type II least squares multiple regression was used to determine the combination of independent variables that best predicted values for biodiversity among the 485 study sites (Sokal and Rohlf 1995). Independent variables were selected using an all-regressions method that performed a multiple regression on every possible combination of dependent and independent variables (Kachigan 1986), with the constraint that the number of independent variables included in the regression could be no more than 10% of the number of observations. Multiple regression was first performed using either water quality or contaminant variables, after which all variables were considered in combination. Regression models that yielded the best fit with the smallest number of variables and most equitable distribution of residuals were selected. Statistical significance of regression models was confirmed through multivariate analysis of variance (α=0.05). Correlation matrices were subsequently calculated for all independent variables utilized in the multiple regression models to ensure results were not biased by high degrees of covariation among independent variables.

Methodological Uncertainties and Limitations

The above methods produce several uncertainties and/or limitations that should be noted. Historical monitoring data were not collected specifically for the purposes for which they were utilized in the current study, and verifying the quality of the data and consistency in analysis methods is problematic. Shannon's index has been criticized due to its assumption that all species are represented in the sample from a population of effectively infinite size, and the value of the index does not vary with sample size (Newman 1995; Weisberg et al. 1997). However, previous studies have demonstrated a strong correlation between Shannon's index and stressors in the Chesapeake Bay (Preston 2002a, 2002c; Preston and Shackelford 2002a). Regardless of the methods used to parameterize and construct the biodiversity models, uncertainty is an unavoidable consequence. Thus, any interpretation of the models must be made with reported variation and uncertainty in mind. Although field data were used as a potential indicator of adverse effects, these effects cannot necessarily be attributed to water quality or contaminant parameters considered in the current study (Havens 1999). Ecotoxicological data were not analyzed to estimate community effect concentrations for stressors that could be compared to environmental concentrations.

Software

Data management and selection were conducted using Microsoft Excel® 98 (Redmond, WA, USA), Statview® 4.0 (SAS Institute, Cary, NC, USA), and the database features of ArcView® GIS 3.2. Mapping of all geographic and environmental monitoring data was performed using ArcView®. Effect modeling and surface interpolation of benthic biodiversity were performed using the Spatial Analyst 2.0 extension of ArcView®. Statistical operations were performed using StatView. Area estimation for the biodiversity models was performed by image analysis using NIH Image 1.62 (National Institutes of Health, Bethesda, Maryland, USA).

Results

Effect Modeling

The basin-wide 15-year mean and standard error for Shannon's index was 2.03±0.02. Based upon the modeling of benthic biodiversity within the basin, a total of 2,015 km^2 was identified as low-impact zones (Fig. 3; Table 1), which contained 16 water-quality monitoring stations and 38 sediment-contaminant monitoring stations. In comparison, a total of 1228 km^2 of the basin was identified as high-impact zones (Fig. 3), which contained 18 water-quality monitoring stations and 72 sediment-contaminant monitoring stations. The means and standard errors for observed values for Shannon's index in the low- and high-impact zones were 3.36±0.02 and 0.59±0.03, respectively, confirming the modeled areas identified as high and low-impact zones represented the observed values for biodiversity within each.

Both empirical data and modeling suggested a potential improvement in benthic biodiversity of Chesapeake Bay between 1997 and 2001. Annual basin-wide means for observed benthic biodiversity displayed a significant upward trend over the 15-year period, despite some variability (Fig. 4). Meanwhile, the three time-series models (1987-1991; 1992-1996; 1997-2001) revealed a substantial retreat of the low-impact areas and expansion of the high-impact areas during 1997-2001, relative to the two prior time periods (Table 1). The total area of high-impact zones (≤20th percentile for 15-year benthic biodiversity) decreased by a factor of 2-3 in the 1997-2001 time period (741 km^2) relative to the two previous time periods (averaging 1331 km^2) (Table 1), while the total area of low-impact zones (≥ 80th percentile for 15-year benthic biodiversity) increased in 1992-1996 and 1997-2001 (averaging 2204 km^2) relative to 1987-1991 (529 km^2).

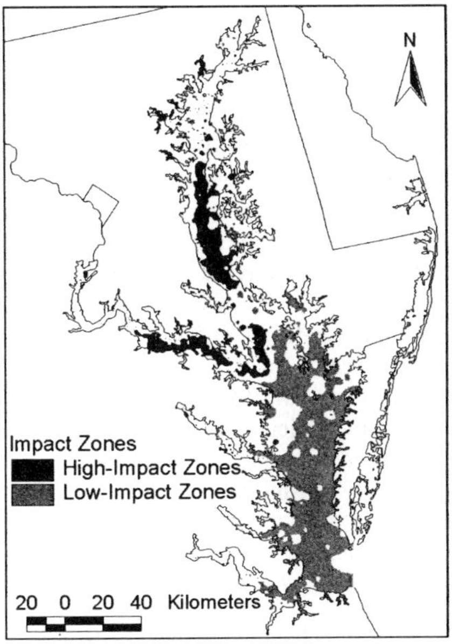

FIG. 3—*Results of effect modeling (1987-2001) based upon observed benthic biodiversity within the study area. Low- and high-impact zones were interpolated from observed values for Shannon's index at each biodiversity sampling location using a second-order inverse distance-weighted (IDW) model.*

Uncertainty analysis of the four biodiversity models indicated that there was low to moderate variability among the observed values of Shannon's index used to interpolate individual grid cells (Figs. 1 and 5). However, the time series models demonstrated differential variability, with models corresponding with the time periods 1987-1991 and 1992-1996 having greater variability than the time period 1997-2001 (Fig. 5). This change in variability was likely due to greater numbers of observations in more recent years.

TABLE 1—*Total area (km^2) associated with high- and low-impact zones in Chesapeake Bay based upon spatial models of benthic biodiversity over various time periods.*

Years	High -Impact Zones	Low -Impact Zones
1987-91	1330	529
1992-96	1302	2240
1997-01	741	2167
1987-01	**1227**	**2014**

Stressor Identification

Aggregate Analyses—Comparison of benthic water quality data from high and low-impact zones aggregated over 1987-2001 indicates several potential factors that might account for observed differences in biodiversity. Dissolved oxygen concentrations in the high-impact zones were less than 75% of those in the low-impact zones, which may be related to the higher concentrations of nitrogen, phosphorus, and dissolved organic carbon in the high-impact zones (Table 2). This increase in nutrients is also reflected in the higher chlorophyll-a concentrations observed in the high-impact zones (Table 2). Collectively, this suggests eutrophication as a potential stressor to benthic biodiversity. However, significant differences were also observed for salinity, conductivity, turbidity, and pH, which may reflect fundamental differences in water quality characteristics between high- and low-impact zones, independent of anthropogenic factors (Table 2). Salinity, in particular, was twice as high in the low-impact zones relative to the high-impact zones, which may have a significant influence on the spatial patterns of biodiversity (Kennish 1990).

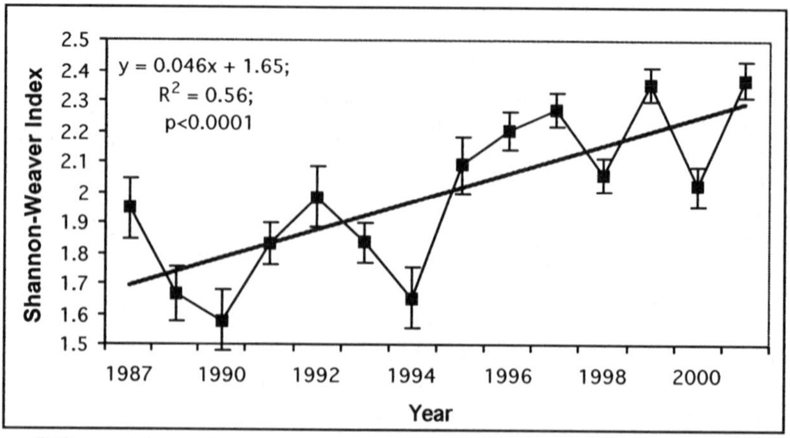

FIG. 4—*Observed annual mean values and linear trend for the Shannon-Weaver index in the Chesapeake Bay basin (1987-2001). The observed trend is statistically significant ($p<0.0001$) by analysis of variance. Error bars represent standard error of the mean.*

Of the ten metals considered in the current study, all were significantly elevated in the high-impact area (Fig. 6). Cadmium concentrations were an order of magnitude higher in sediments from the high-impact zones (1304 µg/kg) compared to the low-impact zones (132 µg/kg), and near order of magnitude disparities were observed for other metals including copper, mercury, and lead. Concentrations of all pesticides considered in the current study were elevated in the high-impact zones. However, observed differences were only statistically significant for dieldrin and trans-nonachlor. All PAHs were significantly elevated in the high-impact zones compared to the low-impact zones, many by an order of magnitude or more. However, these differences were only significant for, benzo[k]fluoranthene, benzo[e]pyrene, benzo[b]fluoranthene, benzo[b+k]fluoranthene, dibenzothiophene, naphthalene, and acenaphthylene. All PCB congeners were elevated in the high-impact zones, several by at least one order of magnitude. However, none of these differences were statistically significant.

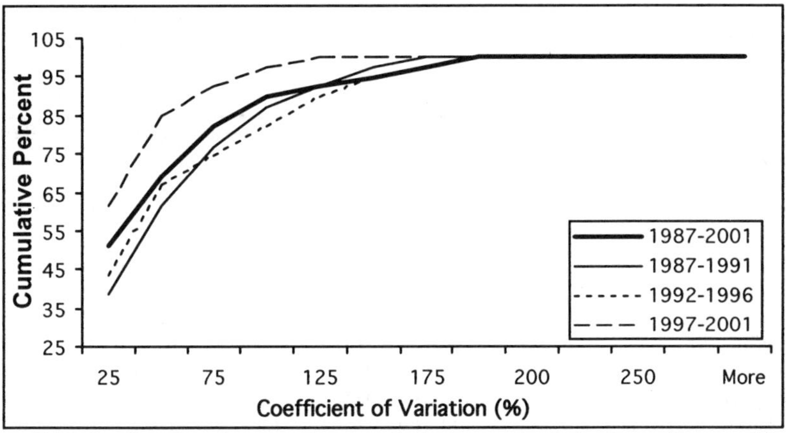

FIG. 5—*Results of uncertainty analysis for each of the four effect models, conducted on the subset of grid cells associated with the 40 calibration points used in sensitivity analysis (See Materials & Methods; Figs. 1 and 3). Uncertainty in the interpolated value for each cell was expressed as the coefficient of variation (CV) among the ten observed values for Shannon-Weaver index used to interpolate each grid cell. CVs are presented as the cumulative percentage of sites with CVs less than the value indicated on the x-axis.*

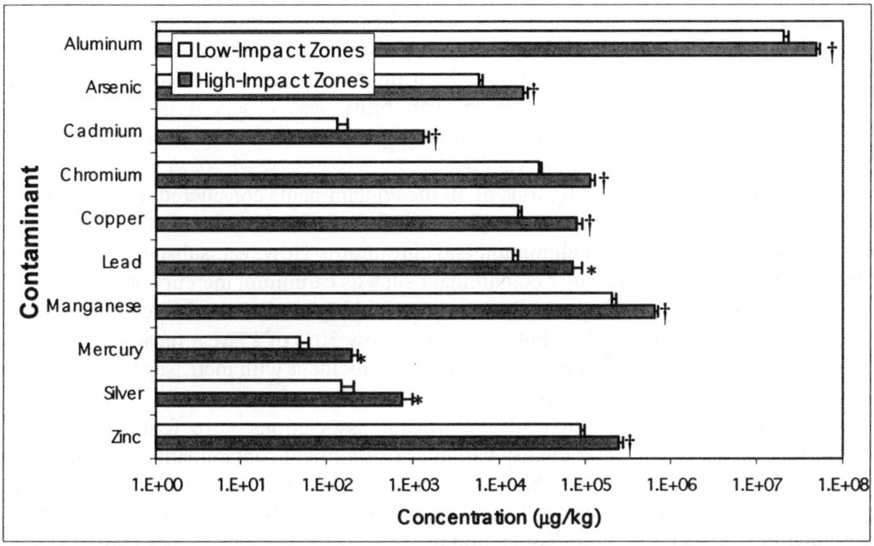

FIG. 6—*Comparison of sediment concentrations for metals in the low- and high-impact zones. Error bars represent the standard error of the mean. * and † indicate significant difference (p<0.05 and p<0.0001, respectively) by two sided t-test.*

TABLE 2—*Comparison of mean benthic water quality parameters between high-impact and low-impact zones of Chesapeake Bay basin during summer months (May-September; 1987-2001). n represents the number of observations per parameter per zone. SE represents the standard error of the mean. High/Low indicates the ratio of mean concentration in the high-impact zones to the low-impact zones. * and † indicate significant difference (p<0.05 and p<0.0001, respectively) by two sided t-test.*

	High-Impact Zones		**Low-Impact Zones**		**Ratio**
Water Quality Variable	n	Mean ± SE	n	Mean ± SE	High/Low
Chlorophyll-a (mg/L)	863	14.38 ± 1.01	441	6.347 ± 0.34	2.27†
Conductivity (μmhos/cm)	1291	21022.2 ± 294.10	1004	34689.219 ± 85.47	0.61†
Dissolved organic carbon (mg/L)	764	3.44 ± 0.07	337	2.644 ± 0.04	1.30†
Dissolved oxygen (mg/L)	1028	4.19 ± 0.11	786	5.808 ± 0.07	0.72†
pH	1289	7.60 ± 0.01	968	7.891 ± 7.89	0.96†
Salinity (g/L)	1027	12.41 ± 0.21	799	21.925 ± 0.15	0.57†
Total nitrogen (mg/L)	1118	0.96 ± 0.02	853	0.52 ± 0.01	1.85†
Total phosphorus (mg/L)	1144	0.07 ± 0.002	853	0.06 ± 0.001	1.20†
Total Suspended Solids (mg/L)	1296	14.73 ± 0.42	1002	22.129 ± 22.13	0.67†
Turbidity (NTU)	39	7.15 ± 1.08	166	16.049 ± 0.94	0.45†
Water Temperature (°C)	1292	22.23 ± 0.12	1005	22.589 ± 0.12	0.98*

Time-series analysis of water quality and contaminant concentrations in the Chesapeake Bay basin identified a number of temporal heterogeneities. For water quality parameters, dissolved oxygen, water temperature, and pH remained relatively constant among the three time intervals (Fig. 7). Chlorophyll-a and dissolved organic carbon concentrations generally increased in the basin over the three time intervals, relative to the 15-year mean. Concentrations of total suspended solids, total nitrogen and phosphorus, and basin conductivity displayed parabolic temporal variability, with the time period 1992-1996 having anomalous high or low values relative to the other two time periods. Thus, although significant variability was observed in water quality parameters over the 15-year period, this variability did not appear to correlate with the overall upward trend in benthic biodiversity.

In contrast, sediment concentrations of all the contaminants considered in the current study decreased between 1987 and 2001 (Fig. 8). These reductions were relatively small for some contaminants, such as aluminum and chromium. However, substantial educations were observed for other contaminant such as cadmium, mercury, silver, and all organic compounds, which generally were two-fold higher than the 15-year means during the 1987-1991 time period, but were at or below 50% of 15-year means by the 1997-2001 time period. These decreases occurred coincident with increases in benthic biodiversity, although one cannot assume cause and effect.

Site-Specific Analyses—Statistically significant associations were found between a wide variety of individual water quality variables or sediment contaminant concentrations and interpolated values for benthic biodiversity among the 485 study sites (Table 3). The strongest correlations were observed for log minimum PCBs (r=-0.47), log minimum chlorophyll-a (r=0.41), log minimum cadmium (r=-0.37), minimum copper (r=-0.35) and log water temperature (r=0.35), and generally water quality and contaminant variables correlated with biodiversity equally well. Those independent variables with the poorest correlation ($r<\pm0.2$) with biodiversity were log maximum lead, log maximum total phosphorus, log minimum PAHs, and log minimum DOC. Generally, the ability of any one water quality or contaminant variable to predict modeled biodiversity values was poor, accounting for only 1-22% of the observed variation (based upon r^2 values, Table 3) in Shannon-Weaver biodiversity.

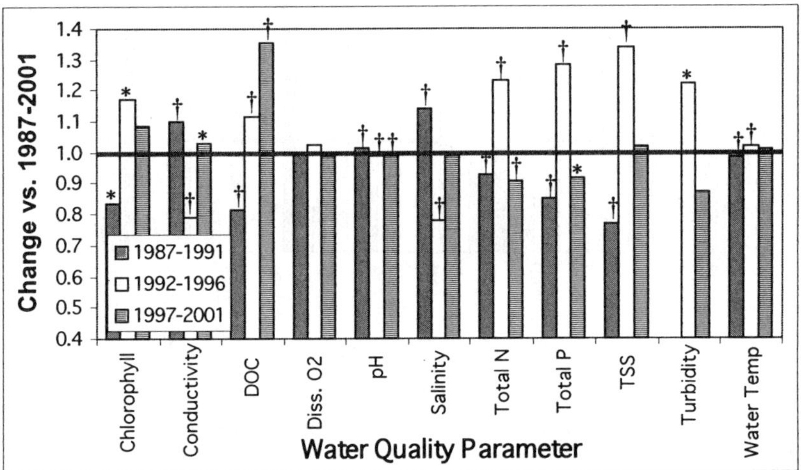

FIG. 7—*Comparison of mean benthic water quality conditions among three different time periods (1987-1991; 1992-1996; 1997-2001) relative to the 15-year mean (1987-2001). Temporal comparisons for each variable are expressed as the ratio of the mean concentration for each time period to the 15-year mean. Values greater than 1 indicate higher concentrations than the 15-year mean, and values less than 1 indicate lower concentrations. * and † indicate significant difference (p<0.05 and p<0.0001, respectively) from the 15-year mean by two sided t-test.*

As multiple potential stressors were significantly associated with benthic biodiversity, multiple regression and analysis of variance (α=0.05) were used to assess the extent to which patterns of biodiversity could be explained by the cumulative and/or net effects of multiple stressors, both anthropogenic and natural, acting in concert. Multiple regression among water quality variables alone yielded a maximum correlation coefficient of 0.72, suggesting combinations of these variables were able to predict biodiversity values better than any single variable. However, water quality variables alone could only account for approximately half of the observed variation in biodiversity among the 485 study sites (r^2=0.52), turbidity was excluded as a variable due to insufficient sample size, and water temperature was a nonsignificant variable. Meanwhile, multiple regression among sediment contaminant variables alone yielded a maximum correlation coefficient of 0.64, suggesting combinations of anthropogenic stressors are reasonable predictors of benthic biodiversity, accounting for approximately 40% of the observed variation in Shannon-Weaver biodiversity (r^2=0.40). The best-fit multiple regression model was obtained when both water quality and contaminant variables were considered (Table 4). However, a number of variables had to be excluded to maintain sufficient sample size for the analysis. Thus, the overall correlation obtained with both water quality and contaminant variables (r=0.78) was not substantially different than that obtained with just water quality variables and only accounted for 61% of the observed variation in biodiversity (r^2=0.61).

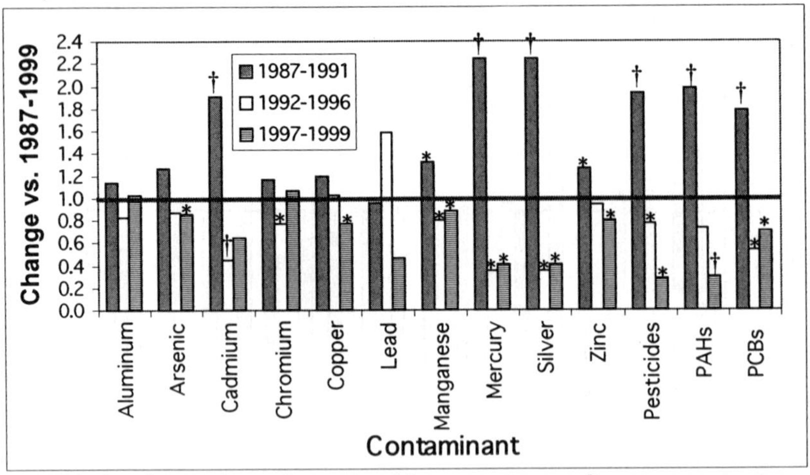

FIG. 8—*Comparison of mean benthic contaminant sediment concentrations among three different time periods (1987-1991; 1992-1996; 1997-1999) relative to the 13-year mean (1987-1999). Temporal comparisons for each contaminant are expressed as the ratio of the mean concentration for each time period to the 13-year mean. Values greater than 1 indicate higher concentrations than the 13-year mean, and values less than 1 indicate lower concentrations. * and † indicate significant difference ($p<0.05$ and $p<0.0001$, respectively) from the 13-year mean by two sided t-test.*

The independent variables used in the multiple regression models were also assessed for covariation that may have invalidated the regressions (data not shown). Generally, poor to moderate correlation among water quality variables was observed. In addition, water quality variables generally correlated poorly with contaminant variables, although a moderate degree of correlation was observed between salinity and conductivity and metals. Metals generally correlated well with other metals, although poor correlation was observed among metals and organic contaminants. The correlation among different classes of organic contaminants was generally poor.

Analysis also indicated that residuals from the multiple regressions were reasonably evenly distributed. A bias was observed among the residuals for both the water quality and contaminant regressions, with more positive residuals at more positive values of biodiversity. This trend was diminished, but remained upon inspection of the residuals from the multiple regression that incorporated both water quality and contaminant variables (Fig. 9).

Discussion

The various analyses in the current study collectively provide a strong indication of anthropogenic impairment of benthic habitat over extensive areas within Chesapeake

Bay, consistent with prior reports of adverse anthropogenic impacts (U.S. EPA 1983; Helz and Hugget 1987). However, it is clear that spatial patterns of biodiversity cannot be attributed to one, or even a small suite, of environmental factors. The aggregate analyses identified associations between a broad range of contaminants and spatial patterns of benthic biodiversity, indicating high spatial autocorrelation among stressors. Water quality variables behaved similarly, although attributing water quality differences to anthropogenic activities is more problematic. Correlation analyses among specific study sites indicated that single factors performed poorly at accounting for observed variability in biodiversity, whereas multiple regression among a broad range of water quality and contaminant variables was capable of accounting for the vast majority of observed variation in biodiversity. This indicates that benthic biodiversity in Chesapeake Bay is driven not by any particular stressor, but by a broad range of anthropogenic and

TABLE 3—*Results of Pearsons's product-moment correlation analysis between 13 contaminants and 11 water quality variables and Shannon-Weaver (SW) biodiversity among 485 study sites. n indicates the sample size. r represents the correlation coefficient. p represent the probability level from Fisher's test. Rank represents the rank correlation with biodiversity among the independent variables. Only those parameters for each variable that yielded the highest r value are reported.*

Dependent	Independent	n	r	r^2	p	Rank
Contaminants						
SW	Aluminum (Maximum)	258	-0.23	0.05	0.000	19
SW	Arsenic (Maximum)	310	-0.26	0.07	<0.0001	15
SW	Chromium (Mean)	312	-0.30	0.09	<0.0001	10
SW	Log Cadmium (Minimum)	227	-0.37	0.14	<0.0001	3
SW	Copper (Minimum)	315	-0.35	0.12	<0.0001	4
SW	Log Lead (Maximum)	299	-0.10	0.01	00.083	24
SW	Log Manganese (Mean)	268	-0.23	0.05	<0.0001	18
SW	Mercury (Mean)	316	-0.22	0.05	<0.0001	20
SW	Log Silver (Mean)	201	-0.29	0.08	<0.0001	13
SW	Zinc (Mean)	319	-0.29	0.08	<0.0001	12
Log SW	Total pesticides	331	-0.30	0.09	<0.0001	11
SW	Log PAHs (Minimum)	230	-0.16	0.03	0.013	22
SW	Log PCBs (Minimum)	45	-0.47	0.22	0.001	1
Water Quality Variables						
SW	Log Chlorophyll-a (Minimum)	342	0.41	0.17	<0.0001	2
SW	Conductivity (Minimum)	463	0.27	0.08	<0.0001	14
Log SW	Log Dissolved organic carbon (Minimum)	443	0.18	0.03	0.0002	21
SW	Dissolved oxygen (Mean)	463	0.33	0.11	<0.0001	6
SW	Salinity (Minimum)	463	0.30	0.09	<0.0001	9
SW	Log pH (Mean)	463	0.30	0.09	<0.0001	8
SW	Log Total nitrogen (Mean)	462	-0.31	0.10	<0.0001	7
SW	Log Total phosphorus (Maximum)	462	0.14	0.02	0.002	23
Log SW	Log Total suspended solids (Mean)	456	0.25	0.06	<0.0001	16
SW	Turbidity (Minimum)	129	-0.23	0.05	0.008	17
SW	Log Temperature (Mean)	469	0.35	0.12	<0.0001	5

natural environmental factors that interact to yield a net ecological effect. However, on a relative basis, it was still possible to prioritize some stressors as particularly problematic, such as cadmium, copper, and PCBs, which also have been previously identified as contaminants of concern in Chesapeake Bay (CBP 1991; Hall et al. 1998). Furthermore, given the spatial autocorrelation among stressors, several of the variables considered in the current study may still serve as useful indicators of anthropogenic disturbance to the benthic environment, even if they explain only a fraction of the observed spatial and temporal patterns of biodiversity.

TABLE 4—*Results of model II multiple regression among independent variables and Shannon-Weaver (SW) biodiversity. n represents the sample size. r represents the correlation coefficient. p represents the probability level from analysis of variance. Only best-fit results are reported.*

Dependent	Independent		n	r	p
	Water Quality Variables	*Contaminant Variables*			
SW	Chlorophyll-a (Maximum)	Arsenic (Mean)	126	0.78	<0.0001
	Conductivity (Minimum)	Chromium (Mean)			
	Dissolved oxygen (Minimum)	Copper (Mean)			
	Log Salinity (Minimum)	Manganese (Mean)			
		Zinc (Mean)			
		Pesticides (Total)			
		PAH (Minimum)			
		PCB (Minimum)			

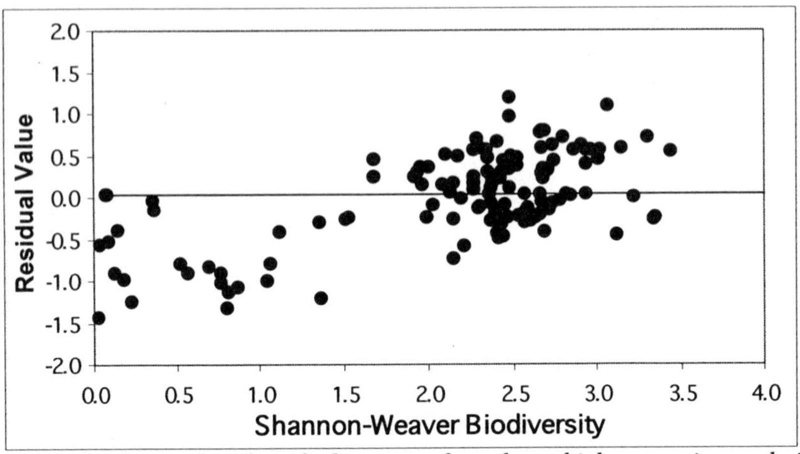

FIG. 9—*Scatterplot of residual statistics from the multiple regression analysis utilizing all water quality and contaminant variables (Table 4).*

Basin-wide models of benthic biodiversity identified significant temporal variability in the benthic community. The three biodiversity time-series models in the current study document a substantial contraction of those areas estimated to lie within the lower 20th percentile for Shannon's Index coincident with an expansion of those areas estimated to lie within the upper 20th percentile. Furthermore, analysis of annual biodiversity observations over the 15-year study period also indicated a significant positive trend for benthic biodiversity in the bay (Fig. 4). This general increase in biodiversity over the 15-year study period was coincident with a reduction in sediment contaminant concentrations in Chesapeake Bay, while the association of biodiversity with water quality variables was more ambiguous (Figs. 7 and 8). This suggests that reductions in sediment contaminant burdens may be the principle driver of observed temporal changes in biodiversity, but the lack of ecotoxicological data in the current study makes it difficult to definitively identify this reduction as a driver of ecological change in the bay.

The analyses performed in the current study are similar to those of Preston (2002c) and Preston and Shackelford (2002a), which examined spatial patterns in benthic biodiversity and multiple stressors over the time period 1984-1999, and yield similar results. However, the current study, which is shifted forward in time relative to the previous studies, identified a slightly larger area as low-impact zones (2015 vs. 1815 km^2) and a smaller area as high-impact zones (1228 vs. 1412 km^2). This provides further validation of the time-series models of biodiversity in the current study, which indicate a positive trend in biodiversity over time. In addition, the current study utilized similar contaminants but a broader range of water quality variables in multiple regression (compared to Preston and Shackelford (2002a)) and benefited from a larger number of study sites (485 vs. 353). Preston and Shackelford (2002a) used multiple regression among select contaminants and water quality variables to account for 73% of the observed variation in benthic biodiversity. Multiple regression among both water quality and contaminant variables in the current study accounted for approximately 61% of the observed variation in biodiversity compared to the 73% obtained by Preston and Shackelford (2002a), due to the current study's constraint on the ratio of independent variables to observations. Although the current regression model would appear to account for the majority of the observed variation in benthic biodiversity (at least among the 126 sites included in the analysis), caution must be used in interpreting this statistical result. Given the limitations of the data (e.g., spatial and temporal aggregation and small sample size for multiple regression) and the complexity of ecological systems, it is unlikely that the factors included in the current study truly capture all of the relevant interactions occurring in Chesapeake Bay. However, it does suggest that multivariate modeling is a potentially valuable tool for elucidating stressor-response relationships in natural systems (Fairbrother and Bennet 2000).

The current study also serves to demonstrate the utility, if not the necessity, of comprehensive environmental monitoring for the development of ecologically relevant conceptual as well as quantitative models of anthropogenic effects at the ecosystem level. The quality and volume of data available for the Chesapeake Bay are largely unmatched by other watersheds or ecosystems, which allows one to investigate stressor-response relationships within the bay without a priori assumptions about which factors/stressors are the most influential (Preston 2002c). In addition, the availability of field data on biotic responses can potentially be used to alleviate the infamous "lab-to-field dilemma" (dos Santos et al. 2002; Ringwood and Keppler 2002) whereby the use of laboratory data in the estimation of field responses introduces uncertainties into environmental assessments. However, the current study also demonstrates some of the challenges associated with the analysis and interpretation of environmental monitoring. The frequent use of ad hoc data collection in monitoring programs, whereby data are collected inconsistently through both space and time, creates challenges for both modeling and statistical analysis. Strategic planning of monitoring programs which focuses not only on the acquisition of data, but also on the manner in which the data can or will be used may enable more streamlined monitoring programs that are simultaneously more amenable to robust analysis.

Lastly, the high degree of spatial autocorrelation among multiple stressors indicated by the current study has important implications for risk assessment. Although risk assessment and management at the ecosystem level is increasingly advocated, the inherent complexity of ecological systems combined with the diversity of possible anthropogenic perturbations creates significant challenges for conducting ecologically relevant risk assessments. Not only must risk assessors be able to prioritize and estimate the net effects of multiple contaminants, they must also be able to distinguish such effects from those caused by natural spatial and temporal variability in physical/chemical conditions (Kennish 1990; Therriault and Kolasa 1999). Clearly there is a need for developing frameworks for integrated ecological risk assessment that can account for the individual and net effects of a diverse range of factors. One possibility for addressing these issues is to shift risk assessment approaches from strictly binary comparisons of contaminated and pristine systems to the analysis of ecological change along environmental gradients (Wickham et al. 2000), where ecosystem responses can be tested in response to variability in both natural and anthropogenic factors. The integration of such geographic approaches (including GIS and multivariate statistics) into traditional ecotoxicology and risk assessment may promise to yield risk assessments that are more representative of ecological complexity and more information-rich for improved environmental management (Johnson 1990).

References

Allan, J. D., Erickson, D. L., and Fay, J., 1997, "The Influence of Catchment Land Use on Stream Integrity Across Multiple Spatial Scales," *Freshwater Biology,* Vol. 37, pp. 149-161.

Barnthouse, L. W., Marmorek, D. R., and Peters, C. N., 2000, "Assessment of Multiple Stresses at Regional Scales," In *Multiple Stressors in Ecological Risk and Impact Assessment: Approaches to Risk Estimation*, S. A. Ferenc and J. A. Foran, Eds., SETAC Press, Pensacola, FL, pp. 1-26.

Boesch, D. F., 1996, "Science and Management in Four U.S. Coastal Ecosystems Dominated by Land-Ocean Interactions," *Journal of Coastal Conservation,* Vol. 2, pp. 102-114.

CBP (Chesapeake Bay Program), 1991, *Chesapeake Bay Toxics of Concern List*, U.S. Environmental Protection Agency, Washington, DC.

CBP (Chesapeake Bay Program), 1993, *Guide to Using Chesapeake Bay Program Water Quality Monitoring Data*, U.S. Environmental Protection Agency, Washington, DC.

CBP (Chesapeake Bay Program), 1999, *Targeting Toxics: A Characterization Report. A Tool for Directing Management and Monitoring Actions in the Chesapeake Bay's Tidal Rivers. A Technical Workplan*, U.S. Environmental Protection Agency, Washington, DC.

CBP (Chesapeake Bay Program), 2000, *The 2000 User's Guide To Chesapeake Bay Program Biological and Living Resources Monitoring Data*, U.S. Environmental Protection Agency, Washington, DC.

dos Santos, M. M., Moreno-Garrido, I., Goncalves, F., Soares, A. M. V. M., and Ribeiro, R., 2002, "An in situ Bioassay for Estuarine Environments Using the Microalga *Phaeodactylum tricornutum*," *Environmental Toxicology and Chemistry*, Vol. 21, pp. 567-574.

Dyer, S. D. and Wang, X., 2002, "A Comparison of Stream Biological Responses to Discharge from Wastewater Treatment Plants in High and Low Population Density Areas," *Environmental Toxicology and Chemistry* Vol. 21, pp. 1065-1075.

Dyer, S. D., White-Hull, C. E., Shephard, B. K., 2000, "Assessments of Chemical Mixtures via Toxicity Reference Values Overpredict Hazard to Ohio Fish Communities," *Environmental Science and Technology,* Vol. 34, pp. 2518-2524.

Fairbrother, A. and Bennett, R. S., 2000, "Multivariate Statistical Applications for Addressing Multiple Stresses in Ecological Risk Assessment," In *Multiple Stressors in Ecological Risk and Impact Assessment: Approaches to Risk Estimation*, S. A. Ferenc and J. A. Foran, Eds., SETAC Press: Pensacola, FL, pp. 67-116.

Fischer, M., Scholten, H. J., Unwin, D., 1996, *Spatial Analytical Perspectives on GIS*, Taylor and Francis, Washington, DC.

Folt, C. L., Chen, C. Y., Moore, M. V., and Burnaford, J., 1999, "Synergism and Antagonism among Multiple Stressors," *Limnology and Oceanography* Vol. 44, pp. 864-877.

Hall, L. W. Jr., Scott, M. C., and Killen, W. D., 1998, "Ecological Risk Assessment of Copper and Cadmium in Surface Waters of Chesapeake Bay Watershed," *Environmental Toxicology and Chemistry,* Vol. 17, pp. 1172-1189.

Harding, J. S., Benfield, E. F., Bolstad, P. V., Helfman, G. S., and Jones III, E. B. D., 1998, "Stream Biodiversity: The Ghost of Land Use Past, *Proceedings of the National Academy of Sciences,* Vol. 95, pp. 14843-14847.

Havens, K. E., 1999, "Correlation Is Not Causation: A Case Study of Fisheries, Trophic State and Acidity in Florida (USA) Lakes," *Environmental Pollution*, Vol. 106, pp. 1-4.

Helz, G. and Hugget, R. J., 1987, "Contaminants in Chesapeake Bay: The Regional Perspective." *In Contaminant Problems and Management of Living Chesapeake Bay Resources*, S.K. Majumdar, L.W. Hall, and H.M. Austin, Eds., Pennsylvania Academy of Sciences, Philadelphia, PA, pp. 270-297.

Hughes, R. M., Paulsen, S. G., and Stoddard, J. L., 2000, "EMAP Surface Waters: A Multiassemblage, Probability Survey of Ecological Integrity in the USA," *Hydrobiologia*, Vol. 422/423, pp. 429-443.

Johnson, L. B., 1990, "Analyzing Spatial and Temporal Phenomenon Using Geographical Information Systems," *Landscape Ecology* Vol. 4, pp. 31-43.

Kachigan, S. K., 1986, *Statistical Analysis*, Radius Press, New York, NY.

Kennish, M. J., 1990, *Ecology of Estuaries, Vol 2.*, CRC Press, Boca Raton, FL.

Kolpin, D. W., Furlong, E. T., Meyer, M. T., Thurman, E. M., Zaugg, S. D., Barber, L. B., and Buxton, H. T., 2002, "Pharmaceuticals, Hormones, and Other Organic Wastewater Contaminants in U.S. Streams, 1999-2000: A National Reconnaissance," *Environmental Science and Technology*, Vol. 36, pp. 1202-1211.

Lozano, R. B. and Pratt, J. R., 1994, "Interaction of Toxicants and Communities: The Role of Nutrients," *Environmental Toxicology and Chemistry* Vol. 13, pp. 361-368.

Moore, D. R. J. and Bartell, S. M., 2000, "Estimating Ecological Risks of Multiple Stressors: Advanced Methods and Difficult Issues," In *Multiple Stressors in Ecological Risk and Impact Assessment: Approaches to Risk Estimation*, S.A Ferenc and J.A. Foran, Eds., SETAC Press: Pensacola, FL, pp:117-168.

Newman, M. C., 1995, *Quantitative Methods in Aquatic Ecotoxicology*, CRC Press, Boca Raton, FL.

Paerl, H. W., 1997, "Coastal Eutrophication and Harmful Algal Blooms: Importance of Atmospheric Deposition and Groundwater as 'New' Nitrogen and Other Nutrient Sources," *Limnology and Oceanography,* Vol. 42, pp. 1154-1165.

Preston, B. L., 2002a, "Hazard Prioritization in Ecological Risk Assessment through Spatial Analysis of Toxicant Gradients," *Environmental Pollution,* Vol. 117, pp. 431-445.

Preston, B. L., 2002b, "Indirect Effects in Aquatic Ecotoxicology: Implications for Ecological Risk Assessment," *Environmental Management,* Vol. 29, pp. 311-323.

Preston, B. L., 2002c, "Spatial Patterns in Benthic Biodiversity of Chesapeake Bay (1984-1999): Association with Water Quality and Sediment Toxicity," *Environmental Toxicology and Chemistry*, Vol. 21, pp. 151-162.

Preston, B. L. and Shackelford, J., 2002a, "Multiple Stressor Effects on Benthic Biodiversity of Chesapeake Bay: Implications for Ecological Risk Assessment," *Ecotoxicology* Vol. 11, pp. 85-99.

Preston, B. L. and Shackelford, J., 2002b, "Risk-Based Analysis of Environmental Monitoring Data: Application to Heavy Metals in North Carolina Surface Waters," *Environmental Management*, Vol. 30 pp. 279-293.

Ringwood, A. H. and Keppler, C. J., 2002, "Comparative in situ and Laboratory Sediment Bioassays with Juvenile *Mercenaria mercenaria," Environmental Toxicology and Chemistry*," Vol. 21, pp. 1651-1657.

Sala, O. E., Chapin, F. S., III, Armesto, J. J., Berlow, E., Bloomfield, J., Dirzo, R., Huber-Sanwald, E., Huenneke, L. F., Jackson, R. B., Kinzig, A., Leemans, R., Lodge, D. M., Mooney, H. A., Oesterheld, M., Poff, N. L., Sykes, M., Walker, B. H., Walker, M., and Wall, D. H., 2000, "Global Biodiversity Scenarios for the Year 2100," *Science,* Vol. 287, pp. 1770-1774.

Sokal, R.R. and Rohlf, F.J., 1995, *Biometry*, W.H. Freeman and Company, New York, NY.

Therriault, T. W. and Kolasa, J., 1999, "Physical Determinants of Richness, Diversity, Evenness and Abundance in Natural and Aquatic Microcosms," *Hydrobiologia* Vol. 412, pp. 123-130.

Tilman, D., Fargione, J., Wolff, B., D'Antonio, C., Dobson, A., Howarth, R., Schindler, D., Schlesinger, W. H., Simberloff, D., and Swackhamer, D., 2001, "Forecasting Agriculturally Driven Global Environmental Change," *Science,* Vol. 292, pp. 281-284.

U.S. EPA (Environmental Protection Agency), 1983, *Chesapeake Bay: A Framework for Action*, U.S. Environmental Protection Agency, Chesapeake Bay Program Office, Annapolis, MD.

U.S. EPA (Environmental Protection Agency), 1998, *Guidelines for Ecological Risk Assessment*, *EPA/630/R-95/002F*, U.S. Environmental Protection Agency, Washington, DC.

Vitousek, P. M., Mooney, H. A., Lubchenco, J., and Melillo, J. M., 1997, "Human Domination of Earth's Ecosystems, *Science,* Vol. 25, Vol. 494-499.

Weisberg, S. B., Ranasighe, J. A., Dauer, D. M., Schaffner, L. C., Diaz, R. J., and Frithsen, J. B., 1997, "An Estuarine Benthic Index of Biotic Integrity for Chesapeake Bay," *Estuaries,* Vol. 20, pp. 149-158.

Wickham, J. D., O'Neill, R. V., and Jones, K. B., 2000, "A Geography of Ecosystem Vulnerability, *Landscape Ecology*, Vol. 15, pp. 495-504.

Appendix 1. Contaminants included in the current study.

Metals (n=10)	Pesticides (n=12)	PAHs (n=20)	PCBs (n=18)
Aluminum	Aldrin	Acenaphthene	PCB 8
Arsenic	Chlordane	Acenaphthylene	PCB 18
Cadmium	DDT (4,4')	Anthracene	PCB 28
Chromium	DDT (O,P)	Benzo(b+k)fluoranthene	PCB 44
Copper	Dieldrin	Benzo[a]pyrene	PCB 52
Lead	Endosulfan	Benzo[b]fluoranthene	PCB 66
Manganese	Endrin	Benzo[e]pyrene	PCB 101
Mercury	Heptachlor	Benzo[ghi]perylene	PCB 105
Silver	Heptachlor Epoxide	Benzo[k]fluoranthene	PCB 118
Zinc	Lindane	Benzo[a]anthracene	PCB 128
	Mirex	Chrysene	PCB 138
	Trans-Nonachlor	Dibenzo(a,h)anthracene	PCB 153
		Dibenzothiophene	PCB 170
		Fluoranthene	PCB 180
		Fluorene	PCB 187
		Indeno(1,2,3-CD)pyrene	PCB 195
		Naphthalene	PCB 206
		Perylene	PCB 209
		Phenanthrene	
		Pyrene	

Vladimir P. Reshetin[1]

Estimating Receptor Sensitivity to Spatial Proximity of Emissions Sources

REFERENCE: Reshetin V. P., "**Estimating Receptor Sensitivity to Spatial Proximity of Emissions Sources,**" ***Landscape Ecology and Wildlife Habitat Evaluation: Critical Information for Ecological Risk Assessment, Land-Use Management Activities, and Biodiversity Enhancement Practices, ASTM STP 1458,*** L.A. Kapustka, H. Galbraith, M. Luxon, and G. R. Biddinger, Eds., ASTM International, West Conshohocken, PA, 2004.

ABSTRACT: The term of the receptor sensitivity of a territory to siting accommodation of emission sources can be effectively calculated by means of a mathematical apparatus of conjugate problems. Solutions of conjugate equations make it possible to determine the effect of pollution in the form of aerosols or harmful gaseous impurity on the environment and human health. Examples of calculation of the sensitivity term by solving a conjugate problem are given. Mathematical models for typical problems of optimal siting of enterprises are formulated and interpretation of the results obtained is given.

KEYWORDS: receptor sensitivity, risk assessment, conjugate problems, emission

Introduction

Contemporary investigations indicate that atmospheric pollution contributes to morbidity and premature mortality (Reshetin and Arutyunyan 2002; Reshetin and Regens 2002; Dockery *et al.* 1993; Kunzli 2000; Arutyunyan et al. 2001; Reshetin et al. 2001). As an example, mention can be made of unique epidemiological investigations (Dockery et al. 1993) in which a coherent and statistically reliable relationship is established between contamination of the atmosphere by fine suspended particles of size less than 10 μm and mortality. Assessments of a number of attributable deaths indicate that about 6% of all the mortality cases in France, Austria, and Switzerland are due to the pollution of the atmosphere by fine particles of size less than 10 μm (Kunzli 2000). Taking into account a higher exposure level in Russia, the number of attributable deaths caused by

[1] Head of Laboratory, Joint Institute for Power and Nuclear Research (JIPNR) of National Academy of Sciences of Belarus, Minsk; E-mail: reshetin@sosny.bas-net.by

the atmospheric pollution can be much higher and attain 16 – 17 % of the total mortality number (Reshetin et al. 2001).

In this connection, the problem of population health risk reduction becomes especially urgent. One of the ways of its solution is optimization of the sitting of new industrial enterprises and complexes. Of no less importance is also the problem of an optimal decrease in the level of emission of harmful chemicals on pre-existing industrial enterprises. In view of the fact that labor resources have been distributed nonuniformly, these enterprises are usually sited in densely populated regions or in close vicinity of them. The benefit for health and environment derived from reduction of the emission level will depend substantially on the location of emissions sources.

The problem of optimal siting of new industrial enterprises and effective reduction of emissions on pre-existing ones can be solved by calculating the receptor sensitivity functional (Reshetin and Arutyunyan 2002). The functional of the receptor sensitivity of a territory to siting an emission object makes it possible to quantitatively assess a change in the population health risk depending on the location of an emission source. New possibilities of calculating the receptor sensitivity are afforded by the mathematical apparatus of conjugate problems (Marchuk 1992). In the present work, the results obtained in (Reshetin and Arutyunyan 2002; Marchuk 1992) were applied to investigation of the problem of optimal siting of enterprises and mathematical models of the most typical situations are considered.

Statement of the Problem

Suppose it is required to site a new industrial enterprise near populated localities or directly on the territory of a large populated area so that the population health risk over the entire region Σ_0 due to the pollution of the latter is minimal or not higher than certain permissible standards. Let us assume that an aerosol source *f(r)* is located at the point $\hat{\mathbf{r}}_\mathbf{0} = (x_0, y_0, z_0)$ and its intensity is Q:

$$f(\hat{\mathbf{r}}) = Q\delta(\hat{\mathbf{r}} - \hat{\mathbf{r}}_\mathbf{0}) \tag{1}$$

Under the action of the wind, the impurity is transferred by air masses and diffuses as affected by low-scale convention. In the simplest statement, impurity transfer in the atmosphere can be described by the equation:

$$\frac{\partial C}{\partial t} + \operatorname{div}(\overset{\times}{\mathbf{U}} C) + \lambda C - K_x \frac{\partial^2 C}{\partial x^2} - K_y \frac{\partial^2 C}{\partial y^2} - \frac{\partial}{\partial z} K_z \frac{\partial C}{\partial z} = Q\delta(\overset{\times}{\mathbf{r}} - \overset{\times}{\mathbf{r}}_0) \tag{2}$$

where C is the impurity concentration in the atmosphere; $\hat{\mathbf{U}}$ is the wind velocity; K_x, K_y, and K_z are the coefficients of diffusion in the direction of the x, y, and z axes; λ is a constant determining the decomposition of the impurity with time; δ(r) - Dirac delta function.

The solution of the problem will be determined in the cylindrical region, in which the boundary conditions taken are

$$C\big|_{\Sigma} = 0, \qquad \left.\frac{\partial C}{\partial z}\right|_{\Sigma_0} = \alpha C, \qquad \left.\frac{\partial C}{\partial z}\right|_{\Sigma_H} = 0 \tag{3}$$

where Σ is the lateral cylindrical surface; Σ_0 is the section of the cylindrical surface at the level z = 0; Σ_H is the section of the cylindrical surface at the level z = H; α - constant determining the interaction between the impurity and the ground surface.

We also assume that function C is periodic, with the period T

$$C(\hat{\mathbf{r}}, T) = C(\hat{\mathbf{r}}, 0) \tag{4}$$

To assess the health risk for the residents of the region Σ_0, the ground-level aerosol concentration is multiplied by the population density $P(r)$ and the resulting function is integrated over the area of the region and the time period T

$$F = a\int_0^T dt \int_{\Sigma_0} P\, C\, d\Sigma \tag{5}$$

Here $a = b/T$; the constant b reflects the dose - response relationship. The numerical values of this constant were found, e.g., in Dockery et al. (1993). Correct to within a multiplier, equation (5) determines the collective exposure, averaged over the time period T that will affect the population of the region due to the emission of aerosols by source (1). The value of term (5) at a given location of the emission source represents assessment of the effect exerted by the aerosol source on the population. Moreover, if, to obtain this assessment, the dose - response function established earlier in epidemiological investigations (Dockery et al. 1993) is used, it determines the relationship between exposure and premature mortality, whereas term (5) represents the assessment of the number of premature deaths caused by the atmospheric pollution. Carrying out calculations of term (5) for the emission sources located at different points of the region, it is possible to assess in which way the number of premature deaths caused by atmospheric pollution can change depending on the location of the source. In work (Reshetin and Arutyunyan 2002) term (5) is called the receptor sensitivity of the territory to the siting of emissions sources.

We note that at the existing level of exposure for many harmful effects to health and an environment, the relationship between exposure and response is linear. Thus, the distribution of the term (5) over the territory of the region represents the assessment of the receptor sensitivity of the territory to siting of an emission source (Reshetin and Arutyunyan 2002). At the given intensity and location of the emission source, the value of term (5) depends on the wind rose typical of this locality, lay of the ground, and the special features of distribution of population over the territory of the region. Due to the dose - response function linearity, the effect exerted by several emission sources on the population and environment is an additive quantity. Thus, the receptor sensitivity of the territory is independent of the location of the pre-existing emission sources in the region. The distribution of the receptor sensitivity function over the region makes it possible, in particular, to analyze the extent to which a decrease in emission at a certain industrial enterprise will be efficient from the viewpoint of risk reduction; moreover, at the same decrease in the level of emission the risk reduction will be the greatest for the enterprises which are located on the territory with a high value of the receptor sensitivity term (Reshetin and Arutyunyan 2002).

The term similar to that used above (5), can be introduced to assess an environment risk. However, depending on the priorities selected, as the function *P(r)* one should select the distribution of these or other parameters significant for assessing the environment risk. Since, in assessing the environment risk, the effects are usually considered at the level of population of community or of an ecosystem, the dynamics of the population, the structure of the community, and the processes occurring in the ecosystem are those end points on which the risk assessment is usually concentrated. In the absence of universal environmental assessment of end points, the risk assessment and the calculation based on the territory receptor sensitivity must rather be restricted by a particular situation. In determining the function *P(r)*, those resources must be considered in the first place which potentially are exposed to emission products. In identifying the end points of risk assessment and determining the function *P(r)* conceptual models, environmental effects, and other factors must be analyzed. Trough the focus of the work reported here was on human health, the approach developed in this study can be extended readily to ecological receptors to provide better estimates of exposure to airborne discharges across complex landscapes.

With the main term of the problem being selected in the form of (5), the problem conjugate to the principal one is formulated as follows:

$$-\frac{\partial C^*}{\partial t} - \mathrm{div}(\overset{\times}{U} C^*) + \lambda C^* - K_x \frac{\partial^2 C^*}{\partial x^2} - K_y \frac{\partial^2 C^*}{\partial y^2} - \frac{\partial}{\partial z} K_z \frac{\partial C^*}{\partial z} = P(r)\delta(z)$$

$$C^*\Big|_{\Sigma} = 0, \qquad \frac{\partial C^*}{\partial z}\Big|_{\Sigma_0} = \alpha C^*, \qquad \frac{\partial C^*}{\partial z}\Big|_{\Sigma_H} = 0 \tag{6}$$

$$C^*(r,T) = C^*(r,0)$$

By virtue of the fact that the problems are conjugate, equation (5) may be written in the following form (double representation of the term (Marchuk 1992)):

$$F = aQ\int_0^T C^*(r_0,t)\,dt \tag{7}$$

Term (7) depends parametrically on the location of the source of aerosols. When $z = 0$, solution of a conjugate problem determines the time dependence of the collective exposure of the population of the region C^* on the location of the emission source of unit intensity. Thus, the attractive side of the solution of a conjugate problem becomes evident: its solution makes it possible to determine collective exposure, whereas, to calculate the receptor sensitivity of the territory of the region, it is necessary simply to average this exposure for a certain interval of time and multiply by the coefficient which is determined by the dose – effect relation. Unlike the principal problem, where, to calculate term (5), it is required to find the distribution of the aerosol concentration for each location of the emission source, in a conjugate problem term (5) can be calculated by performing only one variant of calculation.

In some cases term (5), which apart from a factor, is equal to the exposure average for the period T, can be calculated as superposition of stationary solutions of conjugate problem (6):

$$F = a\,Q\sum_{i=1}^{n} C_i^{*}\Delta t_i\,,\ \text{where}\quad \sum_{i=1}^{n}\Delta t_i = T \tag{8}$$

where Δt_i is the time of a stable regime of air masses.

Stationary solutions of the conjugate problem can be used to average the exposure C^* over the wind directions with allowance for the wind rose in the region. The averaging over stationary solutions allows one to calculate, with a sufficient accuracy, the receptor sensitivity for a situation in which the contribution of transient processes is insignificant.

Mathematical Models of Typical Situations

As the simplest example we consider the following problem. Suppose it is required to site an industrial enterprise emitting harmful aerosols into the atmosphere, between two populated localities A_1 and A_2 with the number of residents P_1 and P_2; A_1 and A_2 are located at the points with coordinates (0,0,0) and (L,0,0). We assume for simplicity that in the time Δt_1 the wind blows with the velocity U_1 in the direction of the locality A_1 and in the time Δt_2 it blows with the velocity U_2 in the direction of the locality A_2, with $T = \Delta t_1 + \Delta t_2$. It is required to determine the distance at which the enterprise is to be sited from the populated localities that the health risk to the population could be minimal. For this case, a stationary solution of the conjugate problem has the form:

$$C^{*} = \frac{P_1\,\Theta(x_0)}{4\pi(K_y K_z)^{1/2} r_1}\exp\left[-\frac{U_1}{4r_1}\left(\frac{y_0^2}{K_y}-\frac{z_0^2}{K_z}\right)\right] + \frac{P_2\,\Theta(L-x_0)}{4\pi(K_y K_z)^{1/2} r_2}\exp\left[-\frac{U_2}{4r_2}\left(\frac{y_0^2}{K_y}-\frac{z_0^2}{K_z}\right)\right] \tag{9}$$

where $r_1 = (x_0^2 + y_0^2 + z_0^2)^{1/2}$ and $r_2 = ((L-x_0)^2 + y_0^2 + z_0^2)^{1/2}$, $\Theta(x)$ –Heavyside step function. The distribution of the receptor sensitivity term F along the straight line which connects the populated localities is described by the following expression:

$$F\,(x_0,0) = \frac{a\,Q\,P_1\,\Theta(x_0)}{4\pi(K_y K_z)^{1/2} x_0}\frac{\Delta t_1}{T} + \frac{a\,Q\,P_2\,\Theta(L-x_0)}{4\pi(K_y K_z)^{1/2}(L-x_0)}\frac{\Delta t_2}{T} \tag{10}$$

By virtue of the linearity of the problem, the receptor sensitivity F represents, apart from a factor, superposition of collective exposure for each populated locality. The contribution of each locality to the collective exposure is inversely proportional to the distance to the emission source. Each term in (10) can be interpreted as the risk potential for the given populated locality, with the receptor sensitivity of the territory representing a sum of the risk potentials for all the populated localities. The minimum value of the term F is attained when the ratio of the distances from the emission source to the populated localities is equal to

$$\frac{x_1}{x_2} = \left(\frac{P_1 \Delta t_1}{P_2 \Delta t_2}\right)^{1/2} \tag{11}$$

For example, when $\Delta t_1 = \Delta t_2$ and the ratio between the residents of the localities is $P_1/P_2 = 9$, the least risk to health is when the emission source is located at the distance $L/4$ from the point A_1. The distribution of the receptor sensitivity term for the example considered is given in Fig. 1.

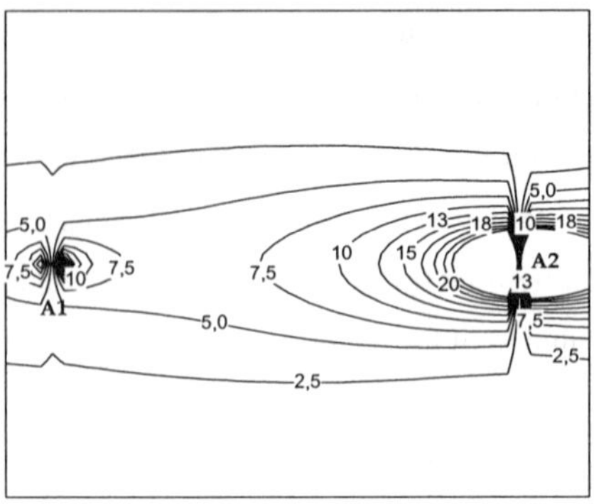

Figure 1 - *Distribution of receptor sensitivity term (relative units) between two populated localities. Computational domain is 1x1 km.*
$P_2/P_1=10$, $K_Y=10^6$ sm^2/s, $K_Z= 10^5$ sm^2/s, $U_1=3$ m/s, $U_2=5$ m/s, $\Delta t_1/\Delta t_2=2/3$.

When, in the area under study, one other populated locality A_3 appears, the distribution of the receptor sensitivity already reflects the contribution of three risk potentials (see Fig. 2), each of which can be calculated in the same way as it was done in the problem with two populated localities.

When the speed of the wind is arbitrarily distributed over the directions, then, to calculate the sensitivity equation in relation (7), it is necessary to sum collective exposure for each direction of the wind, taking into account the data on the wind rose in the region considered. With the wind being directed at the angle φ to the abscissa axis, the following transformation of the coordinates makes it possible to go over into the coordinate system (x', y', z), where the speed of the wind is directed, as before, along the foregoing expressions for the risk potential (see equation (10)):

$$x = x' \cos \varphi - y' \sin \varphi , \quad y = x' \sin \varphi + y' \cos \varphi$$

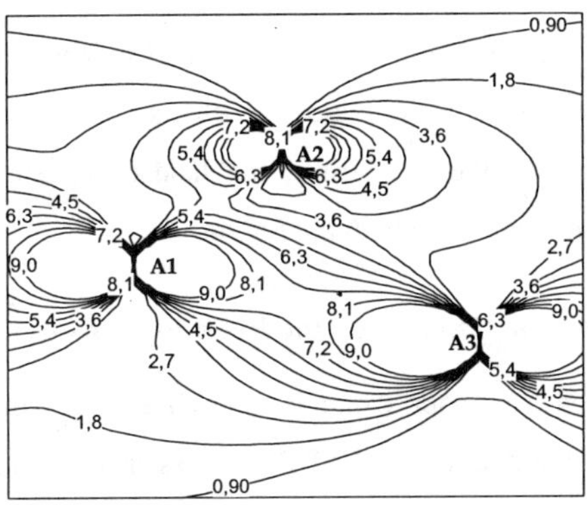

Figure 2 - *Distribution of receptor sensitivity term (relative units) in the region with three populated localities. The term was obtained by using semiempirical model ASME for neutral atmosphere (IAEA 1986).*
Computational domain is 1x1 km. P_{A1}: P_{A2}: P_{A3} = 8:5:10,
$K_Y=10^6$ sm^2/s, $K_Z= 10^5$ sm^2/s, $U_1=3$ m/s, $U_2=5$ m/s, $\Delta t_1 / \Delta t_2 = 2/3$, $h = 0$.

Note should be taken of the fact that in the present work the simplest model of the transfer of aerosols in the atmosphere is used. This model disregards the dependence of the coefficients of diffusion on the height of the emission source and distances from it, and also the effect of plume rise due to the initial momentum and the buoyancy force. The influence of these effects can be investigated with the use of the well-known computer codes intended for describing impurity transfer in the atmosphere. In particular, such investigation could be made with the use of the "Nostradamus" code (Arutyunyan *et al.* 1995) which is based on the present-day understanding of the boundary-layer structure and which makes it possible to describe impurity transfer in the given wind field using the data of regional meteorological stations. Due to the similarity of the equations for the direct and conjugate problems (2), (3) and (6), the tuning of computer programs to solving a conjugate problem is quite an easy task. In the present work, we assumed for simplicity that the populated localities are point objects. Even though this assumption is of help in a number of cases, the problem can easily be generalized also for areal objects. When a populated locality cannot be considered as a point object, equations (7) and (8) result from integration over all the residential territories of the region.

It is important that the receptor sensitivity term is independent of the location of the pre-existing emission sources in the region. A change in the value of this term over the territory is determined by the wind rose, the landscape, and special features in the distribution of the population density in the region.

Semiempirical Models for Calculating the Receptor Sensitivity Term

The receptor sensitivity term can be calculated with the use of semiempirical models, in which standard deviations of the Gaussian distributions are parametrically dependent on *x*, *y*, and *z*. Thus, for $\hat{\mathbf{U}} = (U,0,0)$ and a continuous source of intensity Q located at the point $r = r_0$, the distribution of concentration can be represented as

$$C(x,y,z,x_0,y_0,z_0) = \frac{Q\,\Theta(x-x_0)}{2\pi U\sigma_x(x-x_0)\sigma_y(x-x_0)}\exp\left(-\frac{(y-y_0)^2}{2(\sigma_y(x-x_0))^2}\right)$$
$$\times\left[\exp\left(-\frac{(z-z_0)^2}{2\sigma_z(x-x_0))^2}\right)-\exp\left(-\frac{(z-z_0)^2}{2(\sigma_z(x-x_0))^2}\right)\right]$$

where for the standard deviations σ_x, σ_y, and σ_z the following empirical parametric dependences on a distance are usually used:

$$\sigma_x(x) = \sigma_y(x) = q\,x^{\nu}, \qquad \sigma_z(x) = s\,x^{\gamma} \tag{12}$$

The values of the coefficients q, s, γ, and ν for different states of the atmosphere are presented in (McElroy 1969; Vogt *et al.* 1974; Pasquill 1974). In the foregoing problem with two populated localities, the semiempirical term of the receptor sensitivity has the following form:

$$F(x_0,y_0,z_0)=\frac{aP_1Q\,\Theta(x_0)}{\pi U_1\sigma_x(x_0)_x\sigma_y(x_0)}\frac{\Delta t_1}{T}\exp\left(-\frac{y_0^{\;2}}{2(\sigma_y(x_0))^2}-\frac{z_0^{\;2}}{2(\sigma_z(x_0))^2}\right)$$

$$+\frac{aP_2Q\,\Theta(L-x_0)}{\pi U_2\sigma_x(L-x_0)\sigma_y(L-x_0)}\frac{\Delta t_2}{T}\exp\left(-\frac{y_0^{\;2}}{2(\sigma_y(L-x_0))^2}-\frac{z_0^{\;2}}{2(\sigma_z(L-x_0))^2}\right) \tag{13}$$

When an emission source acting near the earth surface is to be sited on the straight line which connects the populated localities A_1 and A_2, the minimum value of the functional F is attained at the point

$$\frac{x_1}{x_2}=\left(\frac{P_1U_1\Delta t_1}{P_2U_2\Delta t_2}\right)^{1/(\nu+\gamma-1)} \tag{14}$$

where x_1/x_2 is the ratio between the distances from the populated localities to the source.

An example of calculation of the receptor sensitivity term for the territory of the city of Minsk with the use of the "Nostradamus" code is presented in Fig. 3. The data (Gidromet 1987) on the wind rose averaged over the period of many years of observation were used. As expected, different regions of Minsk are not equivalent from the viewpoint of their sensitivity to the emission sources. It is especially interesting that the sensitivity term differs for different regions of the city by more than an order of magnitude. This means that if there are two identical emission objects, the impaired health of the residents may differ more than tenfold depending on the location of a source. For a source located near the ground the value of the receptor sensitivity term at the given point on the territory of Minsk depends substantially on the number of people living in the region of size σ_y^{-1}, where the concentration of aerosols decreases due to convective diffusion. The higher the elevation of the emission source, considerably greater is the contribution of more distant territories. In this case, the asymmetry in the directions and strengths of winds substantially influences the distribution of the term (5).

The distribution of the equation over the territory of one region of the Minsk district is presented in Fig. 4. The characteristic distribution of the populated localities along the roads is manifested in higher values of the receptor sensitivity term in the singled-out directions. From the viewpoint of the population health risk, this layout of the roads inflicts the greatest damage to health.

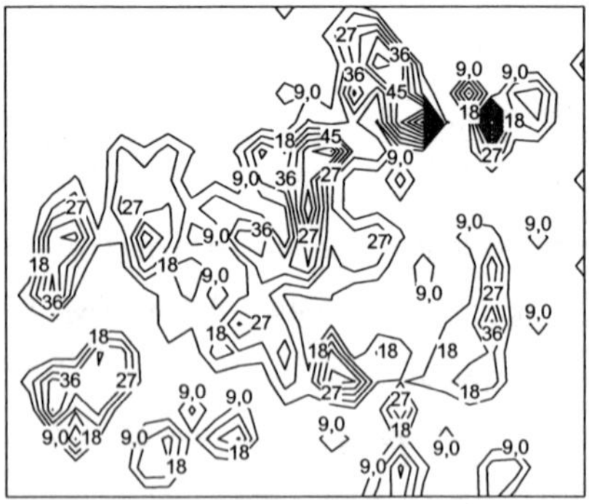

(a)

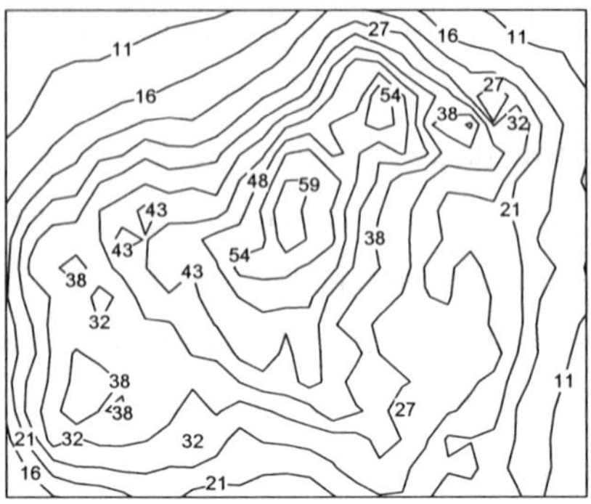

(b)

Figure 3 - *Population density distribution, thousand people/km^2 (a) receptor sensitivity functional (b) (relative units) for Minsk. Computational domain is 21 x 16 km, h=0 m.*

Just as for Minsk, here the receptor sensitivity functional is also distributed non-uniformly, Fig. 4, with the maximum and minimum values differing more than tenfold.

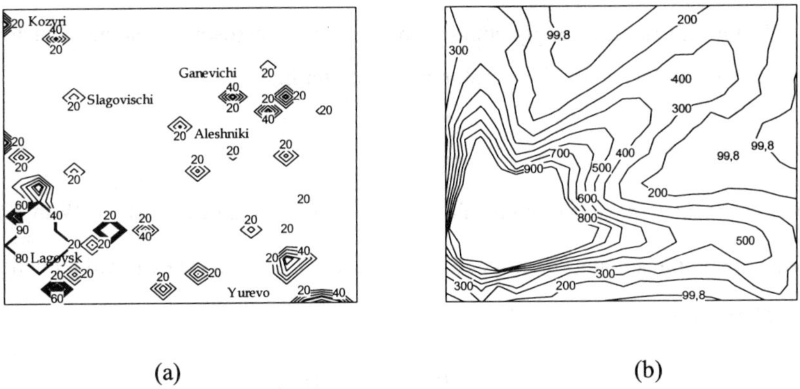

Figure 4 - *Population density distribution, people/km^2 (a) receptor sensitivity term (b) (relative units) for Logoisk region of Minsk district. Computational domain is 20 x 20 km, h=50 m.*

Effective Reduction of Emission at Pre-existing Enterprises

Attention in the previous sections was mainly paid to siting new industrial objects emitting deleterious aerosols into the atmosphere, with allowance for the minimum risk for the population health and environment. It is of interest to consider another aspect of the problem. We will assume that all industrial enterprises are already available and that they emit a certain amount of harmful aerosols into the atmosphere. The task is to determine such an amount of emitted aerosols for each enterprise that their total health risk for people could not exceed a certain established maximum value. It is evident that the total emissions must not be reduced substantially, since this may entail reduction in the economic performances of operating industrial objects. Thus, such emission constraints should be found which will eventually ensure maximum of the economic efficiency at the imposed restrictions.

Let in a given region G at the points $\hat{\mathbf{r}}_{i0}$ $(i = 1, 2, \ldots n)$ n industrial objects are located which emit $Q_i^{(0)}$ aerosols per unit time into the atmosphere; we consider the composition of these aerosols identical.

The health risk caused by emission of deleterious aerosols to the atmosphere can be evaluated with the aid of the receptor sensitivity term:

$$R = \sum_{i=1}^{n} Q_i^{(0)} F(\check{\mathbf{r}}_{i0}) \tag{15}$$

provided that term $F(\hat{\mathbf{r}})$ is calculated for a source of unit intensity. With a planned reduction in the amount of emissions to the level $Q_i \leq Q_i^{(0)}$, the expected reduction in the health risk is expressed:

$$\Delta R = \sum_{i=1}^{n} (Q_i - Q_i^{(0)}) F(\check{\mathbf{r}}_{i0}) \tag{16}$$

Naturally, the present problem should involve the term which will make it possible to evaluate the economic expenditures connected with technological reconstructions of the enterprises which will maintain industrial output at reduced emissions:

$$I = \sum_{i=1}^{n} \zeta_i (Q_i - Q_i^{(0)}) \tag{17}$$

where ζ_i defines capital investment to the technology which ensures the same industrial output on reduction in the emissions (per unit rate of emission). The term I represents the total expenditures needed to improve the technologies of all n enterprises.

The problem is reduced to finding such amounts of emissions Q_i, at which the term I could take a minimum value provided the following conditions are satisfied:

$$I = \sum_{i=1}^{n} \zeta_i (Q_i - Q_i^{(0)}) - \min$$

$$\Delta R \geq \Delta R_0 \tag{18}$$

$$0 \leq (Q_i - Q_i^{(0)}) \leq \Delta Q_i^{(\max)}$$

The above inequality represents the restrictions that may appear in planning reduction in the emissions due, e.g. to the technical feasibility of an appreciable reduction in the emissions; ΔR_0 is the planned decrease in risk.

The problem concerning minimization of the term I can be solved by the methods of linear programming. In the space R^n of inequality (18), a region of admissible values of Q_i is established, and grad($I(Q_i)$) assigns the direction of the most rapid increase in the term I. In the simplest case, where there are two industrial objects, the problem on the minimization of the term can be solved graphically (see Fig. 5). As is seen from Fig. 5, for the optimum solution there always corresponds one of the angular points of the space of solutions. In the general case of n industrial objects, the problem can be solved by the simplex - method. This method is not as pictorial as the geometric one. In its computation procedure, the computational process is implemented, in which, beginning from a certain initial admissible angular point; there occur successive transitions from one admissible angular point to another until the point corresponding to the optimum solution is found.

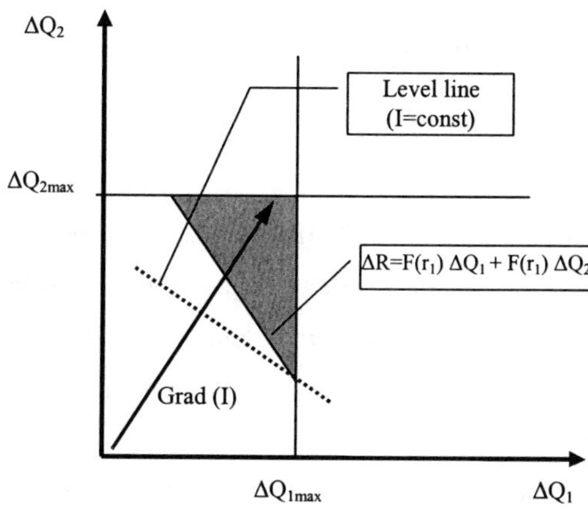

Figure 5 - *Graphical solution of the problem of the expenditure minimization in the case of reduction of emission level at two industrial objects. Optimal solution is the lower apex of the triangle.*

Conclusions

1. The term of receptor sensitivity of the territory to siting of an emission source can be calculated efficiently with the aid of the mathematical apparatus of conjugate problems.

2. To reduce the risk to health and environment, it may well be that for each region a program can be composed for siting of industrial enterprises that release harmful aerosols and gases into the atmosphere. For each region, with allowance for climatic conditions, fields of winds, and specific features of the terrain, maps could be prepared which would reflect the distribution of the receptor sensitivity over their territory. This work should be done in the first place when planning the building of objects in economic development regions, where decisions rational from the viewpoint of protecting the environment and population could be made.

3. Mapping of the territory of the region by the magnitude of the receptor sensitivity would also be useful in drawing a plan of measures intended for decreasing the emission by industrial enterprises and vehicles. The distribution of the receptor sensitivity term over the region would make it possible to assess the efficiency of the measures suggested from the viewpoint of health risk reduction.

The authors are grateful to E. T. Batalova and V. I. Kazazyan for the help in preparing this article.

References

Arutyunyan R. V., Belikov V. V., Reshetin V. P., et al., 2001, "Evaluation of the Effect of the Ryazan' Electric Power Plant on the Environment and Human Health," *Seventh International Conference "Harmonisation within atmospheric dispersion modeling for regulatory purposes,"* Belgirate, Italy, 28-31 May 2001, Ispra, Italy, pp. 163-170.

Arutyunyan R. V., Belikov V. V., Belikova G. V., et al., 1995, "Computer Code "Nostradamus" for Supporting Decision- Making in Emergency Release at Radiation Hazard Objects," *Izvestiya Akademii Nauk, Energetika*, No. 4, pp. 23-29.

Dockery D., Pope A., Xu X., et al., 1993, "An association between air pollution and mortality in six U.S. cities," *New England Journal of Medicine*, Vol. 329, pp. 1753-1759.

International Atomic Energy Agency, 1986, Atmospheric Dispersion Models for Application in Relation to Radionuclide Releases. *IAEA-TECDOC-379*, Vienna, IAEA.

Kunzli N., Kaiser R., Medina S., et al., 2000, "Public-Health Impact of Outdoor Air Pollution and Traffic Related Air Pollution: a European Assessment," *Lancet*, Vol. 356, pp. 795-801.

Marchuk G.I., 1992, *Conjugate Equations and Analysis of Complex Systems*, Moscow, pp.1-336.

McElroy J. L., 1969, "A Comparative Study of Urban and Rural Dispersion," *Journal of Applied Meteorology*, Vol. 8, pp. 12-19.

Pasquill F., 1974, *Atmospheric Diffusion*. NY, John Wiley and Sons.

Reshetin V. P. and R.V. Arutyunyan, 2002, "Receptor Sensitivity to Special Location of Emissions Sources," *Izvestiya Akademii Nauk, Energetika* , No. 3, pp.234-239.

Reshetin V. P. and Regens L., 2002, "Modeling Mobile Source Emissions During Traffic Jams in Micro Urban Environment," *Environmental Science and Pollution Research*, Vol. 9, No. 5, pp. 333-336.

Reshetin V. P., Lantsov I. E., Khatkovski D. V., et al., 2000, "Sustainable Urban Development Problem: Environmental Risk Assessment as a Tool for Decision-Making," *Final report on ISTC project* № B320d, Minsk.

Reshetin V. P, Arutyunyan R.V., Sorokin V. S., 2001, "How many attributable cases of deaths associated with air pollution occur in Russia?" *Preprint IBRAE* I-2001-17, p.1-14.

Scientific and Applied Handbook on Climate of USSR, 1987, Ser. 3, Multi-year Data, Parts 1-6, issue 7, BSSR (in Russian), Gidrometeoizdat, Leningrad.

Vogt K. J. and Geiss H., 1974, Tracer Experiments on Dispersion of Plumes over Terrain of Major Surface Roughness. Kernforschungsanlage Julich, FRG, *Report No.* Jul–1131-ST.

Session III

Rebecca A. Efroymson,[1] *Tina M. Carlsen,*[2] *Henriette I. Jager,*[1] *Tanya Kostova,*[3] Eric *A. Carr,*[4] *William W. Hargrove,*[1] *James Kercher,*[5] and *Tom L. Ashwood*[1]

Toward an Ecological Framework for Assessing Risk to Vertebrate Populations from Brine and Petroleum Spills at Exploration and Production Sites

REFERENCE: Efroymson, R. A., Carlsen, T. M., Jager, H. I., Kostova, T., Carr, E. A., Hargrove, W. W., Kercher, J., and Ashwood, T. L., **"Toward an Ecological Framework for Assessing Risk to Vertebrate Populations from Brine and Petroleum Spills at Exploration and Production Sites,"** *Landscape Ecology and Wildlife Habitat Evaluation: Critical Information for Ecological Risk Assessment, Land-Use Management Activities, and Biodiversity Enhancement Practices, ASTM STP 1458,* L. A. Kapustka, H. Galbraith, M. Luxon, and G. R. Biddinger, Eds., ASTM International, West Conshohocken, PA, 2004.

ABSTRACT: Brine and petroleum spills may affect terrestrial vertebrates through loss of reproductive habitat or reduced food availability rather than direct toxicity. A proposed ecological framework for evaluating impacts of these spills includes individual-based population models, a site conceptual trophic model, habitat suitability maps, and a stochastic brine spill generator. Simulation results for mammal populations in the Tallgrass Prairie Preserve petroleum exploration and production (E&P) site in Oklahoma are presented. The persistence of simulated American badger (*Taxidea taxus*) populations decreased with increasing brine spill area. The decline in persistence and average final population size was much steeper in highly fragmented landscapes. The simulated time to extinction for prairie vole (*Microtus ochrogaster*) populations showed a threshold at 30% habitat loss from spills; above this threshold the time to extinction decreased with increasing spill area. Vole density was sensitive to the interaction of predation and fragmentation, with fragmentation causing population extinction in the presence of predation, yet stabilizing the population in the absence of predation. We anticipate that our results will aid in future development of "exclusion criteria" for leaving unrestored habitat at E&P sites.

---

[1] Research staff member, Environmental Sciences Division, Oak Ridge National Laboratory, P. O. Box 2008, Oak Ridge, TN 37831-6036

[2] Environmental scientist, Environmental Restoration Division, Lawrence Livermore National Laboratory, 7000 East Avenue, P. O. Box 808 L-528, Livermore, CA 94551

[3] Computational mathematician, Center for Applied Scientific Computing, Lawrence Livermore National Laboratory, 7000 East Avenue, P. O. Box 808 L-561, Livermore, CA 94551

[4] Mathematician, Tennessee Institute of Ecological Modeling, University of Tennessee, Knoxville, TN 37996, formerly of Oak Ridge National Laboratory

[5] Physicist, Environmental Sciences Division, Lawrence Livermore National Laboratory, 7000 East Avenue, P. O. Box 808 L-235, Livermore, CA 94551

KEYWORDS: individual-based model, population, habitat suitability, American badger, prairie vole

Introduction

Petroleum exploration and production (E&P) sites are often located in rural areas with diverse populations of mammals and birds. Terrestrial vertebrates may be exposed to hydrocarbons from petroleum spills or salts from brine spills; however, the more important exposures may be indirect, i.e., the removal of habitat and forage vegetation. Therefore, researchers from Lawrence Livermore National Laboratory and Oak Ridge National Laboratory have investigated the role of disturbance patches on vertebrates at the Tallgrass Prairie Preserve (TPP) in Osage County, Oklahoma, an E&P site (reports at http://gis.llnl.gov/mei/). This research has two, long-term goals: (1) to develop an ecological framework for evaluating impacts of brine and/or oil spills at E&P sites, utilizing population models based on patchiness of landscapes and, in some cases, trophic transfer; and (2) to develop thresholds (if possible) based on size and distribution of spills that would result in *de minimis* impacts on wildlife populations. These "exclusion criteria" could be applied to exclude certain well or spill locations from formal ecological risk assessment. In addition, this ecological approach could be used to inform (1) restoration priorities and strategies for companies that may be undertaken prior to exiting a site, or (2) siting and construction of drilling and road locations and associated E&P infrastructure in newly accessed areas.

An ecological approach may be superior to a toxicological approach for assessing population viability at E&P sites. Few studies have measured direct toxicity to vertebrates at petroleum-impacted sites, and these have typically measured biomarkers within an individual, not abundance or reproduction within a population (Charlton et al. 2001, McBee and Wickham 1988; McMurry et al. 1999). Moreover, several factors mitigate against toxicological risk at E&P sites. Hydrocarbon and salt contaminants are not generally taken up by the components of the wildlife diet. Plant uptake of hydrocarbons is usually low (Chaîneau et al. 1997; Anghern et al. 1999). Moreover, where phytotoxicity is evident, as with brine scars (API 1997), plants are largely absent. Earthworms, a common component of the wildlife diet in many ecosystems, tend to avoid moderate levels of hydrocarbons in soil (Wong et al. 1999) and saline soils (Piearce 1982). Metal constituents of crude oil may not be present at toxic concentrations. Thus, toxicological risk to wildlife at E&P sites may often be negligible.

If forage is absent, habitat suitability is low. Even if invertebrate or vertebrate prey are present, vertebrate consumers often avoid disturbed areas. However, the empirical evidence concerning avoidance relates to infrastructure rather than spills and is equivocal. For example, caribou avoided human developments, including wells, roads and seismic lines, showing maximum avoidance distances of 1000 m from wells and 250 m from roads (Dyer et al. 2001). Elk avoided a recently installed oil well but continued to include it within their home range (van Dyke and Klein 1996). Anecdotal evidence suggests that badgers (*Taxidea taxus*) do not construct burrows on spill sites at the TPP. Moreover, the odor of hydrocarbons may deter some species; food avoidance has been observed for other contaminants, particularly pesticides (Pascual et al. 1999; Kononen et al. 1987). We found as many exceptions to the hypothesis of avoidance: (1) caribou did not avoid oil field infrastructure (Cronin et al. 1998); (2) house mouse (*Mus musculus*) abundance was higher in petroleum-contaminated, disturbed areas than in uncontaminated, disturbed areas in former tallgrass prairie ecosystems (Lochmiller et al. 2000); (3) most black bears did not alter the size or location of home ranges at an E&P site in Alberta, Canada (Tietje and Ruff 1983); and (4) lesser prairie chickens use lek (male group display) sites on abandoned oil pads and soils denuded by herbicide treatments (Haukos and Smith 1999; NRCS and Wildlife Habitat Council 1999). Furthermore, several studies of small mammals suggest that habitat fragmentation can sometimes have beneficial or neutral effects on population densities (Dooley and Bowers 1998; Aars et al. 1999). These vertebrates may not avoid disturbed areas.

Habitat disturbance can have adverse effects on population abundance or reproduction for various reasons. Individuals unable to find territories may emigrate. Movement costs may increase for animals that avoid or do not settle in disturbed areas. Forage vegetation or prey may be less available. Remaining habitat may provide fewer refuges from predators. Population declines due to habitat loss may ultimately lead to local extinction at low population densities because of the inability to find mates or breeding territories ("Allee effect," Allee 1938).

Brine and hydrocarbon spills at E&P sites (as well as wellheads, roads, burned areas, grazed areas, mowed areas, etc.) can be considered islands of disturbance in a sea of good habitat (Fig. 1). As accidental brine spills occur during production, underlying soil becomes saline, and the exposed area of the landscape is usually denuded of vegetation. Denuded soil is exposed to erosion, causing an enlargement of the denuded area with time (API 1997). Restoration of brine scars is possible, but revegetation with nonnative, salt-tolerant species or unpalatable plants (Keiffer and Ungar 2002) does not necessarily restore suitable habitat for native species. In contrast, hydrocarbons degrade rapidly, and fertilization can lead to recovery of production of vegetation within a few years.

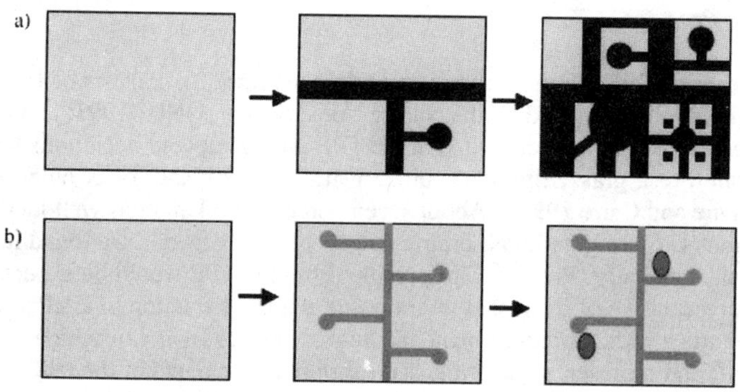

FIG. 1–*Two views of habitat fragmentation: a) "Traditional" habitat fragmentation resulting from industrial or residential development, black representing paved roads and buildings that create a hard barrier to migration and dispersal; and b) fragmentation from chemical releases to the environment, gray representing dirt or gravel roads and well pads that form a "porous" barrier (adapted from Carlsen et al. 2004)*

Data on the spatial and temporal frequencies of spills at E&P sites are not readily available. 567 brine spills were reported in Louisiana between 1990 and the first half of 1998 (Bass 1999), but others may have gone unnoticed. 900 brine spills per year were reported by the state of Oklahoma between 1993 and 2002 (Jager et al. 2004a).

Even though the evidence above suggests that habitat disturbance is more important at E&P sites than chemical toxicity, most spatially-explicit models that have been developed for ecological risk assessment emphasize foraging and chemical bioaccumulation through the food chain, rather than habitat preferences and species life histories (Freshman and Menzie 1996; Clifford et al. 1995; Baveco and de Roos 1996). Other models include habitat preferences but do not incorporate species life histories or address situations in which habitat is removed (Linkov et al. 2001; Henriques and Dixon 1996; Hope 2000). Population models that address habitat fragmentation simulate movement of animals between patches of suitable habitat, rather than population-level effects of unsuitable habitat (Gustafson and Gardner 1996). In contrast, individual-based models (IBMs) can simulate mechanistic linkages between the physical environment, as modified by human activities, and animal populations.

This chapter describes individual-based modeling methodologies and results for two species (American badger, prairie vole) at the TPP, using models emphasizing different aspects of vertebrate ecology (e.g., habitat suitability, predator-prey relationships). We describe a spill generator program that can create permanent or temporary brine or hydrocarbon spills of varying size and number. An ecological framework for evaluating vertebrate population impacts at E&P sites is described which incorporates population models. In the future, results of population models are expected to inform recommendations for no-effect criteria that would exclude E&P sites with particular spill densities or patterns from rigorous ecological risk assessment requirements.

Developing an Ecological Framework: The TPP Case Study

Tallgrass Prairie Preserve

Our case study site, the TPP in northeastern Oklahoma (Fig. 2), consists of 15,200 hectares of prairie grassland owned by the Nature Conservancy (ONHI 1993; Hamilton 1996). Additional terrestrial habitats found at the TPP include upland deciduous forest, deciduous riparian forest, grassy riparian habitat, disturbed areas, and rocky outcrops (ONHI 1993; Payne and Caire 1999). About seven percent of oil and gas well locations in the conterminous U.S are in tallgrass prairie and 32 percent of wells are found in prairie ecosystems, generally (Fig. 3). Thus, results from the TPP would be expected to be somewhat representative of those that might occur at a large fraction of E&P sites.

The TPP supports a wide variety of plant and animal species (many of which are prairie-dependent) and represents one of the last substantial remnants of the tallgrass prairie ecosystem, which historically covered 5.7 million ha of the United States and Canada (Madson 1990). Bison were reintroduced to the TPP in 1993. Since that time, the Nature Conservancy has used fire (median burn rate of 6700 ha/yr) and bison grazing as management practices for prairie restoration, reenacting the natural disturbances that historically functioned to maintain the ecosystem (Hamilton 1996). The TPP is an E&P site with more than 600 historic and 120 active oil and gas wells. The site contains five large, historic brine scars; several recent (within past two years) spills of brine, oil, or both; and older spill sites (8 to 15 years). The total brine spill area is approximately 17 ha (about 0.1% of the total preserve area), with a median spill size of 0.02 ha and a maximum spill of 4.9 ha. Additional wells are located outside of the TPP in Osage County. Approximate total areas of roads, well disturbance, vegetation classes, pastures, bison paddocks, and streams are available from the authors.

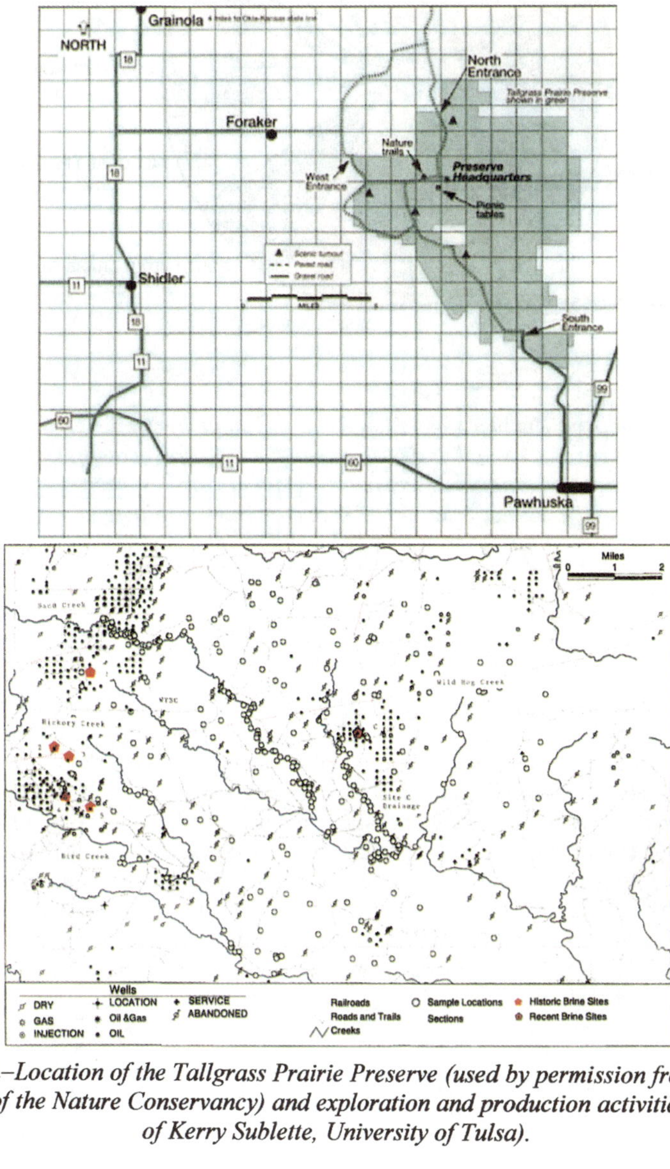

FIG. 2–*Location of the Tallgrass Prairie Preserve (used by permission fromBob Hamilton of the Nature Conservancy) and exploration and production activities (courtesy of Kerry Sublette, University of Tulsa).*

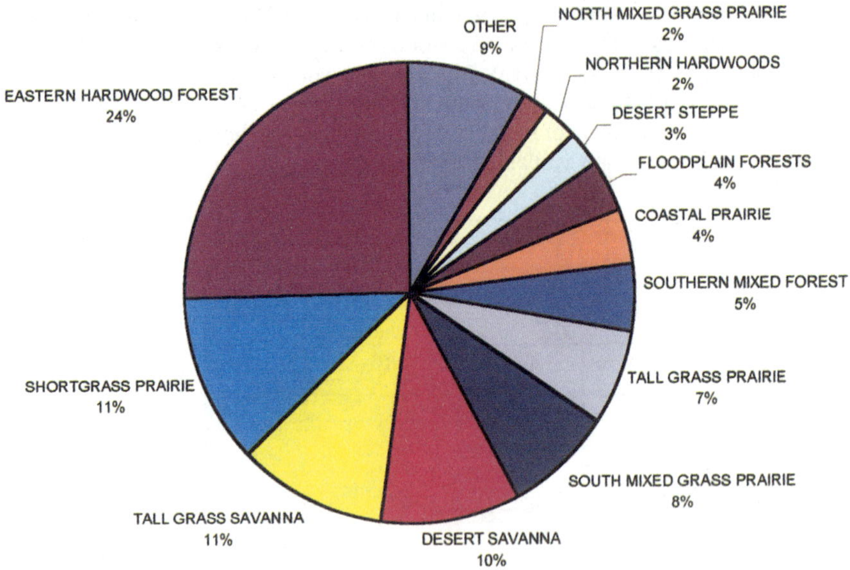

FIG. 3–*Predominant Kuchler vegetation forms in ¼ mile by ¼ mile cells in which productive and unproductive oil and gas wells in the U.S. are located. Data on well locations from the 1995 National Assessment of Oil and Gas Resources were obtained from David Ferderer at USGS and are now available at http://energy.cr.usgs.gov/oilgas/noga/.*

Geographic Information Systems (GIS)

Both the development of this ecological framework and the use of spatially-explicit IBMs require spatial data. We developed a GIS data collection protocol to investigate and manage ecological impacts at E&P sites (Hall et al. 2001). The protocol describes recommended data for assessing ecological impacts and their sources, as well as procedures for 1) assessing the quality, accuracy, precision and applicability of the data; 2) establishing a common projection system; 3) associating tabular data with spatial locations where useful; and 4) preprocessing or correcting the data when necessary. The GIS for the TPP was developed using Arc IMS, Arc INFO and associated modules as the primary GIS analytical engine. Table 1 shows the uses of these spatial data for modeling spills and vertebrate populations at E&P sites. National, state and site-specific sources of these data are identified in Hall et al. (2001).

TABLE 1—*GIS layers for use in modeling vertebrate populations.*

Data layer	Use
Digital elevation model	Predicting probability of pipeline rupture, flow of brine and oil, soil erosion potential, slopes unsuitable for animal movement
Raster coverage of vegetation categories	Depicting forage, predator refuges; contributing to habitat suitability designations
Digital Orthophoto Quarter-Quadrangles, Landsat Images	Depicting changes in spill boundaries and habitat suitability over time, with ground-truthing
Vector coverages of roads, fence-lines	Depicting potential barriers to movement or contributing to habitat suitability designations
Vector coverage of site boundary	Depicting boundary of local population of concern
Raster coverages of well locations, tank farms and other structures	Depicting potential barriers to movement or contributing to low habitat suitability designations
Raster coverages of vegetation disturbances (e.g, prescribed burns, grazing)	Contributing to habitat suitability designations
Raster coverage of soil taxonomy	Providing soil texture information relevant to burrowing mammals

Conceptual Trophic Model

A conceptual trophic model describes interactions among ecological receptors at a site and identifies the important populations that may be the focus of ecological evaluations
and management and significant trophic relationships that may be included in an individual-based, predator-prey model. Three steps are involved in the construction of a conceptual terrestrial trophic model (Stevenson et al. 2001): (1) creating a list of species expected to occur at the site, (2) assigning the species to guilds, and (3) constructing the food web through an analysis of the relationships between guilds. The conceptual model focuses on guilds with high societal interest or that are representative of many species.

Species data can be gathered from a variety of sources. For the TPP, these included a report of the Oklahoma Biological Survey (ONHI 1993), species lists obtained from the Nature Conservancy (1996) and Oklahoma State University (Palmer, unpublished data), and open literature references. Species data from similar, well-studied ecosystems (e.g., National Science Foundation Long Term Ecological Research sites) are also useful.

The procedure for assigning species to alpha and beta guilds is described in Stevenson et al. (2001), based on the guild definitions of Wilson (1999). Alpha guild members use a class of resources in a similar way. Beta guild members share similar space along environmental gradients (i.e., occupy the same or a similar niche). Alpha guilds are the most important for creating a conceptual food web, but beta guilds provide information about species interactions such as competition.

At the TPP, we determined relationships between the guilds based on dietary information collected in the creation of the alpha guilds. Sixteen alpha animal guilds and six beta guilds are found at the TPP (Stevenson et al. 2001). Table 2 shows the community food web at the TPP, describing the binary feeding relationships between the alpha guilds in the community. Figures in Stevenson et al. (2001) show the source webs (Pimm et al. 1991) of the guilds of vertebrates of the TPP.

TABLE 2—*Food relationships between the various alpha guilds at the Tallgrass Prairie Preserve, OK. An "x" denotes that a given prey species along the y-axis may be consumed by the corresponding predator on the x-axis.*

	Predators															
Prey	AI	BI	HM1a	HM1b	HM2	OM1	OM2	OB	OH	CH1	CH2	CB1	CB2	CM1a	CM1b	CM2
Plants	X	X	X	X	X	X	X	X	X							
Detritus	X	X				X	X	X	X				X	X		
AI	X					X	X	X	X	X	X	X	X		X	
BI		X						X	X	X		X		X		
HM1a						X										X
HM1b						X										X
HM2						X		X			X		X			X
OM1						X							X			X
OM2						X		X			X		X			X
OB						X		X	X		X		X			X
OH						X										X
CH1						X	X	X			X		X			X
CH2						X					X		X			X
CB1						X		X	X		X		X			X
CB2																
CM1a						X					X		X			X
CM1b						X							X			
CM2																

Key to Alpha Guilds

Plants	Plants and Fungi	OB	Omnivorous Birds
Detritus	Detritus and Carrion	OH	Omnivorous Herptiles
AI	Aboveground Invertebrates	CH1	Invertivorous Herptiles
BI	Belowground Invertebrates	CH2	Other Carnivorous Herptiles
HM1a	Herbivorous Mammals - grazers	CB1	Invertivorous Birds
HM1b	Herbivorous Mammals - browsers	CB2	Raptors
HM2	Small Herbivorous Mammals	CM1a	Fossorial Invertivorous Mammals
HB	Herbivorous Birds	CM1b	Other Invertivorous Mammals
OM1	Large Omnivorous Mammals	CM2	Large Carnivorous Mammals
OM2	Small Omnivorous Mammals		

Artificial Landscapes

A map of the brine spills at the TPP and a discussion of structures (e.g., pipelines, well heads, tank batteries) that affect spill probabilities is presented in Jager et al. (2004a). However, the TPP case study reported here relied on artificial maps of disturbance features. Artificial spills are necessary to identify potential spill area or fragmentation thresholds that result in Allee effects. Maps with different spill patterns also aid in understanding causes of declines. Three methods were used to develop artificial landscapes. For the trophic (vole) model described below, we used a heuristic method where the spill area, size, number and placement were based on knowledge of existing or potentially possible conditions present at the TPP. For this method we conducted simulations that distributed various numbers of 0.09-ha spills (30 m by 30 m cells) randomly across the TPP and simulations that distributed impenetrable structures (such as roads and fences) across the landscape, resulting in fragmentation. For the badger model two spill generators were developed, one theoretical, and one more realistic and dependent on pipeline distribution (Jager et al. 2004a). The theoretical model distributes spill centers randomly in two-dimensional space. A Dirichlet distribution is used to allocate the total spill area across spill centers, and this area is distributed using a random walk algorithm to simulate diffusion into neighboring cells. The well-complex model simulates spills along gathering lines that connect each well in a rectangular grid (based on many of the well arrangements at the TPP) with a tank battery located at one corner. The user specifies the number and dimensions of well complexes. The model assumes that the likelihood of encountering a spill along any segment of pipe of a specified length is constant, so that the likelihood of a spill within a cell increases with the length of pipe located within its boundaries. The distance along the pipe to the next brine spill is a gamma variate, and area of each brine spill is simulated as a Dirichlet variate, which ensures that the specified total area of spill is exactly met. Badger simulations presented in this manuscript were performed using landscapes created with the theoretical spill generator.

Spatially Explicit Individual-based Models

Two spatially-explicit IBMs were developed for terrestrial vertebrates. The structure of the template for the two models is described in Fig. 4. Each spatial cell, as well as its immediate surroundings, is conceived as the source of food resources and shelter for individual animals. The models can simulate population changes over time in response to disturbances by fire, petroleum spills, and brine spills, though only static brine scars and other static habitat disturbances are presented here. Modeled events include local biological processes that influence individual animals (e.g., mortality, reproduction, aging, mating choice) and external or landscape-wide events (e.g., disturbances, redistribution of organisms).

Habitat Model

Habitat IBMs are well suited for studying the differential susceptibility of species with different life histories and habitat requirements to habitat loss from brine spills. We implemented a habitat-based model for the American badger (*Taxidea taxus*), a voracious, solitary predator with low tolerance for other individuals. Here, we provide a brief overview of the model, which is described more completely by Jager et al. (2004b). We assign habitat suitability indices to various vegetation categories based on known compatibilities with the presence of small, fossorial mammals or burrowing requirements. Brine spills, structures, and streams are designated as unsuitable habitat. Habitat quality of cells influences reproduction simulations through acquisition of territory used for breeding and survival via movement costs and habitat-related mortality. This model does not explicitly represent foraging or predation.

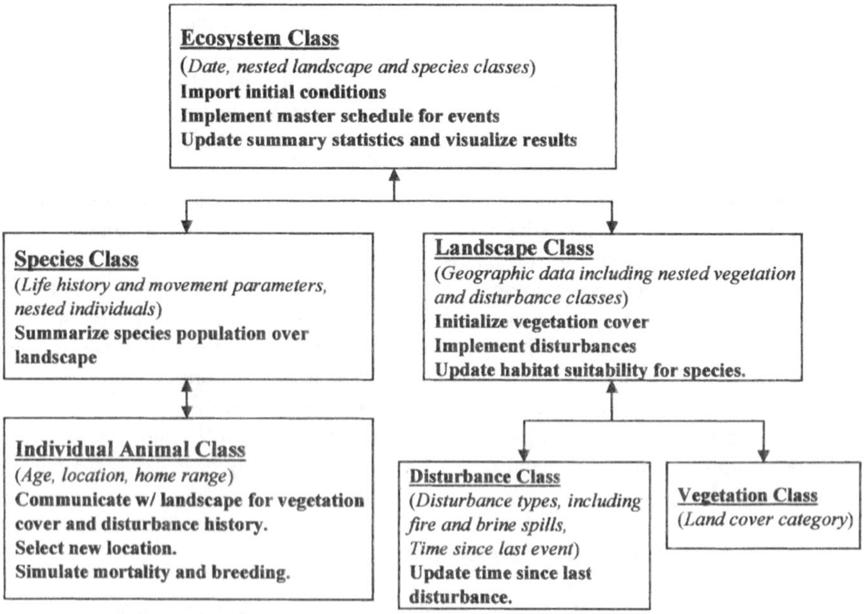

FIG. 4–*Diagram of the general model template. Objects in the model are defined by classes that include data members (italics) and member functions (bold). Each class is represented by a box in the diagram.*

Individuals pass through five periods of pre-breeding, mating, post-mating, birthing, and rearing of young. Juveniles seek to establish a permanent home range, equivalent to the breeding territory, once they leave the care of their mother. Once this range is established, a badger's movements are restricted to cells within its range. Movement depends on season and gender. Mating is assumed to occur for any mature female with a home range overlapping the home range of at least one mature male. Reproduction timing and survival of young are also described in Jager et al. (2004b). Sources of mortality include: baseline, age-related mortality; habitat-related mortality; mortality due to intraspecific aggression; mortality based on movement; and emigration from the study area. Sensitivity analyses are in progress.

We conducted a simulation experiment to investigate the effects of loss of habitat area and fragmentation (represented here by increasing numbers of spills). We used the statistical model described above and in Jager et al. (2004a) to generate spill landscapes with a specified target percentage of area covered by spills (0%, 1%, 10%, 20%, 30%, 40% and 50% and a specified number of spills (100 or 1000). Effects of spills on final average population sizes and the proportion of potential breeding females that successfully mated were compared. In addition, we quantified Allee effects.

Results showed a decrease in the average size of badger populations at the end of a 100-yr simulation with increasing area of habitat loss (Fig. 5a). This decrease was accompanied by a similar decline in the proportion of replicate populations that persisted (Fig. 5b). Results of the Jager et al. (2004b) study suggest that thresholds, defined as steeper declines in persistence with increasing habitat loss, occur when the habitat is highly fragmented by disturbances. Parameter explorations show that steeper, threshold-like declines occur when the mortality risk in poor habitat is high and when poor habitat is not excluded from the animal's territory. The decline in persistence associated with habitat loss was caused by a combination of elevated habitat-related mortality and increased difficulty in finding mates (Allee effects) (Fig. 5c). Fragmentation increased the difficulty in carving out high quality territories and increased mortality during the dispersal phase. The good news is that the likelihood of persistence is high for landscapes with fragmentation characteristics similar to those found at the TPP, that is 0.1% of the area covered with brine scars (and less than 1% of the area directly disturbed by wells, roads or spills). However, empirical verification of these modeling results is necessary before any conclusions can be drawn.

Comparing minimum habitat requirements for a social and asocial prairie species, Wolff (2001) identified behavioral attributes that influence species response to habitat loss for mammals, including (1) habitat specificity, (2) social structure, (3) dispersal ecology, and (4) trophic level. In our framework, we view these attributes as part of the spatial life history of a species. Future research with the habitat IBM will focus on how differences in social structure influence species responses to fragmentation and habitat loss due to brine spills. The badger represents one extreme: an asocial animal that is solitary and highly intolerant of same-sex conspecifics. We hope to contrast our results for the badger with a social breeder, such as a prairie chicken. During the breeding season, male prairie chickens aggregate into lekking displays on bare, elevated areas surrounded by grasslands. Large breeding aggregations benefit from group defenses against predation, as well as access to mates. Simulations may show that social breeders are more susceptible to habitat loss than asocial species because of strong Allee effects. This result would be consistent with field observations that suggest a threshold lek size. Alternatively, simulations may show that social species are better at packing into small habitat areas, and benefit from a brine spill because it creates lekking sites.

Trophic Model

Trophic IBMs focus on interactions that may cause indirect, vertebrate population-level effects associated with habitat loss (e.g., vegetation growth and reduction due to grazing, herbivory, and bioenergetics). The trophic approach captures the interdependence between population density and environmental characteristics such as vegetation density, unsuitable vegetation, and climatic dependence. Trophic concepts were the leading principles of a model that was implemented for the prairie vole (*Microtus ochrogaster*) (Kostova and Carlsen 2003, Kostova et al. 2004), a monogamous herbivore that feeds on grassland vegetation and is preyed upon by predators such as owls, badgers, and snakes. A large number of well-established experimental values are available for parameterization of the prairie vole model. Depending on the availability of

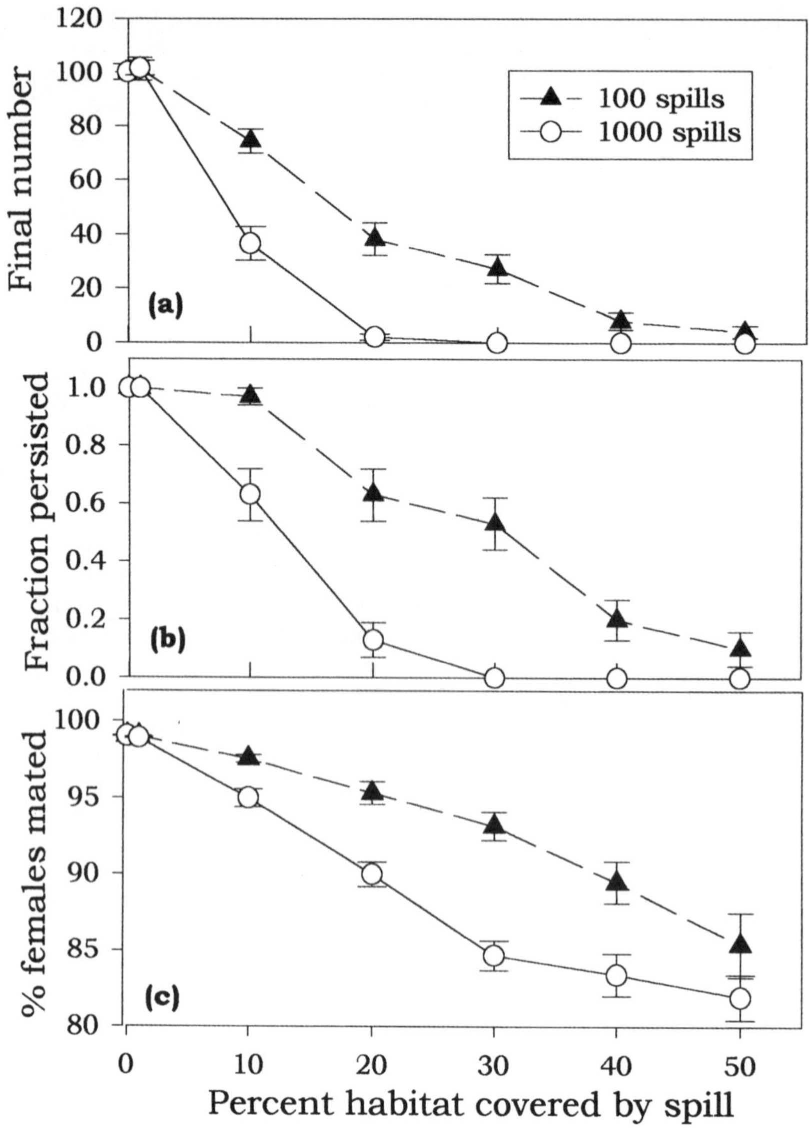

FIG. 5–*Response of simulated (a) final population size and (b) fraction of replicate populations that persist, and (c) fraction of females eligible to breed that find mates to habitat loss. Error bars show 1 standard error surrounding the mean of 30 replicate simulations for landscapes with 100 and 1,000 spills (Jager et al. 2004b)*

sufficient species data, the model is extendable to other monogamous rodents and can be adapted for polygamous species as well.

The model uses growth rates of grasses obtained from a 30-year simulation of CENTURY v. 4.0 (Parton et al. 1993) with historical temperature and precipitation data. The vegetation growth rates are combined with the grazing rate of the cumulative vole population in each cell. Body mass and metabolic status of each individual determine behavioral characteristics, such as territorial competition, mating success, and dispersal. Metabolic status is adjusted for the pregnancy or weaning status of females. Voles do not produce offspring unless a pair is formed from two dispersing animals (floaters) that occupy the same cell.

The survival of an individual vole depends on the availability of vegetation and the individual's physiological status. Starvation, age and predation are factors that contribute to the death rate. Predation is incorporated into the model by removing a density-dependent fraction of the voles.

The spatial structure of the model is based on the notion of home range. A simulated landscape is represented as a collection of cells whose size is equal to the home range of the vole. Voles are residents of a cell or floaters. The status changes over time depending on vegetation availability, age, body size, presence of a potential mate, etc. Floaters choose new cells based on vegetation suitability and quantity as well as on opportunities for mating. In cases when the current cell is on the border of the region and the floater cannot find an unpopulated cell into which to move, it is forced to leave the modeled region.

A series of runs with random initial animal distributions and spill locations were performed in order to establish the dependence of population density and average time-to-extinction (ATE) on factors such as predation level, available habitat size, fragmentation caused by barriers dividing the landscape into connected patches and fragmentation caused by spills. Simulations were performed on artificial, square habitats with uniformly growing tallgrass vegetation as well as on a landscape representing the TPP, using geospatial and vegetation data.

Effect of Area and Predation—Patch size and predation were found to have a combined effect on population density. The reduction of habitat area led to higher vole population density in the absence of predation and dispersion, which destabilized the vole population and decreased the ATE (Kostova et al. 2004). However, the reduction of habitat area had little effect on the maximum population densities in simulations if predation and dispersion were taken into consideration (Fig. 6). Increasing the predation level had the clear effect of decreasing population density. However, the shapes of the density curves, i.e. the locations of the minima and maxima and periodicity, were not sensitive to either area or predation level change (Kostova and Carlsen 2003). Both area and predation level had a significant effect on the time to extinction. Figure 7 represents the ATE at three predation levels.

A direct relationship was observed between the habitat area and the ATE; the larger the patch, the higher the ATE. On the largest patches for the low and intermediate predation levels, almost all simulations produced populations that persisted for the whole 30-year period.

Low predation levels led to dramatically decreased persistence, which can be explained by high vole densities leading to overgrazing in the months of low vegetation.

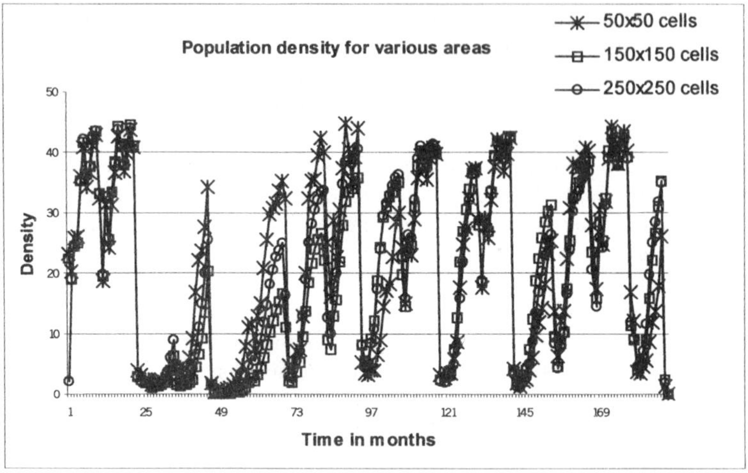

FIG. 6–*Predicted population density for three different areas of artificial square habitat;* a_A*=0.02,* a_J*=0.04, where* a_A *and* a_J *are fraction of adults and juveniles removed, respectively.*

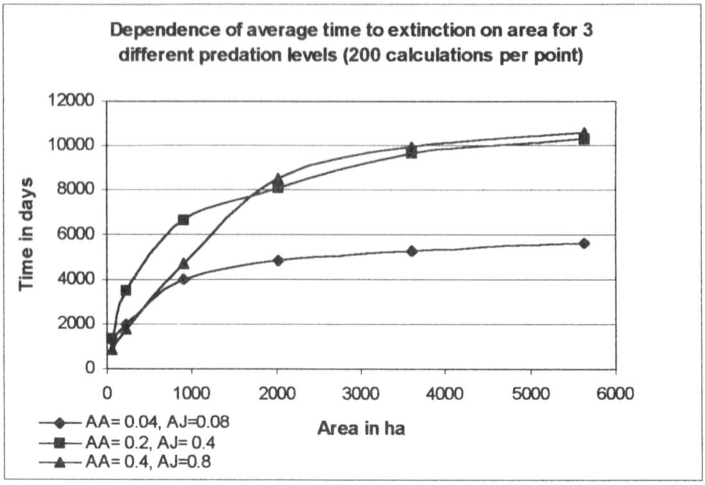

FIG. 7–*Average time to extinction (ATE) as a function of habitat area and predation, averaged over 200 simulations. If the population persisted for the whole simulation period, ATE was taken to be 30 years (10800 days). AA and AJ are fraction of adults and juveniles removed, respectively.*

However, patches of small area displayed low ATE at all predation levels. Because the maximum densities were similar for various areas, the high extinction risk on the smallest simulated patches is not connected with density effects. One possible explanation is that small patches provide lower numbers of surviving individuals to restore the population after a period of insufficient vegetation and no births (e.g., in winter months). The simulations reveal that for patches of practically all sizes, there is an "optimal" predation level for which the ATE is highest (Kostova and Carlsen 2003). Using the "optimal" predation coefficients also resulted in vole population densities characteristic for tallgrass prairie, i.e., 5-7/ha (Getz et al. 2001).

Effect of Non-spill Fragmentation—Habitat fragmentation contributes to the reduction of available habitat and would be expected to lead to a lower ATE. However, the effect of fragmentation on population persistence appears to depend on the population density of voles. We modeled enclosures (no dispersal across boundaries) with no predation (Kostova et al. 2004). The simulations yielded high population densities with mortality mainly due to winter starvation. Fragmentation had a positive effect on population persistence, as it reduced population density and stabilized the populations.

Effect of Spills—The introduction of predation and dispersal as well as more accurate trophic calculations produced lower population densities (Kostova and Carlsen 2003). The effect of introducing randomly distributed "spills" consisting of separate polluted cells in artificial vegetation grids was investigated by performing simulations with an increasing percentage of spills on the patch. The area of the grids was increased so that the inhabitable (unpolluted) area was kept constant at 10,000 cells. Figure 8 presents the results of simulations with two predation levels. Each point represents the result of 200 simulations performed by either fixing an initial animal distribution and varying the number of spills or fixing the spill distribution and varying the initial animal distributions. The fragmentation actually had a beneficial effect, increasing the ATE, in the case with "low" predation if as much as 60% of the area was covered by random spills. Fragmentation had a similar effect as predation in decreasing population densities, and this explains the observed phenomenon. In the case of the "optimal" predation level, fragmentation did not have any effect on the ATE if up to 50% of the area was covered with spills. For spill areas above these levels, the ATE decreased with increasing percentage of spill area.

Effect of Spills and Other Sources of Fragmentation at the TPP—We investigated the combined role of spills and other sources of fragmentation on the persistence of vole populations at the spatial scale of the TPP. Runs with different initial vole distributions inevitably resulted in extinction of the vole population in the southern and western portion of the TPP, which are very fragmented by roads (not considered as barriers in the experiments) and in rivers and patches of non-grass vegetation (considered as barriers). Voles persisted in the northeastern part of the preserve, which is not as fragmented (Fig. 9a). In other experiments, hypothetical random distributions of 1000 spills (or development sites) of the size of one cell (0.09ha) were placed in the northeastern portion of the preserve, and simulations were carried out at the "optimal" predation level. (Fig. 9b). However, due to the large scale of the simulation on the TPP landscape, the number of simulation repetitions was insufficient for a valid prediction. The relatively small number of simulations resulted, in some of the cases, in the extinction of the population

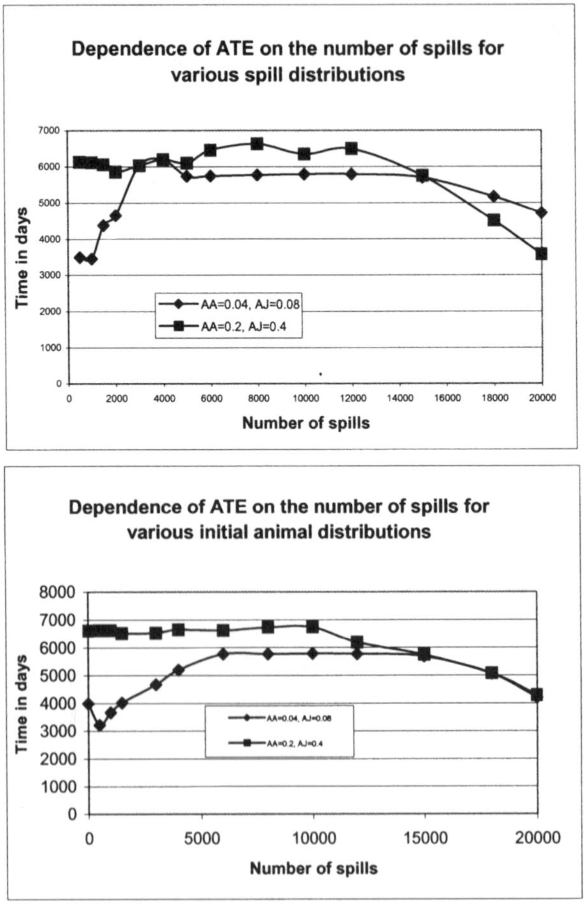

FIG. 8–*Dependence of ATE on the number of spills in an area with 10000 inhabitable cells for various spill and animal distributions. AA and AJ are fraction of adults and juveniles removed, respectively*

in the areas with spill fragmentation even before this happened in the naturally fragmented areas, while the same populations persisted for the whole 30-year period in the absence of spills.

Model Validation

Rigorous model validation has not yet been performed. However, the results of simulations with the trophic model for voles were compared with time series density data from the literature (Krebs et al. 1969; Getz et al. 2001), and the model was adjusted accordingly. The model predicted correctly the average vole density, multiyear

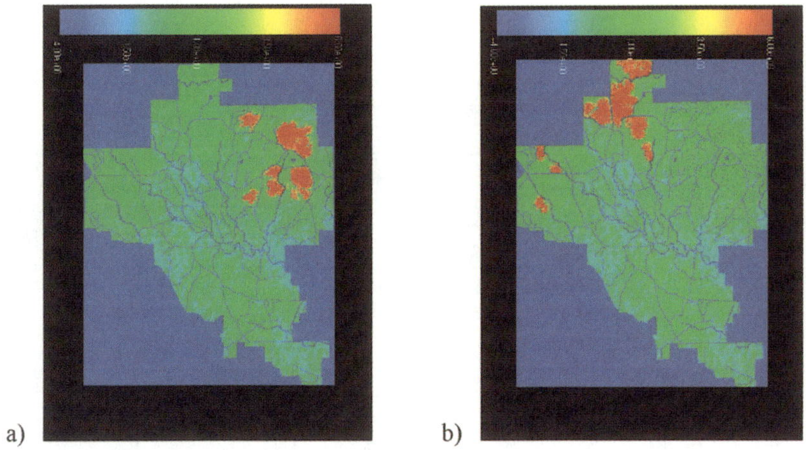

FIG. 9–*Two instantaneous plots of simulations of vole density dynamics at the scale of the TPP, using a) a realistic representation of roads (blue), rivers (dark blue) and patches of non-grass vegetation (light blue), and b) a distribution of 1000 artificial spills in the Northeast. Red represents areas of high vole density that change in location and size during the year.*

fluctuations at higher densities, annual fluctuations at lower densities, and annual dynamics with peaks in October to December and minima in February to March. Studies of impacts of brine and E&P sites that would be useful for validation are rare, and even studies of habitat loss are rare for some vertebrate species. Moreover, multiple stressors that are present at E&P sites can confound field results. For example, Cronin et al. (1998) cite several investigators who recognize the difficulty of distinguishing human impacts from environmental stochasticity affecting caribou herds. Field verification of model results is planned.

Toward an Ecological Framework for E&P Sites

Framework Components

A preliminary ecological framework for evaluating terrestrial vertebrate populations at E&P sites is presented in Fig. 10. Assessment endpoint populations are chosen using a site conceptual trophic model and other management criteria. The framework includes two parallel paths for determining risk from toxicity or habitat loss. The potential for exposure to contaminants is determined by contaminant bioavailability and animal behavior. The threshold for conducting a toxicological risk assessment may be lower for threatened and endangered populations than for other populations. The spatial exclusion criteria (contaminated area thresholds) that determine whether an exposure to habitat disturbance may be significant and may require a spatial ecological assessment as described below. Species life history information, trophic relationships, and habitat

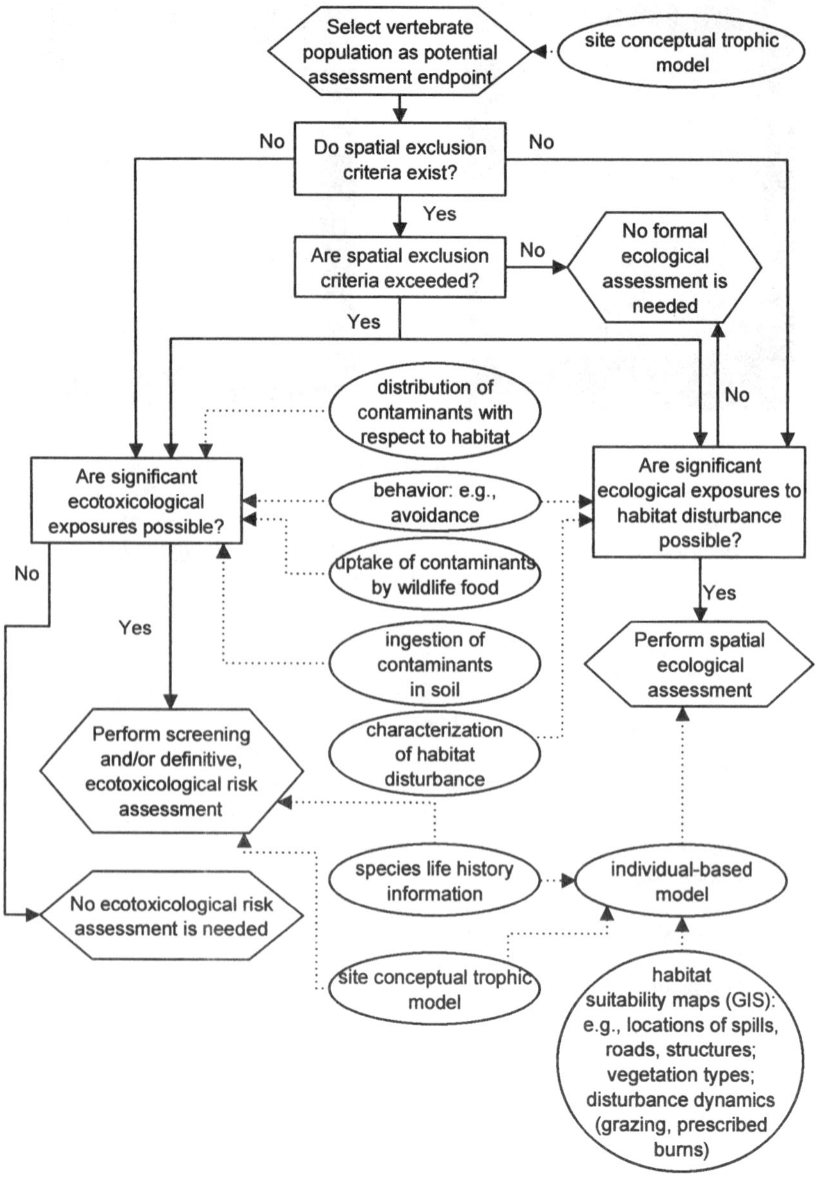

FIG. 10–*A preliminary ecological framework for evaluating terrestrial vertebrate populations at E&P sites.*

suitability may be explicitly or implicitly modeled with an IBM. If an ecological risk assessment is performed, the level of effort should be proportional to the magnitude of the risk management decision.

Spatial Exclusion Criteria

The large number of small brine and oil spills on E&P sites of high habitat value prompts the question of whether simple field criteria (e.g., threshold total area or particular distributions of spills associated with *de minimis* population-level effects) may be used to exclude the spills from formal ecological risk assessment. In the past, this question has been treated as a cost-effectiveness issue, with small spills simply being excavated, restored, or allowed to recover. Sorensen and Margolin (2002) review spatial scale ecological screening criteria for contaminated sites in various states. For example, the Pennsylvania Department of Environmental Protection assumes that two acres of surface soil contamination does not pose risk to vertebrate populations (PADEP 1998). One American Petroleum Institute paper suggests that a petroleum release to surface soil is not of environmental concern if it is farther than 500 ft from the nearest receptor or habitat (Claff 1999). However, these values are not based on a landscape approach to ecological assessment that specifically considers vertebrate populations.

Some guidance regarding habitat loss from spills might be distilled from existing ecological literature. Carlsen et al. (2004) review minimum patch size requirements (e.g., areas below which species are never found or which are associated with unsustainable populations) of several species and taxonomic groups for potential use in screening-level ecological risk assessments at E&P sites. Similar information may be available on the number of territories required to support sustainable populations. A caribou avoidance distance from wells is derived in Dyer et al. (2001). Massey (2001) notes that the Bureau of Land Management has the regulatory authority to move drill pads 200 m away from known lesser prairie chicken lek sites. A rule of thumb for carnivore density states that 10,000 kg of prey supports about 90 kg of carnivore, and this relationship "provides a basis for identifying species that require conservation measures" (Carbone and Gittleman 2002).

In this study, insufficient species, ecosystems, and model structures have been tested to recommend general criteria for excluding E&P sites from formal ecological assessment. However, relevant results are available for American badger and prairie vole. Based on our limited modeling of the American badger in grasslands, this species shows a decline in final population size with increasing habitat loss. If the modeling results were confirmed with field studies, a risk manager could set spatial exclusion criteria in the following manner. If a risk manager wanted an 80% likelihood of population persistence and 1000 spills were anticipated, then a spatial exclusion criterion of greater than 1% and less than 10% spill area could be chosen. One would choose a similar spatial exclusion criterion if a population of at least 50 badgers were desired at the TPP.

Our limited modeling of the prairie vole suggested a threshold at 30% habitat loss due to spills. Below this threshold, the average time to extinction was not affected. Above this threshold, the average time to extinction decreased with increasing spill area. Vole density was sensitive to the interaction of predation and fragmentation, with fragmentation causing population extinction in the presence of predation and stabilizing the population in the absence of predation. Where threshold spill areas for population-level effects are observed, acceptable levels of effects are relatively easy for risk managers to specify.

Conclusions and Caveats

Modeling results from the TPP indicate that vertebrate populations may decline as the area of brine spills at E&P sites increases. However, the spill area associated with detrimental effects is probably much larger than the actual fractional landscape area directly disturbed by spills at the TPP (0.1%). The impacts of increased habitat fragmentation caused by spills, structures, and/or roads can range from beneficial (simulation of vole populations with no predation) to adverse (simulation of vole populations with predation and simulation of badger populations). Simulations of simplified ecosystems with only one explicitly modeled species at a time, on a relatively homogenous grassland landscape, yield complex results. Until sensitivity analyses are performed, the relative importance of life history parameters, habitat suitability designations, bioenergetics, territory acquisition algorithms, impenetrable barriers, predation, edge behavior algorithms, and other factors will be unknown. The dynamic nature of brine and petroleum spills, including chemical degradation, active restoration, or natural recovery time frames was not considered in these IBMs. Moreover, the modeled results have not been verified in field studies. Studies such as ours may help focus scientific and regulatory attention on potential ecological impacts and potentially away from potential toxicological impacts. Conceptual trophic models can be useful in focusing an assessment on appropriate species. IBMs may incorporate many realistic variables, and sensitivity analyses may identify those that are most important. The habitat model applied to the American badger identified situations leading to the existence of steep threshold responses to increasing disturbance areas. Results from both models can be used as qualitative guidance for land managers and regulatory agencies, although field experiments should be designed to check model predictions for quantitative accuracy.

Acknowledgments

This research was sponsored by the National Petroleum Technology Office, U.S. Department of Energy, and performed at Oak Ridge National Laboratory (ORNL) and Lawrence Livermore National Laboratory (LLNL). ORNL is managed by UT-Battelle, LLC, for the U.S. Department of Energy under contract DE-AC05-00OR22725. LLNL is managed by the University of California for the U.S. Department of Energy, and work was performed under the auspices of Contract W-7405-Eng-48. We thank Nancy Comstock, project manager, for her support of this project. We thank Bryan Tapp and Elizabeth Word of the University of Tulsa and Mary Sorensen of Arcadis, Inc. for providing useful information or data. Jim Loar of Oak Ridge National Laboratory, Will Gala of ChevronTexaco Corp., and anonymous reviewers selected by ASTM provided comments on earlier drafts of this manuscript.

References

Aars, J., Johannesen, E., and Ims, R. A., 1999, "Demographic Consequences of Movements in Subdivided Root Vole Populations," *Oikos*, Vol. 85, pp. 204-216.

Allee, W. C., 1938, *The Social Life of Animals*, W.W. Norton & Co., New York.

Anghern, D., Schluep, M., Gälli, R., and Zeyer J., 1999, "Movement and Fate of Residual Mineral Oil Contaminants in Bioremediated Soil," *Environmental Toxicology and Chemistry*, Vol. 18, pp. 2225-2231.

API, 1997, *Remediation of Salt-affected Soils at Oil and Gas Production Facilities*, API Publication Number 4663, American Petroleum Institute, Washington, DC.

Bass, A. S., 1999, "Magnitude, Distribution and Impacts of Accidental Brine Spills in the Louisiana Coastal Wetland Conservation Plan Area," *AAPG Bulletin*, Vol. 83, pp. 1347-1348, as cited in Colgan et al. (2002).

Baveco, J. M. and deRoos, A. M., 1996, "Assessing the Impact of Pesticides on Lumbricid Populations: An Individual-based Modelling Approach," *Journal of Applied Ecology* Vol. 33, pp. 1451-1468.

Carbone, C. and Gittleman, J. L., 2002, "A Common Rule For The Scaling Of Carnivore Density," *Science*, Vol. 295, pp. 2273-2276.

Carlsen, T. M., Coty, J. D., and Kercher, J. R.. 2004 in press. "The Spatial Extent of Contamination and the Landscape Scale: An Analysis of the Wildlife, Conservation Biology and Population Modeling Literature," *Environmental Toxicology and Chemistry*, Vol. 23.

Chaîneau, C. H., Morel, J. L., and Oudot, J., 1997, "Phytotoxicity and Plant Uptake Of Fuel Oil Hydrocarbons," *Journal of Environmental Quality*, Vol. 26, pp. 1478-1483.

Charlton, K. G., Hird, D. W., Griffey, S. M., Zinkl, J. G., and Spiegel, L. K, 2001, "Clinical Findings in Kit Foxes and Deer Mice from an Oil Field Environment," 2001, *Transactions of the Western Section of The Wildlife Society* Vol. 37, pp. 61-65.

Claff, R., 1999, "Primer for Evaluating Ecological Risk at Petroleum Release Sites," *Drug and Chemical Toxicology*, Vol. 22, pp. 311-316.

Clifford, P. A., Barchers, D. E., Ludwig, D. F., Sielken, R. L., Klingensmith, J. S., Graham, R. V., and Banton, M. I, 1995, "An Approach to Quantifying Spatial Components of Exposure for Ecological Risk Assessment," *Environmental Toxicology and Chemistry*, Vol. 14, pp. 895-906.

Colgan, W., III, Vavrek, M. C., and Bolton, J., 2002, "Re-Vegetation of an Oil/Brine Spill: Interaction Between Plants and Mycorrhizal Fungi," 9th Annual International Petroleum Environmental Conference, October 22-25, 2002, Albuquerque, NM.

Cronin, M. A., Amstrup, S. C., Durner, G. M., Noel, L. E., McDonald, T. L., and Ballard, W. B., 1998, "Caribou Distribution During the Post-Calving Period in Relation to Infrastructure in the Prudhoe Bay Oil Field, Alaska," *Arctic*, Vol. 51, pp. 85-93.

Dooley, J. L. and Bowers, M. A., 1998, "Demographic Responses To Habitat Fragmentation: Experimental Tests at the Landscape and Patch Scale," *Ecology*, Vol. 79, pp. 969-980.

Dyer, S. J., O'Neill, J. P., Wasel, S. M., and Boutin, S., 2001, "Avoidance of Industrial Development by Woodland Caribou," *Journal of Wildlife Management*, Vol. 65, pp. 531-542.

Freshman, J. S. and Menzie, C. A., 1996, "Two Wildlife Exposure Models to Assess Impacts at the Individual and Population Levels and the Efficacy of Remedial Actions, *Human and Ecological Risk Assessment*, Vol. 2, pp. 481-498.

Getz, L. L., Hofmann, J .E., McGuire, B., and Dolan, T. W., 2001, "Twenty-Five Years of Population Fluctuations of *Microtus ochrogaster* and *M. pennsylvanicus* in Three Habitats in East-Central Illinois," *Journal of Mammalogy*, Vol. 82, pp. 22-34.

Gustafson, E. J. and Gardner, R. H., 1996, "The Effect of Landscape Heterogeneity on the Probability of Patch Colonization," *Ecology*, Vol. 77, pp. 94-107.

Hall, C., Neher, L. Wilder, L., Carlsen, T., Ashwood, T., Washington-Allen, R., and Efroymson, R., 2001, *Data Collection Protocol for Developing Geographical Information Systems to Investigate and Manage Ecological Impacts at Petroleum Exploration and Production Sites*, UCRL-ID-142371, Lawrence Livermore National Laboratory, January 2001.

Hamilton, R.G., 1996, "Using Fire and Bison to Restore a Functional Tallgrass Prairie Landscape," *Transactions of the 61st North American Wildlife and Natural Resources Conference*," Wildlife Management Institute, New York.

Haukos, D. A., and Smith, L. M., 1999, "Effects of Lek Age on Age Structure and Attendance of Lesser Prairie-Chickens (*Tympanuchus pallidicinctus*)," *American Midland Naturalist*, Vol. 142, pp. 415-420.

Henriques, W. D., and Dixon, K. R., 1996, "Estimating Spatial Distribution of Exposure by Integrating Radiotelemetry, Computer Simulation, and Geographic

Information System (GIS Techniques)," *Human and Ecological Risk Assessment* Vol. 2, pp. 527-538.

Hope, B. K., 2000, "Generating Probabilistic Spatially-Explicit Individual and Population Exposure Estimates for Ecological Risk Assessments," *Risk Analysis*, Vol. 20, pp. 573-589.

Jager, H. I., Carr, E. A., and Efroymson, R. A., 2004b submitted, "A Simulation Study to Evaluate the Effects of Habitat Loss and Fragmentation on the American Badger (*Taxidea taxus*)," *Landscape Ecology*

Jager, H. I., Efroymson, R. A., Sublette, K., Carr, E. A., and Ashwood, T. L., 2004a manucript, "Unnatural Landscapes in Ecology: Generating the Spatial Distribution of Brine Spills," *Biometrics*.

Keiffer, C. H., and Ungar, I. A., 2002, "Germination and Establishment of Halophytes on Brine-affected Soils," *Journal of Applied Ecology* Vol. 39, pp. 402-415.

Kononen, D. W., Hochstein, J. R., and Ringer, R. K., 1987, "Avoidance Behavior of Mallards and Northern Bobwhite Exposed to Carbofuran-Contaminated Food and Water," *Environmental Toxicology and Chemistry*, Vol. 6, pp. 41-50.

Kostova, T. and Carlsen, T., 2003 submitted, "The Effect of Area Size and Predation on the Time to Extinction of Prairie Vole Populations. Simulation Studies Via SERDYCA: a Spatially Explicit Individual-Based Model of Rodent Dynamics," Proceedings of the International Conference on Mathematical Ecology, Alcala, Spain, September 2003.

Kostova, T., Carlsen, T. and Kercher, J. 2004 in press, "Individual-Based Spatially-Explicit Model of an Herbivore and Its Resource: The Effect of Habitat Reduction and Fragmentation," Proceedings of the International Conference on Mathematical Population Dynamics, Bedlewo, Poland, June 2002, *Comptes Rendus de l'Académie des Sciences, Biologies.*.

Krebs, C., Keller B. and Tamarin, R., 1969, "*Microtus* Population Biology: Demographic Changes in Fluctuating Populations of *M. Ochrogater* and *M. Pensilvanicus* in Southern Indiana," *Ecology*, Vol. 50, pp. X-Y.

Linkov, I., Grebenkov, A., Baitchorov, V. M., and Zibold, G., 2001, "Spatially Explicit Exposure Models: Application to Military Sites," *Toxicology and Industrial Health*, Vol. 17, pp. 230-235.

Lochmiller, R. L., Rafferty, D. P., McBee, K., Basta, N. T., and Wilson, J. A., 2000, "Disruption Of Rodent Assemblages In Disturbed Tallgrass Prairie Ecosystems Contaminated With Petroleum Wastes," *Environmental Contaminants and Terrestrial Vertebrates: Effects on Populations, Communities, and Ecosystems*,

Albers, P. H., Heinz, G. H., and Ohlendorf, H. M., Eds., SETAC Press, Pensacola, FL, pp. 277-301.

MacArthur, R. H., 1957, "On the Relative Abundance of Bird Species," *Proceedings of the National Academy of Sciences*, Vol. 43, pp. 293-295.

Madson, J., 1990, On the Osage, *Nature Conservancy*, Vol. 40, pp. 7-15.

Massey, M., 2001, "Long-range Plan for the Management of Lesser Prairie Chickens in New Mexico: 2002-2006, Division of Wildlife, New Mexico Department of Game and Fish, Santa Fe.

McBee, K., and Wickham, J. W., 1988, "Petrochemical-Related DNA Damage in Wild Rodents Detected by Flow Cytometry," *Bulletin of Environmental Contamination and Toxicology*, Vol. 40, pp. 343-349.

McMurry, S. T., Lochmiller, R. L., McBee, K., and Qualls, C. W., 1999, "Indicators of Immunotoxicity in Populations of Cotton Rats (*Sigmodon Hispidus*) Inhabiting Petrochemical-Contaminated Environments," *Ecotoxicology and Environmental Safety*, Vol. 42, pp. 223-235.

Nature Conservancy, 1996, *Animals of the Tallgrass Prairie Preserve*, Unpublished data.

NRCS (Natural Resources Conservation Service) and Wildlife Habitat Council, 1999, *Lesser Prairie-chicken* (*Tympanuchus pallidicinctus*), Fish and Wildlife Habitat Management Leaflet Number 6, www.westgov.org/wga/initiatives/HighPlains/leaflet.html

ONHI (Oklahoma Natural Heritage Inventory), 1993, *Part II. Species Inventory for Significant Species at the Tallgrass Prairie Preserve*, Oklahoma Biological Survey, Norman, OK.

PADEP, 1998, *Ecological Screening Process*. Pennsylvania Department of Environmental Protection, November 10. www.dep.state.pa.us/dep/subject/advcoun/cleanup/attachmentve3.doc

Parton, W. J., Scurlock, J. M. O., Ojima, D. S., Gilmanov, T. G., Scholes, R. J., Schimel, D. S., Kirchner, T., Menaut, J. C., Seastedt, T., Moya E. G., Kamnalrut, A., and Kinyamario, J. I., 1993, "Observations and Modeling of Biomass and Soil Organic-Matter Dynamics for the Grassland Biome Worldwide," *Global Biogeochemical Cycles,* Vol. 7, pp. 785-809.

Pascual, J. A., Fryday, S. L., and Hart, A. D. M, 1999, "Effects of Food Restriction on Food Avoidance and Risk of Acute Poisoning of Captive Feral Pigeons From Fonofos-Treated Seeds," *Archives of Environmental Contamination and Toxicology* Vol. 37, pp.115-124.

Payne, T, and Caire, W., 1999, "Species Diversity of Small Mammals in the Tallgrass Prairie Preserve, Osage County, Oklahoma," *Proceedings of the Oklahoma Academy of Science*, Vol. 79, pp. 51-59.

Piearce, T. G., 1982, "Recovery of Earthworm Populations Following Salt-Water Flooding," *Pedobiolgia*, Vol. 24, pp. 91-100.

Pimm, S.L., Lawton, J.H., and Cohen, J.E., 1991, "Food Web Patterns and Their Consequences," *Nature*, Vol. 350, pp. 669-674.

Sorensen, M., and Margolin, J. A., 2002, *Evaluating Potential Ecological Impacts At Exploration & Production Sites. A Compilation of Screening Criteria*, Topical Report. GRI-02/0006, GRI, Des Plains, IL.

Stevenson, M., Coty, J., Stewart, J., Carlsen, T., and Callaham, M., 2001, *Developing Terrestrial Trophic Models for Petroleum and Natural Gas Exploration Sites: The Oklahoma Tallgrass Prairie Preserve Example*, Lawrence Livermore National Laboratory, January 26, 2001, UCRL-ID-142165, Livermore, CA.

Tietje, W. D., and Ruff, R. L., 1983, "Responses of Black Bears to Oil Development in Alberta," *Wildlife Society Bulletin*, Vol. 11, pp. 99-112.

Van Dyke, F., and Klein, W. C., 1996, "Response of Elk to Installation of Oil Wells," *Journal of Mammalogy*, Vol. 77, pp.1028-1041.

Wilson, J.B., 1999, "Guilds, Functional Types and Ecological Groups," *Oikos*, Vol. 86, pp. 507-522.

Wolff, J. O., 2001, "Behavioral Model Systems," *Landscape Ecology of Small Mammals*, Barrett, G. W., and Peles, J. D., Eds., Springer, NY, pp. 11-40.

Wong, D. C. L., Chai, E. Y., Chu, K. K., and Dorn, P. B., 1999, "Prediction of Ecotoxicity of Hydrocarbon-Contaminated Soils Using Physicochemical Parameters," *Environmental Toxicology and Chemistry*, Vol. 18, pp. 2611-2621.

Igor Linkov[1], Alexandre Grebenkov[2], Anatoli Andrizhievski[2], Alexei Loukashevich[2] and Alexander Trifonov[2]

Risk-Trace: Software for Spatially Explicit Exposure Assessment

REFERENCE: Linkov, I., Grebenkov, A., Andrizhievski, A., Loukashevich, A., and Trifonov, A., **"Risk-Trace: Software For Spatially Explicit Exposure Assessment,"** *Landscape Ecology and Wildlife Habitat Evaluation: Critical Information for Ecological Risk Assessment, Land-Use Management Activities, and Biodiversity Enhancement Practices, ASTM STP 1458,* L. A. Kapustka, H. Galbraith, M. Luxon, and G. R. Biddinger, Eds., ASTM International, West Conshohocken, PA, 2004.

ABSTRACT: Large areas of valuable habitats have been physically disturbed and/or contaminated by hazardous pollutants as a result of industrial activities. This paper presents a methodological approach and a software prototype for spatially explicit risk assessment of contaminated terrestrial ecosystems, to be implemented as a part of a risk-based decision protocol to support the assessment of ecological value and site reuse options. Exposure estimates for wildlife in areas containing spatially localized contaminants are functions of spatial factors, such as the receptor's average foraging area, the size of the habitat being assessed, and the distribution of contamination. Species exhibiting different foraging strategies may experience significantly different chemical exposures from the same site, even if their foraging areas overlap. Currently, exposure estimates and subsequent human health and ecological risk projections usually assume a static and continuous exposure of an ecological receptor to a contaminant concentration represented by some descriptive statistic, such as the mean or maximum concentration. These assumptions are generally overly conservative and ignore some of the major advantages offered by advanced risk assessment techniques, such as the ability to account for site-specific conditions and to conduct iterative analyses. We developed a spatially explicit foraging model that provides a time series estimation of soil and food contamination that receptors may encounter in their daily movements. The model currently inputs information on: the geospatial parameters of the contaminated area, surrounding land, and the habitat types found in each; density and distribution of ecological receptors; receptor home range; maps of contamination concentrations and habitat disturbance; and receptor's foraging range. The model also employs habitat quality factors that account for differential attraction to various habitat types within the site. This paper presents a software prototype that calculates chemical accumulation by a receptor foraging in areas having specified contamination patterns and habitat parameters.

KEYWORDS: Ecological risk assessment, military sites, decision support, modeling, risk analysis

Introduction

Industrial activities create both acute and chronic disturbances in ecosystems surrounding industrial facilities and infrastructure. In the case of the military and some industries, facilities have frequently been inaccessible to the public. As a result, many of these sites are actually relatively undisturbed ecologically, and harbor high biodiversity

[1] ICF Consulting, Inc., 33 Hayden Avenue, Lexington, MA, USA. ilinkov@icfconsulting.com, ph: 781-676-4053, FAX: 781-676-4072

[2] IPEP, 99 Akademik Krasin St., Minsk, Belarus. greb@sosny.bas-net.by, ph: 375 17-246-7542

and large expanses of habitat. Many countries, including developing countries, face the enormous challenge of planning the reincorporation of these sites into the local ecological, economic and cultural fabric while assuring their safe reuse for civilian, industrial and ecological purposes. Another challenge is to conduct limited ongoing military activities in active sites and/or industrial operation in a manner having minimal impact on the environment.

In all cases, activities to remediate affected sites must result in the protection of biodiversity, the reduction of present and future pollution, and the restoration of habitats in surrounding ecosystems. These actions will be effective only with an integrated site management approach, which will further support economic development in a manner that is sensitive to the parallel goal of natural resources conservation. In order to accomplish these often dichotomous goals, management specialists and relevant institutions would benefit from a guiding framework that would lead them through a systematic process for planning and decision-making, explicitly integrating both remedial investigation and ecological restoration goals, while considering the socio-economic context.

Our previous study (Linkov et al. 2001) presents a framework that integrates a number of risk and habitat assessment techniques into a systematic protocol for assessing and managing natural ecosystems at military sites. By integrating proven methods and principles of ecological impact assessment, risk assessment, habitat evaluation and habitat restoration, the protocol is designed to help managers develop creative solutions to the problem of cumulative stresses to the ecosystem from continuing and past military activities. Linkov et al. (2002), present a model that incorporates spatial scales into exposure assessment and risk characterization for a hypothetical aquatic site.

This paper presents both the methodological approach and a software prototype for spatially explicit risk exposure assessment of contaminated ecosystems. Currently, exposure estimates and subsequent human health and ecological risk projections usually assume a static and continuous exposure of an ecological receptor to a contaminant represented by some descriptive statistic, such as the mean or maximum concentration. These assumptions are generally overly conservative and ignore some of the major advantages offered by advanced risk assessment techniques, such as the ability to account for site-specific conditions and to conduct iterative analyses. The results of this study show that a simple model could explain the contaminant accumulation in ecological receptors foraging in heterogeneously contaminated sites with patchy landscapes.

Methods

The analysis employs a spatially explicit foraging sub-model that provides a time series of contaminant concentrations in soil and forage that a receptor may encounter within its habitat. The approach used to design the spatial sub-model is an extension and modification of a prior method (Linkov et al. 2002). The habitat is divided into a grid of one-meter by one-meter cells. Contaminant concentrations are then assigned to each cell based on site-specific measurements and/or a GIS coverage. The spatial sub-model uses the habitat grid to calculate exposure point concentrations for a receptor via soil and plant pathways. The probabilistic receptor migration sub-model then generates random receptor movements to model which exposure concentrations the receptors will encounter. In general, receptors are modeled to prefer areas with high habitat quality; i.e., they move in preferred directions that are determined by location, volume and attractiveness of habitat and forage resources. The rate of receptor migration within a habitat is inversely proportional to the forage volume and habitat quality of the surrounding cells. A probability of random movements is also assigned: at specified time periods, each individual receptor in the simulation is modeled foraging in randomly selected areas within the habitat.

We developed an illustrative example to represent a predominantly soil-driven food web that is common at contaminated terrestrial military sites. The conceptual model is a simple food chain in which the contaminant of concern is the radionuclide Cs-137. Although the current analysis addresses only Cs-137, the general methodology and conclusions are applicable to a wide range of contaminants. The exposure media are soil and forage (grass and mushrooms). The ecological receptors are roe deer (*Capreolus capreolus*). The roe deer is a reasonably representative mammal species because it: 1) is an important recreational species; 2) occurs abundantly in many countries; and 3) is a resident species with a relatively small foraging area. Thus, this species is likely to more frequently encounter localized contaminated sites ("hot spots") than other species that forage over larger areas.

The exposure point concentration for each time step is the average concentration across the cells that a deer encounters within its foraging area for a specified time period. The current simulation uses a daily time step, but different time intervals could be implemented.

The model inputs include information on: habitat size for a species; size and location of contaminated zones within the species habitat; size of the species' foraging area; and size and location of zones with different habitat quality within the species habitat. In the future, spatial distribution of radionuclide contamination within habitats and habitat quality will be entered as GIS coverages. The output of the spatial sub-model consists of combinations of radionuclide concentrations in soil and forage that the roe deer population may encounter while foraging in this habitat over time.

To characterize risk, the analysis then applies bioaccumulation factors (BAFs) for transfer of radionuclides from soil and plants to tissues. Within this conceptual model, we assumed that the deer feed solely on mushrooms and grass. The output of the risk characterization sub-model is a time series of radionuclide concentrations in deer tissues. The tissue concentration can be converted into doses (mg/kg/day) and estimated risks.

Software Implementation

This software was developed within Microsoft Office and functions as a Microsoft Excel macro (a subprogram). It uses Visual Basic and FORTRAN to perform calculations and data processing. The software package includes the following modules:

- User interface
- Modeling module
- Database module

The **user interface** was developed using Visual Basic and is compatible with Microsoft Office. Through this interface, a user can develop scenarios and specify model parameters. The visual interfaces developed for the prototype version include:

- General site information
- Information on site contamination, habitat and foraging resources
- Receptor selection window, containing links to the receptor database
- Model and parameter selection window
- Visual depiction of results

The **modeling module** uses FORTRAN to perform calculations of receptor exposures and risks. The appropriate calculation algorithms are automatically selected depending on how the user describes the scenario *via* the interface. The modeling module includes the following submodules:

- *Probabilistic receptor migration submodel.* Generates receptor movement in random directions, as well as movement towards receptor-preferred directions (determined by location, volume and attractiveness of local habitat and forage resources).
- *Spatially explicit exposure assessment submodel.* Calculates internal dose resulting from ingestion of contaminated food, as well as any other applicable routes of exposure (*e.g.*, soil).
- *Risk characterization submodel.* Calculates Hazard Quotients (HQs) for each contaminant; these are equal to the ratio of the exposure estimate to the selected safe benchmark dose for ecological receptors (toxicity reference values, TRVs).

The **database module** uses Microsoft Access as a data management platform. The libraries of receptor characteristics (*e.g.*, body weight, habitat size, forage resources) are stored as separate tables. The chemical libraries include physical characteristics, data on ecotoxicology, *etc*. Finally, site characteristics are stored as raster geographic maps.

Model Testing

To test the model performance, roe deer migration in an artificial landscape was modeled. Several landscapes with the same level of contamination and total area of high habitat were generated. In one extreme case, the areas of attractive habitat consisted of three squares (Fig. 1a), in the other extreme, three thin strips were modeled (Fig. 1b). Four additional landscapes with varying width/length ratios were also studied.

Figure 1 displays a sample graphic output showing contamination and habitat maps for the two extreme cases of the modeled artificial landscapes. The contaminated zones are shown as solid lines. These zones are also assumed to be of a high habitat quality. Each dot in the figure presents the location of one of the 20 modeled receptors at a different time. In both the square and rectangular habitat presented, receptors migrate extensively within the zones with high habitat quality. For the time of simulation (180 days) receptor forage extensively in the northern portion of all three zones and in the middle zone (Fig. 1a), while more extensive foraging in the eastern zone was modeled for square patches (Fig. 1b). Patches with good habitat quality are connected by corridors where the receptor moves (see more details about corridors in Hargrove et al., this volume).

Figure 2 shows radionuclide accumulation resulting from receptor migration in an artificial landscape with patches of different shape. The radionuclide accumulation is high in the landscape with square patches and is lower in the landscape with more fragmented, thinner patches. In the landscape with square patches, the receptor migrates continuously in the area of high contamination and accumulates significant amount of radioactivity. In fragmented habitats, the likelihood of migration in the contaminated field is smaller. Because more animals could be exposed to different contamination levels (distributed nature of the contaminants) the overall uncertainty in the mean accumulation level (expressed as confidence interval) is higher for the patchy landscape.

The influence of the landscape patchiness on the contaminant accumulation by receptors depends on the receptor's migration rate (Fig. 3). If the receptor moves fast, it covers more area and reaches steady-state concentration that is determined by the landscape structure. In this case, the difference among individual receptors is much smaller than in the situation with low migration rate. A slowly moving animal covers only a fraction of the habitat and thus interindividual variability may be high.

Initial Model Validation

Even though the presented model incorporates simplified assumptions about the nature of the spatial behavior of ecological receptors, it is useful for capturing some of the major components of an exposure and risk analysis for contaminated sites. Since the current paper illustrates the general framework for artificial landscapes, a rigorous model validation cannot be presented here and will be the subject of a subsequent publication. Nevertheless, in this paper we show that the model can be calibrated to predict radionuclide accumulation in deer in a forest in Germany. In addition, we show that the contamination pattern observed in fish collected in the New York Bight area supports the general trends and assumptions that our model predicts.

The terrestrial model was calibrated using the data collected at the prealpine region of Oberschwaben in southern Germany. This area was among the areas in central Europe that were most severely contaminated with cesium radionuclides from the Chernobyl fall-out. The observation of fairly high contamination values among individual roe deer from this area resulted in a surveillance program covering all roe deer shot by the state forest authorities in the first five years after fallout in that region. Since 1987, more than 8500 samples of roe deer shot within an area of about 40 x 40 km^2 have been analyzed with respect to their specific cesium activity (Zibold et al. 2001).

The model calculates ^{137}Cs concentration in roe deer tissues resulting from foraging in a habitat with spatially heterogeneous contamination. The probability of a receptor's presence in the specified areas was assigned based on deer hunting data. Field measurements for soil contamination were used. Figure 4 shows typical accumulation patterns for an individual animal. Initially, the animal is assumed to be uncontaminated. As foraging progresses, it starts to accumulate radionuclides, with amounts accumulated related to its modeled foraging pattern. After about 40 days, a dynamic equilibrium is observed, due to the fact that incorporation of additional dietary radionuclides is balanced by their depuration from the organism.

To model radionuclide accumulation in the roe deer population, 20 simulation runs were performed, and the results were then averaged and compared with field data. Figure 4 presents the annualized geometric means of the field data in comparison with model predictions. Good consistency was achieved.

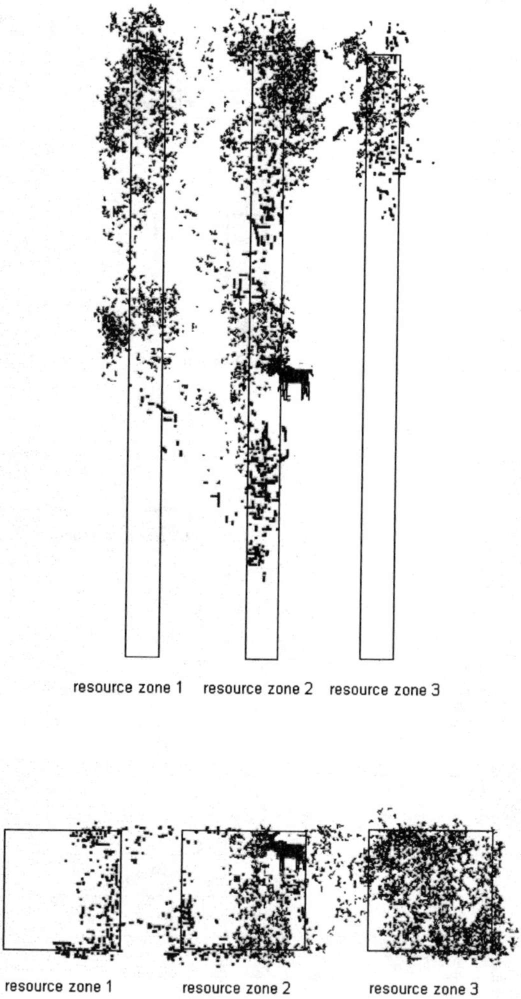

Figure 1. *Receptor Migration Pattern in Artificial Landscape*

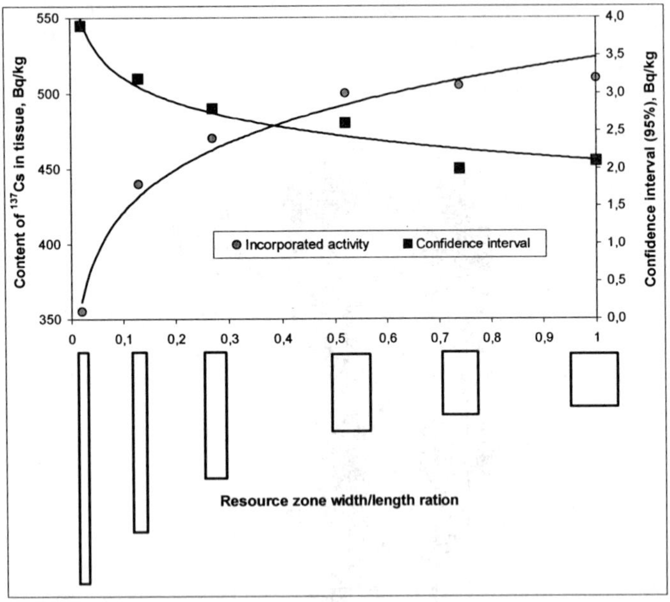

Figure 2. *Cs Accumulation by Roe deer foraging in landscapes with different patch shapes.*

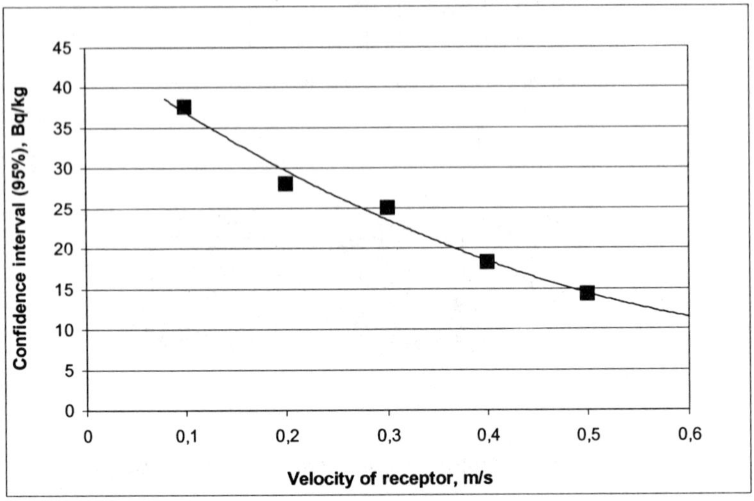

Figure 3. *The influence of the receptor's migration rate on the contaminant accumulation*

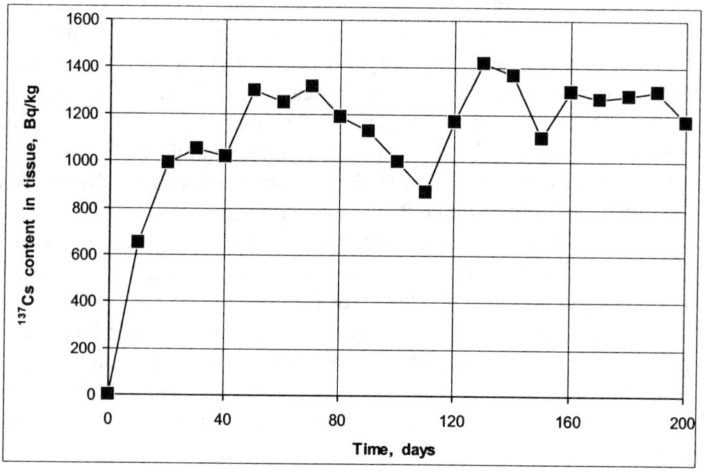

Figure 4 - *^{137}Cs accumulation in roe deer tissues*

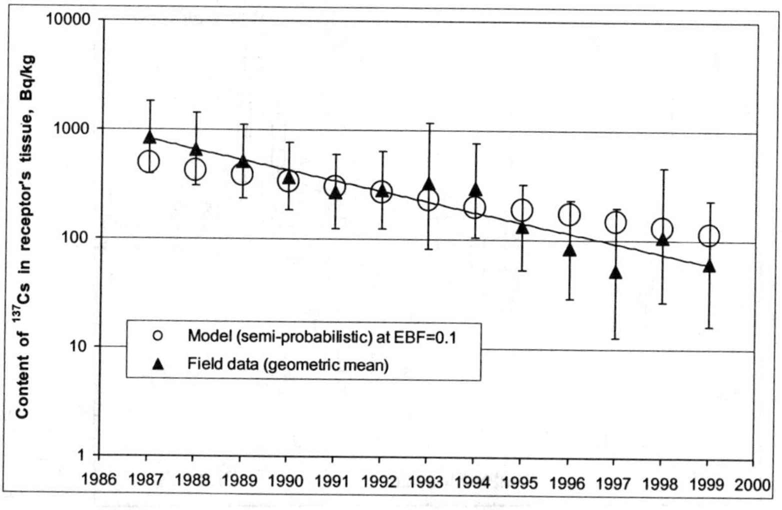

Figure 5. - *Comparison of experimental data and modeling results for mean roe deer contamination (geometric mean).*

Figure 6 presents PCB measurements in eel, winter flounder and blue fish collected in six sampling areas in the NY-NJ harbor estuary in fall 1993 or early winter 1994 (Skinner et al. 1996). The six areas have varying sediment PCB contamination resulting from industrial discharges and disposal of waste materials. The three fish species represent different foraging strategies. Eels spend most of the time foraging in the same area, while bluefish are known to cover large distances within short time periods. Winter flounder is a residential fish; its foraging area is quite large compared to that of eel, but much smaller than that of bluefish. The figure shows that the average PCB concentration in eel varies over three orders of magnitude among sampling areas, while the range for bluefish contamination is less than one order of magnitude. The range of body burdens within the same sampling area shows a similar trend. Winter flounder caught within the same general area exhibit quite a wide range of PCB concentrations, while individual bluefish do not show as much variation in tissue PCB concentrations.

These trends in concentration variation can be explained by the fact that fish with small foraging areas are likely to reflect local sediment contamination. Eel that happen to forage in a contamination hotspot are likely to be heavily contaminated, while other eel collected in a non-contaminated area are likely to be uncontaminated. Since bluefish forage over large areas as well as consume fish that forage over extended areas, they are affected by both clean and contaminated areas. No matter where bluefish are captured, they reflect the average contamination of a large habitat. Figure 6 shows that winter flounder falls between these two extremes.

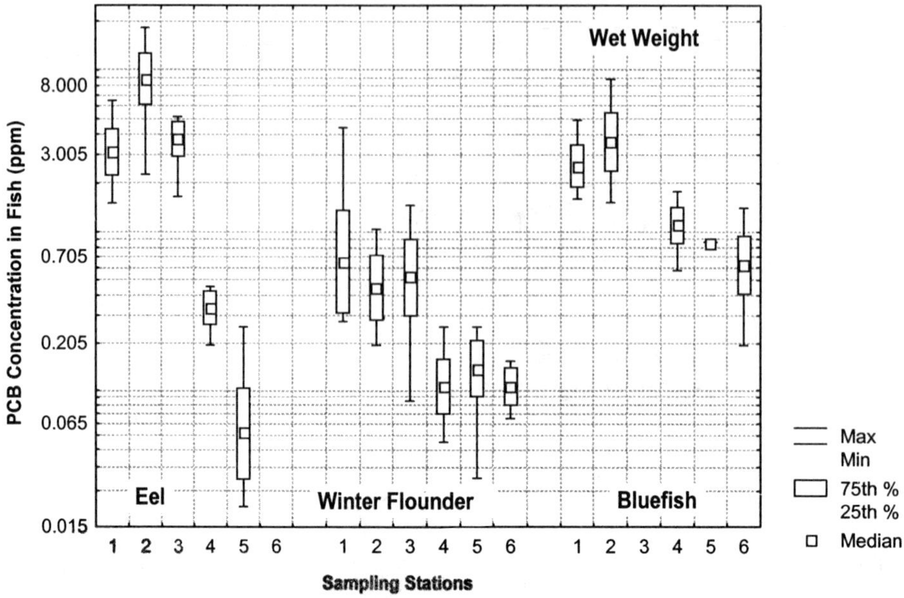

Figure 6 -- *PCB concentration in fish collected within the New York-New Jersey Harbor Estuary by Skinner et al. (1996). Sampling Stations: 1-Upper Bay, 2-East River, 3-Kills, 4-Jamica Bay, 5- Lower Bay, 6-NY Bight) (reproduced from Linkov et. al. 2002, with permission).*

Conclusions

We developed a spatially explicit foraging model that provides a time series of media and forage contamination that receptors may encounter during their daily movements. The model currently inputs information on: geospatial parameters of the contaminated area, surrounding land, and habitat types found in each; density and distribution of ecological receptors; receptor home range; maps of contamination concentrations and habitat disturbance; and size of the receptor's foraging range. The model also employs habitat quality factors that account for differential attraction to various habitat types within the site. The model is developed for both terrestrial and aquatic ecosystems.

The consistency between the modeled and experimental data achieved in this study reveals the utility of simple bioaccumulation and exposure models for site-specific risk assessment. Nevertheless, this consistency is the result of careful model calibration implemented in this study. Since the objective of the study was to test the model, many parameters were based solely on expert opinion. A full-scale implementation of this and similar models will require the collection of species-specific and habitat-specific data. Several papers regarding this issue provide additional details on possible methods and procedures one can use (Akçakaya 2000, Kapustka et al. 2001, Hope 2000, 2001, Freshman and Menzie 1996, Suter 2000).

This paper is a part of our overall effort to incorporate spatially explicit ecological risk assessment into a risk-based protocol to be used in decision-making regarding the reuse and/or sustainable use of disturbed sites. We propose to approach these complex problems by combining the approaches from traditionally disparate schools of assessment. The tools and methodologies to be developed will incorporate concepts from both risk assessment and ecological assessment to simultaneously address the factors (e.g., pollutants) that decision makers need to eliminate or minimize and the factors (e.g., habitat, rare species) decision makers want to maximize. Comparative risk assessment (CRA) is likely to be a key process in assessing the ecological value of contaminated or disturbed military sites and in the development of a reuse decision and site use protocol. CRA is emerging as a methodology that may be applied to facilitate decision-making when various possible activities compete for limited resources. CRA may be an especially valuable tool for prioritization of remediation efforts and for choosing among environmental policies related to military operations (Linkov et al. 2001).

Further development of the risk-based protocols and related prototype software will:

- Further develop risk assessment algorithms;
- Make direct use of geographic information systems (GIS) technology, and further integrate data with GIS;
- Supplement the database with profiles for a wider range of receptors;
- Enhance the current default database of exposure parameters and risk benchmarks;
- Expand functional modeling capabilities to include food chains and other dynamic factors of the specific ecological situation; and
- Link the user to expert decision support systems.

Acknowledgments

This work is partially funded by the United Nations Global Environment Facility and NATO Science for Peace Programme. The data were provided by Gregor Zibold, Fachhochschule Ravensburg-Weingarten, Germany. The authors are grateful to Dr. E. Klemt, Dr. G. Zibold, Dr. L. Kapustka, and A. Varghese for their review and useful suggestions.

References

Akçakaya, H.R. 2000. *Population Ecology.* Vol. 42, pp. 45-53.

Freshman, JS and Menzie CA. 1996. Two wildlife exposure models to assess impacts at the individual and population levels and the efficacy of remedial actions. *HERA*; Vol 2, pp. 481-498.

Hope, B. 2000. Generating Probabilistic Spatially-Explicit Individual and Population Exposure Estimates for Ecological Risk Assessments. *Risk Analysis*; Vol 20, pp. 573-589.

Hope, B. 2001. Consideration of Bioenergetic Factors in Spatially-Explicit Assessments of Ecological Receptor Exposure to Contaminants. *Journal of Toxicology and Industrial Health* Vol. 17, pp. 236-246.

Kapustka, L.A., Galbraith, H., Luxon, M., and Yocum, J. 2001. Using Landscape Ecology to Focus Ecological Risk Assessment and Guide Risk Management Decision-making. *Journal of Toxicology and Industrial Health* Vol. 17. pp. 322-332.

Linkov, I., Burmistrov, D., Cura, J., and Bridges, T.S. 2002. Risk-Based Management of Contaminated Sediments: Consideration of Spatial and Temporal Patterns of Exposure Modeling. *Environ. Sci. Technol.*; Vol. 36, pp. 238-246.

Linkov, I., Grebenkov, A., Baitchorov, V. 2001. Spatially Explicit Exposure Models: Application To Military Sites. *Journal of Toxicology and Industrial Health* **17**: 230-235.

Skinner, L.C., Jackling, S.J., Kimber, G., Waldman, J., Shastay Jr., J., Newell, A.J. 1996. *Chemicals in Fish, Shellfish and Crustaceans from the New York – New Jersey Harbor Estuary: PCB, Organochlorine Pesticides and Mercury.* New York State Department of Environmental Conservation, Division of Fish, Wildlife and Marine Resources, November 1996.

Suter, G.W., Efroymson, R.A., Sample, B.E., Jones, D.S. 2000. *Ecological Risk Assessment for Contaminated Sites.* Lewis Publishers: Boca Raton.

Zibold G., Drissner J., Kaminski S., Klemt E., and Miller R. 2001. Time-dependence of the radiocaesium contamination of roe deer: measurement and modeling. Journal of Environmental Radioactivity Vol. 55, pp. 5-27.

W. T. Wickwire[1], *Charles A. Menzie*[2], *Dmitriy Burmistrov*[3], *and Bruce K. Hope*[4]

Incorporating Spatial Data into Ecological Risk Assessments: The Spatially Explicit Exposure Module (SEEM) for ARAMS

REFERENCE: Wickwire, W. T., Menzie, C. A., Burmistrov, D., and Hope, B. K., **"Incorporating Spatial Data into Ecological Risk Assessments: The Spatially Explicit Exposure Module (SEEM) for ARAMS,"** *Landscape Ecology and Wildlife Habitat Evaluation: Critical Information for Ecological Risk Assessment, Land-Use Management Activities, and Biodiversity Enhancement Practices, ASTM STP 1458*, L.A. Kapustka, H. Galbraith, M. Luxon, and G.R. Biddinger, Eds., ASTM International, West Conshohocken, PA, 2004.

ABSTRACT: Although the tools available to ecological risk assessors have become more sophisticated, the basic questions remain the same. Foremost among those questions is what spatial scale is appropriate from an ecological, toxicological, operational and regulatory perspective for the ecological risk assessment. Once a spatial scale has been defined, the risk assessor needs useful modeling tools with enough power to evaluate exposure at the selected spatial scale and models that include a consideration of not only the physical size of an assessment area, but also the habitat suitability with respect to the needs of a number of wildlife species. To address this need, our team developed a spatially explicit exposure module (SEEM) for the U.S. Army that considers these aspects for some terrestrial wildlife species. SEEM offers the risk assessor the opportunity to improve the ecological relevance of the risk assessment by considering spatial aspects of exposure through an evaluation of heterogeneous habitat use and chemical patterns and a comparison of exposure with the potential for toxicological effects, resulting in a population measure of risk. SEEM predicts and compiles exposures for all individuals within a local population, rather than a single representative individual. In addition, SEEM increases the predictive capabilities of the exposure assessment by incorporating habitat preferences in the determination of daily exposure estimates. The model will track an individual over an ecologically-relevant period of time as it travels across a landscape. The individual will move according to a set of pre-determined rules and exposure for a population of individuals will be tracked over time. The module is being developed for inclusion within the U.S. Army Risk Assessment Modeling System (ARAMS).

KEYWORDS: wildlife exposure, spatial assessment, ARAMS, population risk assessment

Introduction

The tools available for wildlife exposure assessment vary from simple statistical tools applied broadly across an entire site to complex population risk modeling utilizing geographic information systems (GIS). Models that balance analytical power, ease of

---

[1] Senior Scientist, Menzie-Cura & Associates, Inc., Winchester, MA 01890, Email: wickwire@menziecura.com.

[2] Principal, Menzie-Cura & Associates, Inc., Winchester, MA 01890.

[3] Senior Scientist, Menzie-Cura & Associates, Inc., Winchester, MA 01890.

[4] Senior Environmental Toxicologist, Air Quality Division, Oregon Department of Environmental Quality, Portland, OR 97204.

application and utility for more accurately estimating risk from chemical exposures offer the risk assessor the ability to provide better information to risk managers. Wildlife species vary in their habitat preferences. Some species utilize habitats that maximize the chances of acquiring food resources and may even seasonally shift diets, thereby altering habitat preferences. Other species may make habitat use decisions based on predator community composition and focus on cover as well as food resources. Regardless, there are a variety of factors underlying wildlife habitat preferences and these factors influence the likelihood that a species will occur at a site. Any of these factors and assumptions may affect an exposure assessment. Additionally, an analysis at one spatial scale may result in a different conclusion than the conclusion reached from analysis at a different spatial scale (Wiens 1989). Current exposure assessment methods do not consider these aspects and use worst-case or conservative assumptions to provide deterministic exposure estimates that are biased, often resulting in improper interpretation of the risk estimate and misguided decisions. The availability of simple, flexible, yet realistic, wildlife exposure tools is important for increasing the value of wildlife exposure assessment in the risk management process.

Simple spreadsheet models may allow a user to explore the interaction of spatially variable chemical distributions, species-specific foraging strategies, and the resulting exposure. Freshman and Menzie (1996) introduce the Average Concentration with Area Curve (ACAC) model; a simple model to explore the relationship between spatial scale and average concentration. In this model, the analytical data are ranked by magnitude and the area over which they apply. A curve is generated from the ranked data reflecting trends in the average concentration as the overall area increases (Freshman and Menzie 1996). The addition of wildlife-specific foraging areas and screening criteria allow a user to identify species that might be susceptible to exposures based on the interaction of contamination patterns and foraging areas. In addition, this model has utility for hotspot analyses and remedial decision-making. Admittedly, the model relies on a number of simplifying assumptions, including treating the entire site as potential exposure habitat. The simplified approach, however, provides easy access to a useful analytical screening tool.

In the Population Effects Foraging Model (PEF), Freshman and Menzie (1996) expand upon the simple spatial assumptions in the ACAC by implementing concepts of movement over a landscape and including simple individual exposure tracking based on a grid of cells. In the PEF model, wildlife exposures may be restricted by a foraging distance, or an individual may move across an entire landscape. Each individual is randomly placed on the landscape in areas considered to provide habitat. The model calculates exposure averaged over the given foraging area for each individual and compares the exposure to a soil effect level. The model then repeats the exposure for all individuals in a local population. The PEF model computes a percentage of the local population affected and then iterates the process for thousands of different, randomly located local populations (Freshman and Menzie 1996). In combination with foraging area to total habitat ratios and ranges of screening values, the PEF model may be used to identify patterns resulting from the interaction of contamination patterns and foraging area sizes. The resulting foraging area-effect patterns are not always intuitive.

Hope (2001) highlights the overriding goal for spatially explicit exposure models, which is to avoid reaching "misleading" risk management conclusions, e.g. identifying wildlife risk when the risk is driven by chemical concentrations in non-habitat areas of a site. Hope (2000, 2001) has developed a habitat area and quality-conditioned exposure estimator (E[HQ]) and a spatially explicit ecological exposure model (SE^3M) to facilitate this calculation. This individual-based model can employ different, movement-based, foraging strategies including: unrestricted, limited and restricted foraging (Hope 2000, 2001). Each strategy includes different rules regarding individual movement based on the presence or absence of habitat and foraging starting points. In this model, the landscape is organized on the basis of spreadsheet cells, each with defined total and habitat areas. Extent of daily foraging by individuals is a function of cell total area. From an ecological perspective , it assumes that individuals may only forage (and therefore be exposed to chemicals) in cells containing habitat; however, there is an option to allow movement through non-habitat containing cells (Hope 2000, 2001).

The Spatially Explicit Exposure Module (SEEM) is being developed to integrate key analytical components of the ACAC, PEF and SE^3M models within an interface that balances the importance of user accessibility and simplicity with more meaningful estimates of exposure and the development of population risk estimates based on these exposure estimates. SEEM employs rule-based foraging used in previous wildlife exposure models (Freshman and Menzie 1996; Hope 2000, 2001; Marsh and Jones 1988). Ultimately, SEEM will become a wildlife exposure module within the Army Risk Assessment Modeling System (ARAMS) and will draw input information from the databases within ARAMS. In this paper, an introduction to the model environment is followed by a summary of model assumptions and development strategies.

Model Environment

The integration of SEEM into ARAMS will add exposure assessment power to the comprehensive library of assessment models within ARAMS. ARAMS is a "decision support" tool that assembles a number of multimedia fate and transport models, bioaccumulation models, exposure models, effects databases and risk assessment models under a single modular framework (Dortch and Gerald 2002). The assemblage of risk assessment tools within ARAMS relies on the object-oriented, modular platform of the Framework for Risk Analysis in Multimedia Environmental Systems (FRAMES). FRAMES is a platform that provides the capability of drawing information from models that individually may not be able to communicate with one another (Dortch and Gerald 2002; Whelan et al. 1997). Each model within ARAMS adds a unique analytical function that complements the other models and increases the power of the package. Currently ARAMS contains a chemical property and effects database, chemical concentration and release module, fate and transport modules for air, surface water, soil-vadose zone and groundwater, modules for human health risk assessment and ecological effects databases, exposure modules and ecological risk assessment tools (Dortch and Gerald 2002). In addition, ARAMS will offer sensitivity and uncertainty tools and a geographic information system (GIS) (planned) (Dortch and Gerald 2002). A habitat suitability index module is currently under development (Kapustka et al. 2004a, 2004b; Kapustka 2003; Kapustka et al. 2001; Terrell and Carpenter 1997). ARAMS is available on the worldwide web.

Although SEEM will operate as a stand-alone model, user accessibility increases when it becomes a module within ARAMS. Models and databases within ARAMS supply wildlife species information, (e.g. ingestion rates, foraging areas, diet compositions), a GIS-based mapping function (planned), habitat suitability indexing tools, chemical toxicity reference value (TRV) and uptake factor databases and chemical concentration assessment and fate and transport tools.

Model Inputs	Model Outputs
Site base map	Report viewer
Chemical concentrations (soil, sediment, water)	Population ecological hazard quotient curves
Habitat suitability indices	Ecological hazard quotient with area curves
Foraging radius	Interactive map displays
Foraging events per day	Ecological hazard quotient patterns
Toxicity reference values	Habitat suitability patterns
Ingestion rates	Real-time cell status bar

TABLE 1 – Summary of SEEM Inputs and Outputs

Spatially-Explicit Exposure Module (SEEM)

SEEM has been developed to improve the analysis of population risk and increase the realism of wildlife exposure assessment. In general, "population-level" assessments consider the individuals that comprise the "population" within the area of interest. SEEM tracks exposure for each individual within the defined local population. Within the ecological risk assessment framework (USEPA 1992, 1997, 1998), SEEM improves the characterization of wildife exposure. SEEM integrates the influence of spatial foraging strategies, the distribution of suitable habitat and the spatial distribution of chemical concentrations and evaluates the impact each has on population risk. In addition, SEEM increases the realism of the exposure assessment process by incorporating habitat preferences at a finer resolution than the entire site.

Conceptual Overview

The model will track an individual over a specified period of time as it travels across a landscape. The core interface within SEEM consists of a user-selected base map and a user-defined grid. The grid is used to characterize the chemical concentrations and habitat suitability across the landscape. The model user selects a grid size appropriate to the existing chemical information and based on available knowledge about the habitat suitability of the site. The grid cells are not used directly to establish or define foraging areas. From the primary interface, the user can access all data inputs required to run the model (TABLE 1). Once integrated into ARAMS, the inputs will be supplied by existing databases, a GIS or fate and transport/uptake models within ARAMS.

SEEM currently contains three submodels. Two of the submodels represent common wildlife foraging strategies, "random walk" and nesting. While these strategies do not represent the universe of potential movements that wildlife might employ to search for food, they do represent common approaches useful for many analyses. In addition to the two foraging submodels, a third submodel, the plant or biota submodel, is provided for acute exposure assessment and to generate prey item concentrations that can be used in the food chain models for higher order organisms in the random walk and nesting submodels.

Individuals move according to a set of pre-determined rules within each submodel. The probability of locating a starting point in any given position for each individual is proportional to the habitat suitability. As discussed previously, wildlife species may utilize different habitats to meet different goals such as acquisition of food compared to protecting against predation, while also acquiring food. The assessment of the habitat suitability is conducted outside of SEEM. The input parameter for habitat suitability is an index (between 0-1). Ultimately, SEEM will use habitat suitability indices (HSI) derived through application of habitat suitability index models compiled in an independent database (Kapustka et al. 2004a, 2004b; Kapustka 2003; Kapustka et al. 2001; Terrell and Carpenter 1997). In this discussion, habitat suitability refers to specific measurable parameters compiled into the HSI representing a species' specific habitat use preferences. Once a starting point is determined, the model operates based on a foraging radius/distance. This radius describes the maximum distance over which an animal may forage in one day. Applying a foraging distance within the model allows the model to be independent from the grid of cells, i.e., the cell size affects the foraging models indirectly in that it reflects the ability of the modeler to predict contamination and habitat suitability at a certain spatial scale across the landscape/site. Each day an animal forages over an area defined by a foraging radius. Although the radius is fixed per day, the actual number of cells to which a receptor is exposed may vary based on the geometry of the cells and the current position of the receptor. Within the foraging radius, the individual will complete a foraging event a defined number of times; the user selects the number of daily foraging events. The locations of the foraging events are determined using a Markov Chain Monte Carlo (MCMC) method with the habitat suitability determining the probability to forage in any given location. At the end of each day, an average exposure dose is calculated from the foraging events for each individual and tallied for all individuals in the population over the exposure period.

Movement occurs within the foraging submodel and is driven by a combination of relative habitat suitability indices and the foraging radii. As a result, the species' relative habitat preferences are used as weighting factors for daily exposure averaging. Conceptually, it is an expression of the likelihood that an individual will forage within a given cell which is based on the amount of time spent and/or food that may be obtained from the particular cell.

Submodels

Our goal in selecting two general foraging strategies is to provide the user with two choices that are ecologically relevant, yet simple, in concept. A third submodel, the plant and biota submodel, is provided for acute exposure assessment. The foraging submodels do not capture every potential foraging strategy, but do represent general strategies employed by wildlife receptors either throughout a lifetime or during specific developmental stages. From a screening perspective, they represent contrasting techniques that are useful for the identification of exposure patterns. Users are able to modify input factors, e.g., ingestion rates, foraging areas, foraging events per day, in order to test the impact of variations in the foraging strategy. The two foraging strategies include: random walk and nesting.

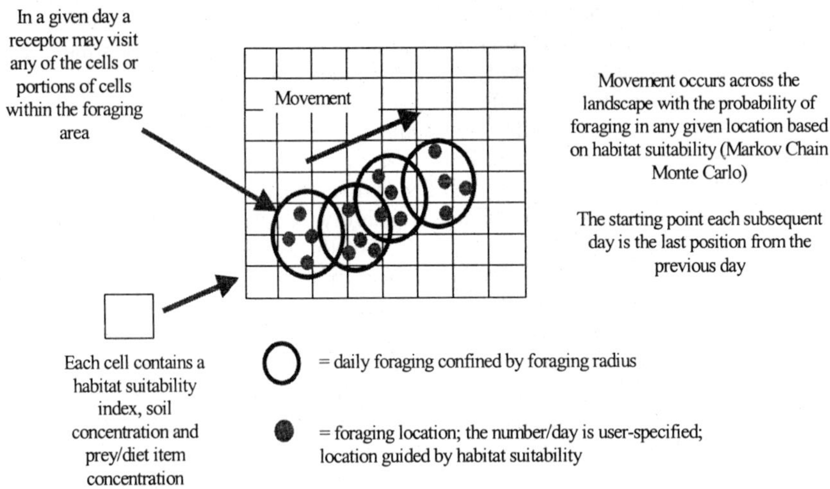

FIG. 1 – Conceptual Overview of the Random Walk Foraging Submodel

Similar to previous wildlife exposure models, movement and exposure in SEEM is guided by a set of rules (Freshman and Menzie 1996; Hope 2000, 2001; Marsh and Jones 1988). Each of the submodels within SEEM operates under a different set of rules and is designed to assess a specific exposure type or to provide information required by the other submodels. The submodels increase the flexibility of SEEM as an analytical tool. The user may choose to explore specific questions through the selection of a set of inputs. For example, one might question how foraging area sizes for different habitat qualities might impact exposure for a group of species. To explore the question, the user might run the model a number of times for different combinations of foraging radii, habitat qualities and wildlife species.

Factors that determine the foraging strategy that a specific organism or species might employ vary from specific metabolic requirements to presence or absence of competitors to temporal factors such as the time of year and age of the individual (Stephens and Krebs 1986). Foraging strategies will also vary based on the suitability/type of available habitat under any given foraging scenario. A review of a few studies reveals the complex factors that influence foraging strategy selection. Remsen and Robinson (1990) begin to explore the complexity of classifying avian foraging strategies. Specifically, they examine five categories/components required to define foraging behavior in birds. They include: search, attack, foraging site, food and food handling (Remsen and Robinson 1990). Within each category of foraging behavior, there are a number of different subcategories of activity, e.g. search: walk, hop, jump, run, climb, glide, flutter, or fly (Remsen and Robinson 1990). Morris and Davidson (2000) studied optimal foraging strategies in mice. In this study, researchers determined that foraging strategy complexity not only is dependent on habitat suitability, but also is influenced by predation threat. For mice, the intensity of mice foraging activity is higher in safer habitats (Morris and Davidson. 2000).

Random Walk — Within the random walk submodel, individuals of a selected wildlife species move across the landscape and are exposed to media and diet items (Fig. 1). For each model run, the probability of an individual beginning in any given cell is proportional to the relative habitat suitability. Habitat suitability also influences movement across the landscape. Movement is guided by a set of pre-determined rules. First, daily foraging is constrained by the foraging radius. An individual may forage in any area within the foraging radius containing a habitat suitability index greater than zero. Foraging area is not defined by a set number of matrix cells. Second, the species will complete a user-defined number of foraging events within the foraging radius each day. Exposure is calculated at the random points within the foraging area and recorded. The exposure point positions are simulated using a MCMC approach in such a way that the probability of a foraging event being placed inside a cell (or part of a cell) is proportional to the local habitat suitability. Third, an exposure dose will be calculated each day as the average over all the exposure events during this day. Since the probability distribution of the exposure points is influenced by the spatial distribution of habitat suitability, this procedure essentially provides a habitat-suitability weighting of the average daily exposure. Fourth, exposure for each subsequent day will begin at the last foraging location from the previous day. Movement under the random walk foraging strategy within SEEM is summarized in Fig. 1. Depending on the spatial distribution of habitat suitability indices and the foraging period, an individual may forage over the entire landscape defined by the base map. A home range is not used to restrict foraging within a landscape. The site boundary is considered a "reflective border" in the sense that an individual cannot migrate offsite, and new individuals cannot enter the site during the modeling period. Individuals also cannot visit cells with a relative habitat suitability rating of 0 (in the current version of SEEM). Specifically, this means that an individual can cross a zero habitat region (as long as the dimensions of that region are less than the foraging radius), but will not be exposed in that area. The large-scale migration of the population, such as seasonal migrations of some bird populations, can be modeled as an option simulating the season-dependent probability to migrate offsite and the inverse probability to return to a random location (weighted by habitat suitability) within the site. The time spent migrating offsite is considered "exposure free".

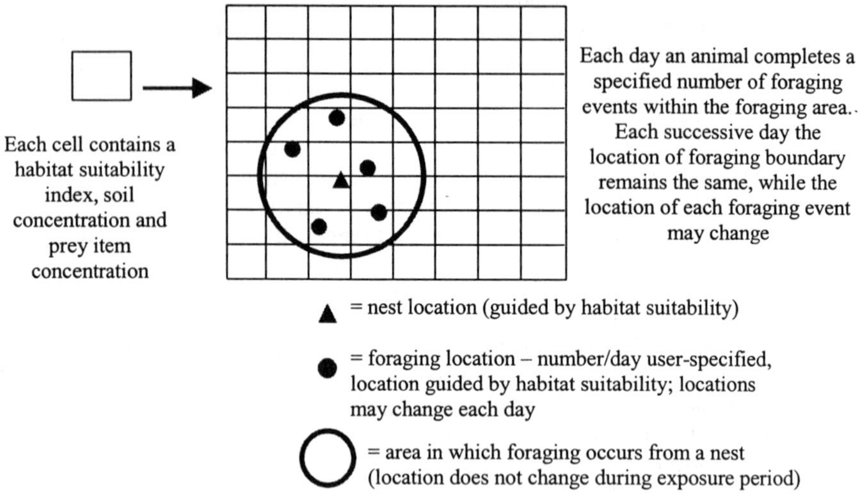

FIG. 2 – Conceptual Overview of the Nesting Foraging Submodel

Nesting – Under the nesting foraging model, daily foraging follows the first three foraging rules under the random walk foraging strategy. In contrast to the random walk foraging submodel, under the nesting submodel an individual remains in a fixed location and forages within the foraging radius for the entire modeling period (Fig. 2). Therefore, in this submodel the foraging area is equal to the home range. While the foraging area defined by the radius remains the same, the location of each foraging event position within this area may vary each day throughout the modeling period. The nesting location for each individual is selected using MCMC simulation providing habitat-weighted randomization. Offsite migration may be selected for this submodel as well.

Plant or Biota Submodel – In addition to the two foraging submodels, SEEM also includes a submodel for estimation of plant or biota (tissue) concentrations employing uptake factors and for acute exposure assessment at single points. In this submodel exposure occurs at a single point. Under the current design, each cell with habitat for a specific receptor also includes a food item representing an important diet component. For example, a fox may consume a small mammal, while a robin may consume a soil invertebrate. SEEM focuses on the vagile terrestrial wildlife species, where we assume that each matrix cell containing species-specific habitat also contains a non-diminishing food item. The food items do not move and are always available to any species or foraging event that intersects a habitat-containing cell. However, a user may choose to evaluate any wildlife species for which a database of foraging areas, ingestion rates, TRVs and habitat suitability indices are available using the plant or biota submodel. This submodel is also useful for identifying potential acute effects based on a one-time exposure to soil (or other contaminated medium) and a diet item within a single cell.

Inputs

SEEM includes a stepwise, guided input interface. The interface introduces users to the inputs and provides valuable guidance regarding the required data. The categories of data required to operate SEEM are described in the sections that follow.

Base Map and Landscape Characterization – The chemical concentrations in site media, habitat suitability indices and prey/diet item concentrations are all organized and compiled on a grid system that overlies the site or landscape map. The model will recognize a number of image file types. Once loaded, the user must define the scale of the map using x- and y-coordinate distances and select units. In addition, the user must select the grid size appropriate to resolution of the site data. SEEM does not provide geospatial averaging tools. However, within ARAMS, a GIS system is planned to assist in developing habitat suitability and chemical concentration data layers. The selection of the grid size is project- and site-specific and will depend on receptor foraging distances, the resolution of habitat suitabilities and the availability of soil chemistry data. A user is required to evaluate the density and spatial distribution of sample points and habitat suitability parameters in order to select the size of the grid cells. The grid is employed to define the chemical and habitat suitability components of the landscape; the grid cells are not used to define foraging areas or to alter the foraging strategies directly. For the model to operate correctly, each cell containing habitat for the species of interest, must contain chemical concentrations in the focal media and a habitat suitability index. Prey/diet concentrations are calculated by SEEM based on the media concentrations.

Foraging Radius and Foraging Events – As discussed previously, the input or constraint on the area foraged per day for SEEM is the foraging radius. Foraging radii are the "distances the animals are willing to travel to potential food sources" (USEPA 1993). The foraging radius within SEEM delineates a circle, however we recognize that in a heterogeneous landscape, the geometry that a "foraging radius" defines will vary (USEPA 1993). Within SEEM, a user enters the foraging radius, but also enters the number of foraging events that occur each day. While the radius defines an area over which an individual will travel to obtain food, it does not necessarily define the number of events completed to obtain food. In SEEM there is no limit on overall area foraged for an individual throughout the modeling period (number of days the model is run). Depending on the number of days selected for a model run, a receptor may cross the entire landscape. However, users retain the flexibility to restrict foraging (e.g., define a home range) through user-defined habitat qualities. While it is not possible to capture the full complexity of foraging theory, SEEM does provide analytical flexibility through modification of inputs.

Habitat suitability – One of the important analytical components of SEEM is the consideration of relative HSIs in the assessment of wildlife exposure. By including habitat suitability indices (based on habitat use preferences) as one of the foraging parameters/"rules", a situation in which a high exposure concentration in a portion of the site that does not contain habitat for wildlife of interest, does not result in a finding of ecological risk. Evaluating relative habitat preferences can be subjective. SEEM will rely on an objective method independent from the model to determine the relative habitat suitability using an evaluation protocol that integrates HSI models (Kapustka et al. 2004a, 2004b; Kapustka 2003; Kapustka et al. 2001; Terrell and Carpenter 1997).

Time-Dependent Model Properties – Exposure calculations within SEEM are completed on a daily basis. The user may define the number of days within a full model run. In addition, the model offers the user the capability to modify exposure assumptions based on the season. Specifically, a user can input information on access to media and time spent onsite versus time spent migrating offsite. The user may enter the total number of days in each season.

Chemical Exposure Parameters – The model currently will assess any chemical for which the necessary concentration and TRV information is provided. In addition to chemical concentration information for all exposure media on a cell-by-cell basis, the other important chemical input required to run the model is a chemical-, species- and endpoint-specific TRV. The user may specify the TRV source (e.g. chronic, acute, subchronic), the target/endpoint effect (e.g.,. reproduction, mortality, development) or the level of observation (No Observed Adverse Effect Level, Lowest Observed Adverse Effect Level, etc.).

Wildlife and Food Chain Exposure Parameters – The model requires each user to provide a number of exposure parameters. First, the user needs to select the species of interest. Selection of the assessment species should occur within the larger context of conceptual model development and incorporate knowledge about the site and habitat, toxicological exposure information for the compound and species of interest, and the interaction of species sensitivity to exposure and spatial mobility. Second, the diet for the species should be selected and uptake factors for any prey/plant diet components should be specified. Third, a user provides the food, soil and water ingestion rates for each species of interest. Finally, before running the model, the user selects the foraging submodel (random walk or nesting) or the plant or biota submodel and the foraging radius.

Assumptions

In order to balance model accessibility with assessment power, a number of simplifying assumptions are made. First, it is assumed that every cell with a relative habitat suitability index greater than 0, contains a prey or food item and the prey item does not diminish when consumed or move across neighboring cells (i.e., food is not density dependent). As a result, an individual may forage multiple times in a given day at points within the same cell. Food availability may be a factor that the user chooses to include within the habitat suitability index, but the habitat suitability index does not vary within a single model run. The tissue concentration in prey is calculated using uptake factors under equilibrium conditions with the corresponding media concentrations in cells where prey is located. Prey are exposed within a single cell and do not migrate to other cells (in the current version of SEEM). Second, the model assumes that a user can populate a landscape with chemical data and relies on the user to select, complete and defend spatial interpolation methods. A GIS module (planned) within ARAMS will facilitate the process. Third, SEEM assumes individual independence in terms of foraging. The foraging paths of individuals may overlap and cross, but we do not assume that inter- or intra-species competition impacts foraging success. Fourth, as discussed previously, we include two foraging strategies that cover two of the primary foraging types, but not the entire spectrum of options.

Model Output

The model generates an exposure dose and ecological hazard quotient (EHQ) on a per day basis for each individual and is run for an ecologically relevant period of time (TABLE 1). The final EHQs consist of an average and maximum of all of the hazard quotients over the period of exposure for the individual. The ultimate goal is to assess exposure and risk to a population of species incorporating the influence of habitat suitability on that exposure. The output provides a summary of the percent of the population at risk. The model can also be applied to identify acute risks, to perform sensitivity analyses using different TRVs, and to evaluate the interaction between chemical concentrations, habitat suitability and foraging behavior parameters. The model is accessible and provides the user with the ability to modify exposure assumptions, modify receptor species, update or modify habitat suitability and can accommodate updated source data (inputs can be modified easily). The development of screening templates within SEEM will increase the exploratory power of the model.

Integration with ARAMS

When integrated into ARAMS, SEEM will draw the required inputs from pre-existing databases and models within ARAMS. The seamless integration will allow a user to prepare a map in which the landscape has been characterized using a GIS, select species-specific exposure data from a comprehensive database, select TRVs from another database and then run the model. Outputs will be provided directly from SEEM, but may also be routed to other modules within ARAMS. A user may also select the standalone SEEM which will retain full operational functionality as an independent wildlife exposure model.

Applications

SEEM offers a flexible and accessible screening platform for the analysis of the interaction of chemical distribution, habitat suitability, foraging strategies and wildlife exposure estimates. Through the review of resulting exposures and hazard quotients, a user can identify specific areas of the site that might significantly impact wildlife. Users may also use SEEM to implement theoretical remedial strategies and observe the resulting impact on population risk. While the calculation of hazard quotients should not represent the only line of evidence in a risk assessment, the analysis can provide useful insights into comparative patterns of impact.

Future work will focus on developing additional screening interfaces for use within SEEM. One interface may provide a user with a broad screening assessment based on body weight differences. For example, using allometric relationships, the interface/template might run a series of exposure assessments for a wildlife species category, e.g., herbivorous, forest-dwelling mammals. The screening template could then be used to run the SEEM model for a series of body weights and display the results as the percent of population affected in each weight class. This screening template would allow a user to identify sensitive weight classes. Previous work with the PEF model illustrated the results from this pattern analysis are not always intuitive and are closely aligned with spatial distribution of contamination and foraging area size.

Additional plans include developing a remedial options template for SEEM. Within this template a user may set specific remedial goals, e.g., "clean" specific areas to a user-defined concentration and then run the model again. In this way, SEEM will be a valuable tool in the remedial cost-benefit assessment process providing insights for balancing risk reduction, habitat loss and acreage remediated.

Currently, the model is designed to allow the user to perform sensitivity analyses. A user may choose to deactivate the randomizing components of the model. This allows the user to run the model multiple times, with variable inputs and obtain comparable results, i.e. results that aren't reflecting inherent randomness from the model. In future versions of the model, we plan on developing templates that can be used for exploratory analyses. The templates could be applied to modify sets of parameters (habitat suitability, foraging areas, foraging events) and produce comparative figures and tables.

Conclusion

Recognition of the importance of spatial scale and habitat suitability in ecological exposure assessment is an important improvement to the process. While SEEM offers an accessible and simple interface for incorporating spatial scale and habitat suitability considerations, the user must ensure that defensible inputs are used. As discussed previously and explored in detail by Wiens (1989), the sensitivity, responses identified and ultimately the conclusions of any study are all influenced by the spatial scale of analysis. Conclusions at one spatial scale, may differ from the conclusions at another spatial scale. Understanding the differences and developing a defensible approach for selecting model inputs will help the user clearly define and begin to address exposure questions. Through the inclusion of habitat suitability information, species specific foraging behaviors and a mechanism for tracking exposure for all individuals within a population, SEEM offers risk managers an accessible, yet powerful wildlife exposure tool.

Acknowledgments

We are grateful to Dr. Mark S. Johnson and Dr. Christopher J. Salice of the U.S. Army for supporting this work and providing model comments throughout the development process. We would also like to acknowledge the support provided by Dr. Mark Dortch of the U.S. Army, and Dr. Gene Whelan and Mitch Pelton of the Pacific Northwest National Laboratory of the U.S. Department of Energy for their input to the development and integration of SEEM into ARAMS. This work is funded through the Environmental Quality Technology Program of the U.S. Army.

References

Dortch, M. S. and Gerald, J. A., "Army Risk Assessment Modeling System for Evaluating Health Impacts Associated with Exposure to Chemicals," *Brownfield Sites: Assessment, Rehabilitation and Development,* C. A. Brebbia, D. Almorza and H. Klapperich, Eds., WIT Press, Southampton, UK, 2002.

Freshman, J. S. and Menzie, C. A., 1996, "Two Wildlife Exposure Models to Assess

Impacts at the Individual and Population Levels and the Efficacy of Remedial Actions," *Human and Ecological Risk Assessment,* Vol. 2, No. 3, pp. 481-498.

Hope, B. K., 2000, "Generating Probabilistic Spatially-Explicit Individual and Population Exposure Estimates for Ecological Risk Assessments," *Risk Analysis,* Vol. 20, No. 5, pp. 573-589.

Hope, B. K., 2001, "A Case Study Comparing Static and Spatially Explicit Ecological Exposure Analysis Methods," *Risk Analysis*, Vol. 21, No. 6, pp. 1001-1010.

Kapustka L.A., Galbraith H., Luxon M., 2001, "Using Landscape Ecology to Focus Ecological Risk Assessment and Guide Risk Management Decision-Making," *Toxicology and Industrial Health*, Vol. 17, pp. 236-246.

Kapustka L.A., 2003, "Rationale for Use of Wildlife Habitat Characterization to Improve Relevance of Ecological Risk Assessments," *Human and Ecological Risk Assessment* (in press).

Kapustka L.A., Galbraith H., Luxon M., Yocum J., Adams B., "Application of Habitat Suitability Index Values to Modify Exposure Estimates in Characterizing Ecological Risk," *Landscape Ecology and Wildlife Habitat Evaluation: Critical Information for Ecological Risk Assessment, Land-Use Management Activities, and Biodiversity Enhancement Practices, ASTM STP 1458*, L.A. Kapustka, H. Galbraith, M. Luxon, G.R. Biddinger, Eds., ASTM International, West Conshohocken, PA, 2004a.

Kapustka L.A., Galbraith H., Luxon M., Yocum J., Adams B., "Predicting Biodiversity Potential Using a Modified Layers of Habitat Model," *Landscape Ecology and Wildlife Habitat Evaluation: Critical Information for Ecological Risk Assessment, Land-Use Management Activities, and Biodiversity Enhancement Practices, ASTM STP 1458*, L.A. Kapustka, H. Galbraith, M. Luxon, G.R. Biddinger, Eds., ASTM International, West Conshohocken, PA, 2004b.

Marsh L.M. and Jones R.E., 1988, "The Form and Consequences of Random Walk Models, *Journal of Theoretical Biology*, Vol. 133, pp. 113-131.

Morris, D.W. and Davidson, D.L., 2000, "Optimally Foraging Mice Match Patch Use with Habitat Differences in Fitness," *Ecology*, Vol. 81, No. 8, pp. 2061-2066.

Remsen, J.V. and Robinson, S.K., 1990, "A Classification Scheme for Foraging Behavior of Birds in Terrestrial Habitats," *Studies in Avian Biology,* No. 13, pp. 144-160.

Stephens, D.W. and Krebs, J.R., *Foraging Theory*, Princeton University Press, New Jersey, 1986.

Terrell J.W. and Carpenter J., *Selected Habitat Suitability Index Model Evaluations*, Information and Technology Report (USGS/BRD/ITR—1997-0005), U.S. Department of the Interior, U.S. Geological Survey, Washington, DC, 1997.

USEPA, *Framework for Ecological Risk Assessment*, Risk Assessment Forum, U.S. Environmental Protection Agency (EPA/630/R-92/001), Washington, DC,1992.

USEPA, *Wildlife Exposure Factors Handbook*, Volume I, Office of Research and Development, U.S. Environmental Protection Agency (EPA/600/R-93/187),Washington, DC, 1993.

USEPA, *Ecological Risk Assessment Guidance for Superfund: Process for Designing and Conducting Ecological Risk Assessments*, Interim Final, Solid Waste and Emergency Response, U.S. Environmental Protection Agency (EPA/540/R-97/006), Washington, DC, 1997.

USEPA, *Guidelines for Ecological Risk Assessment*, Risk Assessment Forum, U.S. Environmental Protection Agency (EPA/630/R-95/002F), Washington, DC, 1998.

Whelan, G., Pelton, M.A., Castleton, K.J., et al., *Concepts of a Framework for Risk Analysis in Multimedia Environmental Systems*, Prepared by, Pacific Northwest National Laboratory, Richland, Washington (PNNL-11748), Prepared for, Office of Research and Development, National Environmental Research Laboratory,

U.S. Environmental Protection Agency, and Office of Environmental Management, U.S. Department of Energy, and Radiation Protection Division, Center for Risk Modeling and Emergency Response, U.S. Environmental Protection Agency, 1997.

Wiens, J.A., 1989, "Spatial Scaling in Ecology," *Functional Ecology,* Vol. 3, pp. 385-397.

Bruce K. Hope, Ph.D.[1]

Approaches to Spatially-Explicit, Multi-Stressor Ecological Exposure Estimation

REFERENCE: Hope, B. K., **"Approaches to Spatially-Explicit, Multi-Stressor Ecological Exposure Estimation,"** ***Landscape Ecology and Wildlife Habitat Evaluation: Critical Information for Ecological Risk Assessment, Land-Use Management Activities, and Biodiversity Enhancement Practices, ASTM STP 1458***, L. A. Kapustka, H. Galbraith, M. Luxon, and G. R. Biddinger, Eds., ASTM International, West Conshohocken, PA, 2004.

ABSTRACT: When assessing risks posed by chemical toxicants, risk assessors must estimate a receptor's level of exposure to them in abiotic (soil, water, etc.) and biotic (tissues, prey items) media. However, any free living terrestrial receptor is constantly challenged to avoid or minimize physical and biological stressors. Thus a receptor at a contaminated site may also face challenges from physical and biological, as well as toxicant, stressors. Toxicant stress may pose a risk on its own, or may add to risk posed by physical and biological stressors that are a part of a receptor's everyday existence. It is generally recognized that the relative spatial positions of receptors and contaminated media can strongly influence estimates of exposure and hence of risk. How a receptor moves with respect to habitat directly influences how it may be affected by one or more stressors. This paper was prepared to provoke further discussions on: (1) the benefits associated with attempting to estimate a terrestrial receptor's exposure to multiple stressors as that receptor moves through both space and time and (2) the challenges posed by attempting such an estimate in the context of a typical production ecological risk assessment.

KEYWORDS: ecological risk, spatially-explicit , bioenergetic, multiple stressors

Introduction

When assessing risks posed by chemical toxicants, risk assessors must estimate a receptor's level of exposure to them in abiotic (soil, water, etc.) and biotic (tissues, prey items) media. Such assessments are typically performed at state or federally regulated hazardous waste sites or for agroecosystems. However, any free living terrestrial receptor is constantly challenged to avoid or minimize physical (lack of habitat) and biological (no or few food items containing high levels of metabolizable energy) stressors. Toxicant stress may pose a risk on its own, or may add to risk posed by physical and biological stressors that are a part of a receptor's everyday existence. Thus a receptor at a contaminated site may face, in addition to toxicant (chemical levels in abiotic habitat (e.g., soils) and food items) stressors, challenges from physical (availability of suitable habitat, degree of structural integrity of extant habitat, access to habitat) and biological (availability and energy content of prey or forage items) stressors. This is a limited definition of a biological stressor due to the focus here on food consumption as a primary toxicant exposure pathway for terrestrial receptors. Biological

---

[1] Senior Environmental Toxicologist, Land Quality Division, Oregon Department of Environmental Quality, 811 SW Sixth Avenue, Portland, Oregon, 97204.

stressors may also include competitive interactions, density-dependent factors, and genetic variation, among others. Stress on an individual receptor or a population of receptors can thus be conceptualized, minimally, as occurring along three axes that vary continuously from low to high: one expressing food energy availability and quantity, one quantifying toxicant levels in the receptor's habitat and food items, and one expressing the availability, accessibility, and integrity of habitat features. Although simultaneous exposure to all three types of stressor is not likely to be beneficial, stressor interactions are not always additive, in that receptors capable of moving through non-habitat may gain relief from toxicant stress at the expense of increased stress from lack of food.

Differing stressor combinations can be grouped into eight exposure scenarios. In the optimal scenario (scenario 1), a receptor's habitat and food energy needs are readily met without risk of exposure to, and effects from, toxicants. An "attractive nuisance" (scenario 2) occurs when both habitat and high-energy, but toxicant-containing, food items are available. Such attractive nuisance situations exacerbate exposure in habitat where the availability of food energy is either unaffected, or worse, enhanced, by co-related toxicants otherwise harmful to a receptor. Examples might be a fertilizer that exerts a toxic effect on the receptor but a beneficial effect on any vegetation that forms its habitat or the increased availability and consumption of prey items ("invertebrate rain") both contaminated and immobilized by organophosphate insecticides aerially applied to a forest (Stehn et al. 1976). However, exposure may be less when toxicant-containing, but high-energy, food is available, since less mass of food (and of contaminant) would need to be consumed to meet basic energy needs.

In scenario 3, where exposure to toxicants is low, poor habitat or food quality may prevent a receptor from meeting it's daily energy needs. Although its ingestion rate can increase to compensate for food with a low energy content, it can only do so up to a physiologically-determined maximum ingestion rate (Shipley et al. 1994). Beyond that an energy deficit may occur, with an attendant loss of foraging capability leading potentially to starvation. Scenario 4 presents a receptor with the dual threats of both not obtaining sufficient food energy and of being exposed to some level of toxicant in its food. As toxicant intake is also a function of the daily rate of food ingestion, exposure to a toxicant could increase if a receptor is forced to consume greater amounts of low-energy, toxicant-containing food to meet its basic energy needs. Because it is a fundamental assumption that foraging and food intake can only occur within suitable habitat, the remaining four scenarios have limited ecological plausibility. When neither habitat or food resources are adequate (scenarios 5 and 6), receptors may simply be absent from the area; in these instances toxicant exposure is of little concern. Because a receptor's prey or food items may have habitat requirements that don't fully overlap with those of the receptor, they can exist absent habitat for that receptor. However, being outside a receptor's habitat is likely to make them inaccessible to that receptor, diminishing the chance for exposure but enhancing the probability of receptor starvation (scenarios 7 and 8).

In addition to the possibility of multiple stressors acting on a receptor, the scientific and regulatory communities also generally recognize that the relative spatial positions of receptors and contaminated media can strongly influence estimates of exposure and hence of risk. How a receptor is assumed to move (i.e., the movement rule to which it typically adheres) directly influences how it may be affected by one or more stressors.

For a receptor that must remain in suitable habitat (percolation movement), lack of habitat itself is a source of physical stress and may exacerbate the impact of other stressors. The absence of habitat can deprive a receptor of food and shelter, potentially leading to its starvation and death, and non-habitat can block movement, potentially confining the receptor to areas with high toxicant levels, exacerbating toxicant stress, or denying it access to food resources, increasing biological stress. For receptors capable of crossing non-habitat areas (nearest-neighbor movement), lack of habitat is also a source of physical stress and may have opposing interactions with other stressors. Because neither energy or toxicants (if present) are consumed in non-habitat, movement through it reduces toxicant stress because of "dilution" (i.e., non-habitat contributes zero toxicant to average intake), but increases the potential for biological (food energy deficit) stress, as evidenced by an increased number of energy deficit days and a decline in average energy level. This could place such a receptor at greater risk of starvation and discourage movement through non-habitat even for receptors capable of such movement. When a toxicant is present in habitat , a receptor may experience both toxicant and biological (food) stresses, but the threat of toxicant stress is potentially lower for a nearest-neighbor disperser than for a percolator.

This paper was prepared to summarize comments prepared for a symposium on interactions between ecological risk assessment and landscape ecology. Its purpose was to provoke further discussions on the ecological realism afforded by attempting to estimate a terrestrial receptor's exposure to multiple stressors as that receptor moves through both space and time and methods for attempting such an estimate in the context of a typical production ecological risk assessment.

Considering Spatial and Temporal Dimensions

Simple individual-based, random walk models, implemented with spreadsheets and Visual Basic® programs, provide a risk assessor an accessible means for exploring the interaction of spatially variable contamination, species-specific foraging strategies, and the resulting exposure. Such models, informed by the rich literature on foraging and dispersal theory and research (e.g., Stephens and Krebs 1986, Turchin 1998), are particularly applicable to the practical problem of ecologically relevant estimates of exposure over space and time. In such models, each individual has an explicit location in space at each time step in the simulation (DeAngelis and Gross 1992; Marsh and Jones 1988). Where an individual is located from one time step to the next depends upon "rules of movement" that individuals follow and how they respond to variations in the landscape. The rules of movement and response to environmental variation have a profound effect on where, when, and how far individuals move across the landscape and where they are exposed to toxicants in various media (Marsh and Jones 1988). These rules (and others) may vary depending on the internal condition, life history stage, and environmental context of an individual (Turchin 1998). These differences help shape how individuals in a particular state are distributed across the landscape, to which environmental stressors they are exposed, and the ultimate population-level consequences. Individual-based models allow for considerable flexibility in the definition of movement rules, thus allowing them to be tailored to better approximate the life history and ecology of a specific species of receptor.

Freshman and Menzie (1996) developed two such spreadsheet models: an average concentration with area curve and a population effects foraging model. Linkov et al. (2002) developed a spatially and temporally explicit model for management of contaminated sediment sites. Hope (2000; 2001a) developed a habitat area and quality-conditioned exposure estimator and a spatially explicit ecological exposure model to facilitate its calculation. In this model, the landscape is organized on the basis of spreadsheet cells, each with defined total and habitat areas. Extent of daily foraging by individuals is a function of cell total area and the receptor's foraging area. Foraging area is defined here as the area a receptor is willing to search to locate potential food sources and/or suitable habitat per day. It is synonymous with home range, the geographic area encompassed by a receptor's activities over a specified time or where a receptor may be located 95% of the time (Minta 1992; USEPA 1993). From an ecological perspective, it assumes that individuals may only forage (and therefore be exposed to contaminants) in cells containing habitat; however, there is an option to allow movement through non-habitat containing cells (Hope 2000; 2001a). This individual-based model can employ different, movement-based, foraging strategies including: unrestricted, limited and restricted foraging. Each strategy includes different rules regarding individual movement based on the presence or absence of habitat and foraging starting points (c.f., Table 1). To include the temporal dimension, the receptor is moved through *d* days of foraging, with *d* approximately equal to the average longevity of the receptor species in the wild. Values for a population of *n* individual receptors are generated by iterating the spatial model, in its entirety, *n* times. After *d* days of foraging by a population of *n* individuals, model execution terminates. Hope (2001a) highlights the overriding goal for spatially explicit exposure models, which is to avoid reaching 'misleading' risk management conclusions, e.g. identifying wildlife risk when the risk is driven by contaminant levels in non-habitat areas of a site.

TABLE 1 - *Movement options available with simple spreadsheet models*

Issue	Option	Consequence
How does habitat influence movement?	With nearest-neighbor rule receptor may move through any cell. Applies to species willing or able to cross areas of unsuitable habitat (e.g., large mammals, birds).	May "dilute" exposure. Time spent in non-habitat cells deducts from foraging capacity without necessarily adding to exposure. No restriction on movement can provide greater access to non-contaminated habitat.
	With percolation rule receptor may move only through habitat-containing cells.	May increase exposure as there is less opportunity for "dilution" through access to "clean" habitat.
	Applies to species unwilling or unable to cross areas of unsuitable habitat (e.g., small mammals).	May decrease exposure because receptor would spend less time in non-habitat.

TABLE 1 (cont'd) - *Movement options available with simple spreadsheet models*

Issue	Option	Consequence
Where does exposure occur?	Food consumption (with incidental soil ingestion) exposures occur only in habitat containing cells with habitat quality > 0.	Assumes that primary exposure is habitat-dependent. Dermal contact exposures could be modeled as habitat independent.
What are the limits on a receptor's movement?	All movement must occur within the model landscape, whose edge is assumed to be a reflective barrier.	Model landscape must encompass both the site and a significant portion of the receptor's forage area.
Where does a given receptor's movement begin on first day?	At a randomly selected habitat-containing cell within a range of cells. From a specific (fixed) habitat-containing cell.	Reinforces assumption of habitat dependency. Range option assumes that not all receptor's in a population are likely to begin movement from the same cell.
	At a randomly selected cell within a range of cells. From a specific (fixed) cell.	Could be used to assess (a) the effect of habitat dependency on a receptor's exposure or (b) a receptor's exposure in an area of missing or marginal habitat.
Where does a given receptor's movement begin on subsequent days?	Movement begins again within the cell range or specific cell used on first day.	Represents a receptor foraging from a specific location (e.g., a nest or den) or specific area (e.g., protecting a territory).
	Movement continues from the location where the previous day's movement ended.	Represents a receptor foraging continuously over an area without needing to return to a specific location.
For how many days does the simulation run?	Value could be selected to approximate a receptor's average lifespan in the wild or in captivity or the duration of chronic toxicity tests on which the toxicity reference value is based.	A receptor with a large foraging area, over many days, is more likely to encounter contamination. The coefficient of variation (CV) in the exposure estimate is higher for short runs by receptors with small forage areas.

TABLE 1 (cont'd) - *Movement options available with simple spreadsheet models*

Issue	Option	Consequence
What determines a receptor's direction of movement?	Movement into any one of 8 adjacent cells is determined at random (With, 1997; Rule #2).	A receptor's movement behavior is not directly affected by landscape or environmental characteristics.
	Movement is "directed" toward the adjacent cell whose habitat quality is highest relative to that in other adjacent cells.	A receptor interacts with its environment to a limited amount. Such directed behavior enhances the possibility of an "attractive nuisance" (i.e., habitat that is both high quality and contaminated).
Does a receptor have to leave its current location on the next move?	Yes	Assumes that a given habitat patch is highly resource limited. A receptor is forced to move because it has quickly exhausted resources in its present location.
	No	Assumes that for any given cell to be resource limited is unlikely.
How many moves will a receptor make each day?	Determined by the receptor's forage area divided by the area of each cell visited.	The CV in the exposure estimate decreases with increases in the forage area/cell area ratio. A ratio of 50 or greater provides a CV < 1 for both nearest-neighbor and percolation rules.
What is the length of each move?	1 cell	The "length" of a move can be adjusted by changing cell size (in terms of cell area).

Considering Physical Stressors

A simplified view of "habitat quality" is as a suite of attributes related to the structural integrity, suitability, attractiveness, and food resource availability of a given habitat for a given receptor (Bowers 1994). Conservation biology research indicates that the primary physical stress facing many free living receptors is the availability (or lack) of suitable quality habitat. This stress can be readily incorporated in spreadsheet models.

First, by controlling a receptor's access to habitat, particularly its ability to move from one habitat patch to another, by applying either a nearest-neighbor movement rule, which assumes a receptor (typically a large mammal or bird) can cross non-habitat or a percolation movement rule, which assumes a receptor (typically a small mammal) is unwilling or unable to cross non-habitat (King and With 2002). Then, second, by assuming that toxicant exposure and food consumption can only occur in habitat-containing cells. A third option, to address food resource limitations and time required for food resource recovery from foraging, requires a receptor to vacate its current cell on each move and, in addition, be restricted from returning to that cell for some length of time (c.f., Table 1).

Considering Biological Stressors

Although biological stressors may also include competitive interactions, density-dependent factors, and genetic variation, among others, the one of interest here is the availability and energy content of prey or forage items. A receptor must, obviously, obtain sufficient energy (in the form of food) to meet its daily energy needs and thus avoid the stress of starvation, which could result from a lower encounter rate with food items (prey mass/time), prey of the wrong size, or lower energy content of food items. Food consumption is also a key link between ecology and ecological risk assessment, where it is often conceptualized as a primary toxicant exposure pathway for terrestrial receptors (Moore et al. 1999). As a receptor consumes food to meet its energy needs, it may also be consuming toxicants contained in or on its food items.

How much food energy a receptor needs to acquire, on a daily basis, is a function of its field metabolic rate (FMR), the total energy cost a wild animal pays during the course of a day. FMR includes the cost of basal metabolism, thermoregulation, locomotion, feeding, predator avoidance, alertness, posture, digestion, food detoxification, reproduction and growth, and other energy expenses that ultimately appear as heat, as well as any savings resulting from hypothermia (Nagy 1987, 1994). Field metabolic rate can be derived allometrically as a function of receptor body weight (USEPA 1993). FMR varies temporally in response to changing environmental, seasonal, and physiological factors. This variation in FMR over time is simulated by sampling a distribution defined by the upper and lower 95% confidence intervals on an allometrically derived value for FMR,

$$FMR = a(BW)^b \tag{1}$$

$$BW \sim Normal(x, s) \tag{2}$$

$$FMR_{95} = \log_{10} FMR \pm c\sqrt{d + e(\log_{10} BW - \omega)^2} \tag{3}$$

$$NFMR \sim \frac{Triangular(-FMR_{95}, FMR, +FMR_{95})}{BW} \tag{4}$$

where:

FMR = Field metabolic rate (kcal/d)
a, b = Receptor-specific coefficients for allometric FMR estimation (unitless)
x = Mean of receptor body weight (g)
s = Standard deviation of receptor body weight (unitless)
$FMR_{\pm 95}$ = Upper and lower bounds of the daily food ingestion rate (kcal/d)
BW = Receptor body weight (g)
c, d, e, ω = Receptor-specific coefficients for FMR bound estimation (unitless)
NFMR = Normalized field metabolic rate (kcal/g·d)

To meet its energy needs as expressed by the FMR, a receptor must obtain metabolizable energy through food consumption. Metabolizable energy is a function of the types of food items potentially available in a given location, the probability that each item is actually present in that location, and the gross energy and assimilation efficiency of each food type potentially present,

$$ME = \sum_{i=1}^{f} GE \cdot AE \cdot DF \cdot PF \tag{5}$$

where:

ME = Metabolizable energy available in a given location (kcal/g, dry wt)
GE = Gross energy content of food item (kcal/g, dry wt)
AE = Assimilation efficiency of food item (unitless)
DF = Fraction of food item in total diet (unitless)
PF = Probability that food item is present in a given location (unitless)
f = Number of food items in a given location (unitless)

To meet its energy needs, a receptor must ingest food at a rate dictated by the metabolizable energy it can obtain from available food items,

$$IR = FMR/ME = FMR/(AE \cdot GE) \tag{6}$$

where:

IR = Daily food ingestion rate available in a given location (g/d, dry wt)

Two assumptions underlying Equation (6) are that neither metabolizable energy or ingestion rate will be limiting factors. Such assumptions are likely implausible under field conditions, where poor habitat quality (structure and extent), restricted availability of food or food with a low energy content, limits on a receptor's ingestion rate, or ingestion of toxicants, which could degrade feeding behavior and increase respiration (Donkin et al. 1989; Nisbet et al. 1996), may singularly or collectively limit energy availability. From a spatially-explicit perspective, the availability and quantity of metabolizable energy is likely to be both limited and uncertain as both habitat and receptor characteristics vary over time and space.

When metabolizable energy is limited due to the presence of food items with a low gross energy content (e.g., dry grasses versus seeds), the expectation of Equation (6) is that a receptor will increase its ingestion rate, and thus its energy intake, in compensation. A receptor's ability to adjust its ingestion rate is not, however, limitless. In mammalian herbivores, for example, maximum intake rate has been shown to scale closely with body weight, $IR_{max} \approx BW^{0.71}$ (Shipley et al. 1994). In habitat with low energy food items, application of Equation (6) may generate very high, physiologically implausible, ingestion rates. Given that a receptor's food ingestion rate can be estimated allometrically as a function of its body weight (USEPA 1993), ecological realism suggests that this rate could be assigned a plausible upper bound, such as the 95th percentile, based on an allometrically derived rate,

$$IR_{95} = \log_{10} IR + c\sqrt{d + e(\log_{10} BW - \omega)^2} \quad (7)$$

$$NIR_{95} = IR_{95}/BW \quad (8)$$

$$NIR = \begin{cases} NFMR/ME & \text{if } NFMR/ME < NIR_{95} \\ NIR_{95} & \text{if } NFMR/ME > NIR_{95} \end{cases} \quad (9)$$

where:

IR_{95} = Upper bound on daily food ingestion rate (g/d, dry wt)
c, d, e, ω = Receptor-specific coefficients for IR confidence interval (unitless)
NIR = Normalized daily food ingestion rate (g/g·d, wet wt)
NIR_{95} = Upper bound of normalized food ingestion rate (g/g·d, wet wt)

Because the field metabolic rate is defined as a receptor's total daily energy requirement, it is to be expected that, if sufficient (or greater) metabolizable energy is available to meet that requirement, then at the end of each day energy supply will meet energy needs, so that: (ME × NIR) - FMR ≈ 0. If FMR/ME > NIR_{95}, then energy intake was potentially limited by physiological restrictions on a receptor's ability to ingest a sufficient mass of food (and energy) and its daily energy balance will be less than zero 0. This is a type of biological stress that a receptor may be able to accept periodically but not sustain indefinitely.

This food energy aspect of biological stress has been explored with a spatially- and bioenergetically-explicit model where: (a) movement rules control access to suitable habitat, (b) habitat quality is expressed in terms of gross energy available from a suite of habitat-specific food types, (c) intake of contaminants is linked to food consumed to meet daily energy needs, (d), fulfillment of a receptor's energy needs as it traverses habitat patches with varying gross energy levels is tracked both daily and over a lifetime, and (e) contaminant doses and resulting tissue residue levels as a receptor moves through habitat patches with differing toxicant levels are also tracked, both daily and over a lifetime (Hope 2001b). This model provided a structured basis for the intuitive conclusion that a free living terrestrial receptor faces a constant demand for food energy, making inadequacies in food resources as potentially as great a stressor as the presence of a toxicant (Hope 2001b).

Considering Chemical Stressors

As a receptor ingests food items, it also has the potential to ingest toxicants contained in or on those items. A receptor's daily intake is linked to the energy quality of the food items available to it. More sophisticated assessments estimate total daily intake of a toxicant with an implicit ingestion rate estimated from data on gross food energy, food assimilation efficiency, and food availability,

$$TDI = NFMR \cdot \sum_{i=1}^{n} \frac{C \cdot DF}{AE \cdot GE} \tag{10}$$

where:

TDI = Total daily toxicant intake (mg/kg·d)
C = Contaminant concentration in food item (mg/kg)
n = Number of food items being ingested (unitless)

As noted previously, food-related biological stress could result from a lower encounter rate with food items, prey of the wrong size, or lower energy content of food items. These have different implications when combined with toxicant stress. For example, both ingestion rate and toxicant intake (at least for that item) would decrease if there were a reduced rate of encounter with a specific food item. Conversely, both ingestion rate and toxicant intake might increase if each food item has a low energy content. With contaminated, but high energy content food, toxicant intake may or may not increase. It may be less if the receptor consumes less food to meet its daily energy needs or it may be more if the receptor seeks to capitalize on the availability of high energy food.

From a spatially-explicit perspective, exposure for ecological receptors is generally assumed not to be a habitat-neutral process (i.e., all areas are not equally, randomly, and completely accessed), but is assumed to occur only in habitat required by a receptor (Hope 2000). Thus the toxicant concentration that a receptor receives after foraging for one day through habitat patches with varying toxicant concentrations in different food items may be represented as,

$$C_{HW} = \sum_{k=1}^{g} (C \cdot (HA/TAHF)) \tag{11}$$

where:
C_{HW} = Habitat area-weighted toxicant concentration (μg/g, wet wt)
C = Chemical concentration in food items in a given habitat patch (μg/g, wet wt)
HA = Habitat area (ha)
TAHF = Total area of habitat foraged in a day (ha)
g = Number of habitat patches foraged in one day (unitless)

One issue not often addressed in ecological risk assessments is that of the changes in a receptor's toxicant tissue residue levels that are likely to occur as it moves through space and time. The toxicant tissue residue level in a receptor at the end of each day's foraging through cells with varying toxicant concentrations in different food items is,

$$\frac{dTC}{dt} = (Dose \cdot \alpha) - (TC \cdot k_e) \quad (12)$$

where:

TC = Average daily tissue residue concentration (μg/g, wet wt)
Dose = Habitat area-weighted average applied daily dose (μg/g·d, wet wt)
α = Toxicant assimilation efficiency (unitless)
k_e = Toxicant elimination rate (d^{-1})
t = Time (d)

In most ecological risk assessments, tissue residue levels are typically estimated to assess bioaccumulation or biomagnification potential, with the view that such levels can only increase. However, a moving receptor may accumulate toxicant residues in contaminated habitat but then depurate a portion of that load if it has access to non-contaminated habitat. Considerable depuration may occur in "clean" habitat if it is occupied for some multiple of the half-life of the toxicant. Linkov et al. (2002) used a probabilistic adaptation of the Gobas bioaccumulation model (Gobas 1993) to account for spatial and temporal variation in fish exposed to concentrations of hydrophobic contaminants in sediment and surface water. Rather than increasing continuously, tissue residue levels rose and fell over time as fish migrated in and out of contaminated areas. As a result, risks to humans from fish consumption were as much as one order of magnitude lower when the spatial and temporal characteristics of the fish (e.g., foraging area, seasonal migration) were considered.

Conclusions

What does this discussion suggest for the typical ecological risk assessment? First, assuming ecological risk is due only to toxicant stress may ignore the actual source of stress, or fail to recognize that it is not due to the toxicant alone but to its interaction with other stressors. In these instances, restoration activities (if planned) may be unsuccessful because remedial actions taken to address toxicants may miss, or worse, exacerbate, the actual sources of stress. Second, there needs to be greater cognizance of the movement rule preferred by each mobile receptor identified as an assessment entity and of the degree of habitat fragmentation on and around a site. For a given contaminated site, percolators (usually small mammals) can experience greater toxicant stress than nearest-neighbor dispersers (usually large mammals and birds) if they are trapped on the site by a lack of habitat-containing escape routes capabilities. Conversely, nearest-neighbor dispersers may be less exposed on average to a toxicant than percolators because of their ability to escape from the contaminated site through non-habitat areas. However, it is rare to find a "production" ecological risk assessment (i.e., one responsive to and constrained by budget, schedule, logistical, and regulatory constraints) that explicitly incorporates spatial factors.

Lastly, several states (e.g., Massachusetts, Texas, Pennsylvania, Louisiana) have included population-level considerations in the screening procedures used to determine if an ecological risk assessment is needed for a chemical release site. These procedures use *de minimis* spatial scale criteria, such as 1 to 2 acres of terrestrial habitat (provided that a variety of other conditions are also met, based on the premise that some sites are too small for population level exposures to occur; therefore, population level impacts are not expected to occur. However, the spatial scale of interest to a receptor is strongly influenced by its forage area capabilities (as evidenced by changes in the coefficient of variation in the exposure estimate) and preferred movement rule. Establishing a "generic" *de minimis* scale for any and all receptors may easily fail to account for significant influence of these species-specific characteristics. Thus, while it requires greater ecological expertise, establishing a "no effect" spatial scale (if any) as a function of a specific receptor's ecological preferences may be preferable to setting a default minimum for any and all sites and receptors.

Acknowledgement

All views or opinions expressed in this paper are those of the author and do not necessarily represent Oregon Department of Environmental Quality policy or guidance. No official endorsement is implied or is to be inferred. Preparation of this paper was greatly assisted by constructive comments from two anonymous reviewers.

References

Bowers. M. A., 1994, "Use Of Space and Habitats By Individuals and Populations: Dynamics and Risk Assessment," In R. J. Kendall and T. E. Lacher, Eds., *Wildlife Toxicology and Population Modeling: Integrated Studies of Agroecosystems*, CRC Press, Inc., Boca Raton, Florida, pp.109-122.

DeAngelis, D. L. and Gross, L. J., 1992, *Individual-Based Models and Approaches in Ecology: Populations, Communities and Ecosystems*, Chapman & Hall, New York.

Donkin, P., Widdows, J., Evans, S. V., Worall, C. M., and Carr, M., 1989, "Quantitative Structure-Activity Relationships for the Effect of Hydrophobic Organic Chemicals on Rate of Feeding By Mussels (*Mytilus Edulis*)," *Aquatic Toxicology*, Vol. 14, pp. 277-294.

Freshman, J. S. and Menzie, C. A., 1996, "Two Wildlife Exposure Models to Assess Impacts at the Individual and Population Levels and the Efficacy of Remedial Actions," *Human and Ecological Risk Assessment*, Vol. 2., No. 3., pp. 481-498.

Gobas, F. A. P. C., 1993, "A Model for Predicting the Bioaccumulation of Hydrophobic Organic Chemicals in Aquatic Food-Webs: Application to Lake Ontario," *Ecological Modelling* Vol. 69, pp. 1-17.

Hope, B. K., 2000, "Generating Probabilistic Spatially-Explicit Individual and Population Exposure Estimates for Ecological Risk Assessments," *Risk Analysis*, Vol. 20, No. 5, pp. 573-590.

Hope, B. K., 2001a, "A Case Study Comparing Static and Spatially-Explicit Ecological

Exposure Analysis Methods," *Risk Analysis*, Vol. 21, No. 6, pp. 1001-1010.

Hope, B. K., 2001b, "A Spatially- and Bioenergetically-Explicit Exposure Model for Terrestrial Ecological Risk Assessment," *Toxicology and Industrial Health*, Vol. 17, No. 5-10, pp. 322-332.

King, A. W. and With, K. A., 2002, "Dispersal Success on Spatially Structured Landscapes: When Do Spatial Pattern and Dispersal Behavior Really Matter?" *Ecological Modelling*, Vol. 147, pp. 23-39.

Linkov, I., Burmistrov, D., Cura, J., and Bridges, T. S., 2002, "Risk-based Management of Contaminated Sediments: Consideration of Spatial and Temporal Patterns in Exposure Modeling." *Environmental Science and Technology*, Vol. 36, pp. 238-246.

Marsh, L. M. and Jones, R. E., 1988, "The Form and Consequences of Random Walk Models," *Journal of Theoretical Biology*, Vol. 133, pp. 113-131.

Minta, S. C., 1992, "Test of Spatial and Temporal Interaction Among Animals," *Ecological Applications*, Vol. 2, No. 2, pp. 178-188.

Moore, D. W. J., Sample, B. E., Suter II, G. W., Parkhurst, B. R., and Teed, R. S., 1999, "A Probabilistic Risk Assessment of the Effects of Methylmercury and PCBs on Mink and Kingfishers Along East Fork Poplar Creek, Oak Ridge, Tennessee, USA," *Environmental Toxicology and Chemistry*, Vol. 18, No. 12, pp. 2941-2953.

Nagy, K. A., 1987, "Field Metabolic Rate and Food Requirement Scaling in Mammals And Birds," *Ecological Monographs*, Vol. 57, No. 2, pp. 111-128.

Nagy, K. A., 1994, "Field Bioenergetics of Mammals: What Determines Field Metabolic Rates?" *Australian Journal of Zoology*, Vol. 42, pp. 43-53.

Nisbet, R. M., Ross, A. H., and Brooks, J., 1996, "Empirically-Based Dynamic Energy Budget Models: Theory and an Application to Ecotoxicology," *Nonlinear World*, Vol. 3, pp. 85-106.

Shipley, L. A., Gross, J. E., Spalinger, D. E., Hobbs, N. T., and Wunder, B. A., 1994, "The Scaling of Intake Rate in Mammalian Herbivores," *The American Naturalist*, Vol. 143, No. 6, pp. 1055-1082.

Stehn, R. A., Stone, J. A. and Richmond, M. E., 1976, "Feeding Response of Small Mammal Scavengers to Pesticide-Killed Arthropod Prey," *American Midland Naturalist*, Vol. 95, No. 1, pp. 253-256.

Stephens, D. W. and Krebs, J. R., 1986, *Foraging Theory*, Princeton University Press, Princeton, New Jersey.

Turchin, P., 1998, *Quantitative Analysis of Movement*, Sinauer Associates, Inc., Sunderland, Massachusetts.

USEPA, 1993, *Wildlife Exposure Factors Handbook, Volume I of II* (EPA/600/R-93/187a), U.S. Environmental Protection Agency, Washington, DC.

With, K. A., 1997, "The Application of Neutral Landscape Models in Conservation Biology," *Conservation Biology*, Vol. 11, No. 5, pp. 1069-1080.

Author Index

Subject Index

P

Q

R

S

T

U

V

W